AF678207

INTRODUCTION TO Occupational Health AND Safety

INTRODUCTION TO Occupational Health AND Safety

Joseph LaDou, M.S., M.D.
Editor-in-Chief

National Safety Council

444 North Michigan Avenue
Chicago, Illinois 60611

Copyright © 1986 National Safety Council
All Rights Reserved

No portion of this book may be reproduced by any process without written permission of the National Safety Council, 444 North Michigan Ave., Chicago, Ill. 60611

International Standard Book Number: 0-87912-133-5
Library of Congress Catalog Card Number: 84-61216
Printed in the United States of America

7M0486 Product No. 122.15

Contents

Council Hardcover Books

The National Safety Council's hardcover book program presents current and vital information intended to help readers establish priorities, gather and analyze data, and develop methods and procedures to eliminate or minimize accident problems that result in injury, illness, and/or economic loss.

Until this book, *Introduction to Occupational Health and Safety,* joined the others on the Council's hardcover bookshelf, little had been published by the Council specifically on the health aspect, although it is well-nigh impossible to separate illness from injury in the work setting. In fact, nearly all contemporary legislation has just about legally married occupational health and safety. Although this "relationship" was recognized in previous Council books, occupational health was not emphasized; similarly, safety is not missing in this present book—it is just not emphasized.

The content of this book overlaps somewhat with that of *Fundamentals of Industrial Hygiene* and the "Administrations and Programs" volume of the *Accident Prevention Manual for Industrial Operations.* The Council recommends that if the reader wants more information on industrial hygiene or on safety techniques, he or she refer to these two sources respectively.

In addition to supplying detailed information about health and preventing occupational illness, this book includes other important subjects—women in the workplace, high-risk workers, information management systems, work scheduling, and understanding legal aspects—that make it an important adjunct to the Council's other books.

Because this book strongly emphasizes health and occupational illness, the word "health" is usually placed before the word "safety" in all places where the two are inseparable. Unless safety is specifically discussed, this text refers to "occupational health and safety."

As with the Council's other books, reader comments and suggestions are welcome.

FRANK E. MCELROY, P.E., C.S.P.
Director, Technical Publications
National Safety Council

Foreword

The field of occupational health and safety has exploded with new career opportunities in recent years. Government regulation for health and safety in the workplace, industry's need to meet those regulations, and universal goals for safer and healthier work environments have created a strong demand for trained professionals in this field. But where are these trained professionals to come from? A number of colleges and universities in the United States offer some form of training in this field. A recent survey by the National Safety Council (1983) indicated that 448 colleges and universities (out of 2,140 that received questionnaires) offered some credit courses in safety education but no degree, 103 offered degree programs at the graduate or undergraduate levels, and 16 offered doctoral degrees in occupational safety. So educational institutions are providing a source of trained health and safety professionals, although the number may be somewhat limited.

Unfortunately, however, many companies—particularly smaller companies—do not or cannot for fiscal reasons seek trained professionals for their health and safety programs. Instead, when faced with the demands of federal and state occupational health and safety agencies, they often assign a staff or production person who has performed well to take over the health and safety program. Most of these individuals will know little about developing or running such a program. Regardless of how efficient they have been in their past positions, they will be mired down in the attempt to start up the new program. At the same time, they will be besieged by government agencies for reports on company injuries/illnesses and plant or premises inspections. There will be no one to help them within the company, because the managers to whom they report will know little more than they do about developing an effective and viable health and safety program.

This book, therefore, has been prepared for those working in occupational health and safety who have not been specifically trained for this field and for the corporate managers responsible for working with them. The authors have attempted to provide a broad base of information on the many aspects of occupational health and safety for those who seek immediate answers, with suggested readings for those seeking greater depth. We hope that the book will prove useful to the audience it addresses.

JOSEPH LADOU, M.S., M.D.
Editor-in-Chief

San Francisco, California

Preface

This book has been prepared to assist both large and small companies in answering questions related to worker health and safety in their plants and establishments. It attempts to answer questions on organizational responsibility and operation, health and safety problems, legal issues, handling of toxic substances, preventive occupational health and safety, and other pertinent questions so that both management and health and safety team members can expand their knowledge and range of activities. Sources for obtaining additional and more in-depth information than is possible to include in this volume are also listed.

The first important question concerns the makeup and requirements of the occupational health and safety team itself. In other words, what should management seek in staffing this team? Part I answers this question by defining the professional skills that are needed and the required educational background and experience. Licensure, ethics, and educational and training opportunities are also discussed.

Whereas the development of the occupational health and safety team is the primary concern of management, the workplace itself is the most important aspect to the occupational health and safety team. These individuals will be asking, in many cases, what they should do and why they should do it. Part II discusses company operations and how they affect programs for workers. In many work environments, workers may be exposed to serious hazards that health and safety professionals must be aware of. These problems range from simple work-related dermatitis to long-term exposure to carcinogens. For example, of 7,000 industrial chemicals tested, 1,500 have been found to be carcinogenic. Unless the members of the team know about these and other hazards in the workplace, they cannot fully and effectively implement a system of preventive occupational health and safety. Toward attaining that objective, Part II discusses general aspects of company health and safety programs, as well as principles of toxicology, carcinogenicity, mutagenicity, teratogenicity, radiation safety, and environmental protection. To protect the worker, the health care professional not only must be skilled in one particular field, but also must have a working knowledge of the other disciplines represented by the team. For example, if an occupational health nurse is unaware of the procedures instituted by the safety engineer, he or she cannot be certain of the source of worker problems that are encountered.

Part III discusses the broad concepts that underlie a healthful and safe work environment. The reader is given detailed information on the treatment of injuries and illnesses and rehabilitation following debilitating injuries and work exposures. Occupational stress and emotional illness have become a growing area of concern among occupational health and safety professionals since stress in the workplace can create problems for both the individual and the company. Such problems as alcohol and drug abuse and the means of detecting and managing them are discussed. Worker education not only in alcohol and drug abuse, but also in hearing conservation, safe handling of toxic substances, and first aid after exposure is a highly important aspect of preventive occupational health and safety and, consequently, a significant responsibility of the health and safety team. Methods for fulfilling this responsibility are outlined. Finally, every company has employees who are at greater risk in the work environment than the general work force. They include women of child-bearing age, older workers, and workers at risk because of genetic disabilities. The final two chapters of Part III discuss the special problems of such workers and point out how they can be managed so that risks are reduced or eliminated.

A comprehensive, interdisciplinary occupational health and safety program with an objective of achieving preventive measures in the workplace is a very large undertaking. If it is to be successful, it must have effective management and administration. Part IV of this book identifies problem areas in operating such a program and offers suggestions for resolving them. The following are considered here: management reporting, medical records and reports, health and safety information systems, counseling of management and employees, work scheduling including effects of different shift methods, plant inspections, and worker protection including disaster management. Also included is a comprehensive list of sources of information on toxic materials and their effects.

Part V addresses the legal aspects of occupational health and safety and answers questions that both company management and team professionals will be asking. What are the impacts of federal regulations and the action of the various responsible agencies? What are the legal issues pertaining to control and disposal of toxic substances, safety in the workplace, environmental hazards, workers' compensation and rehabilitation, and discrimination? Can a company or an individual on the team be liable for malpractice? These and many more questions are answered with examples given from recent legal decisions and the judicial outlook indicated.

For purposes of company and personal planning, a chapter is included indicating what may be expected in future areas of occupational health and safety. It discusses the outlook for programs by both labor and government, industry, and academic institutions, as well as indicating opportunities in this field.

This book, as any volume covering such a broad subject, cannot answer all questions in great detail. Consequently, for additional information, the reader is referred to the detailed sources listed at the end of most chapters. References have been carefully selected by authors to provide sources of information that are of particular value to a wide range of readers.

Unless the information in this book is directly attributed to the National Safety Council, it is to be considered as being the opinion of the individual chapter authors and not necessarily that of the National Safety Council.

Although the information and recommendations contained in this publication have been complied from sources believed to be reliable, the National Safety Council makes no guaranty as to, and assumes no responsibility for, the correctness, sufficiency, or completeness of such information or recommendations. Other or additional safety and health measures may be required under particular circumstances.

Contributors

The chapter authors were carefully chosen on the basis of their expertise and experience in their respective areas of specialization. The editors are grateful to each of them for their valuable contribution to this text, even to making last-minute updates on the galley proofs.

In addition to the chapter authors listed below and the work of the editor-in-chief, Joseph LaDou, in pulling their efforts together, editorial guidance was provided by Frank E. McElroy, director of technical publications of the National Safety Council. Editing, proofreading, and production were handled by Jan Elliott, Carol North, and Nancy Seeger. Cover design was by Jake Kasparian, Council staff artist.

Roy Fisher, editor of *National Safety and Health News,* published some of the chapters as articles; his suggestions are appreciated.

Chapter authors are listed with their affiliations at the time of their contribution. "Former" means that the individual's affiliation has changed between the time of final approval of his typeset contribution and the date of final editorial OK.

Charles E. Becker, M.D.
Professor, Department of Medicine
University of California
School of Medicine
San Francisco, Calif.
Chapter 5

Edward J. Calabrese, Ph.D.
Associate Professor
School of Health Sciences
University of Massachusetts
Amherst, Mass.
Chapter 10

James Cinney
Senior Vice President
Safety and Health
Zenith Insurance Company
Encino, Calif.
Chapter 17

Barbara Cohrssen, M.S., M.L.S.,

C.I.H.
Industrial Hygienist
Department of Health and Safety
Stanford University
Stanford, Calif.
Chapter 15

Sally R. Cowles, M.D., Dr. P.H.
Medical Director, Epidemiology
Corporate Medical Department
Shell Oil Company
Houston, Texas
Chapter 6

Jean S. Felton, M.D.
Clinical Professor
Southern Occupational Health Center
University of California
School of Medicine
Irvine, Calif.
Chapter 3

Charles R. Goerth, J.D.
Attorney-at-Law
Dugan & Goerth, Ltd.
Wilmette, Ill.
Chapter 16

John V. Grimalidi, M.A., Ph.D., C.S.P.
Executive Director and Professor
Institute of Safety and Systems Management
University of Southern California
Los Angeles, Calif.
Chapter 4

Steven D. Herron, C.I.H, C.S.P.
Former Director
Occupational Health Programs
National Safety Council
Chicago, Ill.
Chapter 19

Joseph LaDou, M.S., M.D.
Acting Chief
Division of Occupational Medicine
Department of Medicine
University of California
School of Medicine
San Francisco, Calif.
Chapters 1, 9, 13

Julian B. Olishifski, M.S., P.E., C.S.P
Former Manager,
Loss Control, Hygiene and Health Services
Alliance of American Insurers
Schaumburg, Ill.
Chapters 1 and 2

Jane E. Parker, R.N., M.S.
Associate Professor
College of Nursing
University of Illinois Health Services Center
Chicago, Ill.
Chapter 2

Dan Petersen, M.S., Ed.D., C.S.P.
Management Consultant
Tuscon, Ariz.
Chapter 8

Mary I. Phares, M.S.
Technical Publications Consultant
Palo Alto, Calif.
Chapter 9

Arthur R. Reich, M.S., M.P.H., C.I.H.
Supervisor, Industrial Hygiene and Toxicology
University of California Health Sciences Center
San Francisco, Calif.
Chapter 14

Joyce A. Simonowitz, R.N., M.S.N.
Nursing Consultant
State of California
Department of Industrial Relations
Division of Occupational Safety and Health
Los Angeles, Calif.
Chapter 12

Joyce A. Spencer, R.N., M.P.H.
Consultant in Occupational Health Nursing
Pasadena, Calif.
Chapter 12

Ralph J. Vernon, C.S.P., C.I.H.
Professor, Industrial Engineering
College of Engineering
Texas A&M University
College Station, Texas
Chapter 2

Leon J. Warshaw, M.D., F.A.C.P.
Executive Director
The New York Business Group on Health
and Clinical Professor of Environmental Medicine
New York University
Medical Center
New York, N.Y.
Chapter 7

Adrienne A. Whyte, Ph.D.
Vice President, Program
Development
OHIS Inc.
Falls Church, Va.
Chapter 11

Gerald R. Williams, M.D., J.D., M.P.H.
Consultant in Occupational Medicine
San Diego, Calif.
Chapter 18

Carl Zenz, M.D., Sc.D.
Consultant in Occupational Medicine
West Allis, Wis.
Chapters 1 and 2

Part I

Background

Chapter 1

Occupational Health And Safety Today

by Joseph LaDou, Julian Olishifski, and Carl Zenz

Comprehensive occupational health and safety is a relatively recent idea that got started in the 1930s. The health and safety of workers is seen as a result of the concerted efforts of a team of individuals who have competence in safety, industrial hygiene, and occupational health, with the cooperation of a management dedicated to the idea that an organization's primary asset is healthy and safe workers.

As shown in Chapter 3, "History of Occupational Health and Safety," late in the 19th century, concerned individuals, journalists, and workers' organizations pioneered to improve the hazardous conditions prevalent in the work environment at that time. With government support, conditions improved because many companies began to employ professionals to oversee the safety and hygienic aspects of their work operations. Some companies hired health professionals to treat their ill and injured workers. Although there were significant gains, these health and safety professionals often functioned independently, sometimes with little or no contact. Typically, the hygiene and safety group would be concerned primarily with sanitation, a clean workplace, and equipment safeguards, while the medical group concentrated on treatment of injured and occupationally ill workers. Such an arrangement did little to contribute to the overall health of the work force since it did not include the concept of preventive occupational health.

INTERDISCIPLINARY APPROACH

Over time, health and safety professionals became aware of the relationships between physical illness and the job, as workers in certain industries exhibited a higher than normal incidence of such problems as dermatitis, musculoskeletal problems, pulmonary disease, mental illness, and cancer. In the 1940s, industrial hygienists, safety professionals, occupational health nurses, and physicians began to realize that their programs should be integrated to provide comprehensive health care, emphasizing greater protection for the worker to lessen the incidence of health problems and accidents in the workplace. As one early proponent of integration, H. G. Dyktor, commented, "Placing side by side the purposes of the official industrial hygiene unit and of the plant medical unit, it will be noticed that they are very similar in their objectives and that there is a tremendous amount of common ground upon which they can cooperate to the fullest extent."[1]

The objectives of the proposed integration were to: (*a*) acquire greater knowledge of health problems in the workplace including the effects of toxic substances; (*b*) provide a greater expertise for testing and evaluating problem areas; (*c*) provide a broader data base for assessing and comparing health problems of other establishments in the same industry; (*d*) stimulate the use of preventive measures plantwide and companywide; and (*e*) make workers' continuing health a primary concern of the entire company or plant. The outcome of this integration movement has been the development of occupational health and safety as we see it today—an integrated concept that identifies incipient health and safety problems and hazards and works to prevent or control them.

Government has played an important role in this development, particularly with the passage of the Occupational Safety and Health Act (OSHAct) in 1970 at the federal level and supporting legislation at the state level. The federal law stipulates that every employer must provide a safe and healthful work environment for employees. The act does not specify that he must have an in-plant health service (many companies are too small for such an outlay), but stipulates that there must be some plan for treating ill and injured workers, provision for the periodic examination and testing of workers exposed to toxic substances, and an individual with knowledge of first aid on duty during all working hours. Larger companies, it has been suggested, should provide:[2]

Periodic medical surveys of plant hazards and employee health
In-plant industrial hygiene and safety professionals
In-plant occupational health nurse
Medical examinations for all employees
Full compliance with OSHA regulations
Personal protective equipment for workers
Reduction of injury frequency rates
Rehire of injured workers
Physician input and direct access to top management on health policy

Medical records held in exclusive custody of medical personnel
Physician or nurse participation in regular consultation
Health and safety training of supervisors and workers.

Fulfillment of these objectives for worker health and safety creates the type of comprehensive occupational care envisioned by the American Conference on Industrial Health, which first proposed such an interdisciplinary approach in 1940.[3]

This integrated approach to worker health *and* safety requires the expertise of health professionals, industrial hygienists, and safety professionals, all working together toward a common objective—preventive health and safety care. The size and number of a company's health and safety staff will depend on the size of the company and the complexity of its operations. Large organizations may have most skills in house, but small companies must obtain them from private safety consultants and occupational health care facilities. Laboratory work, including analysis of toxic or suspected substances, is usually performed by commercial or government laboratories because only the very largest companies can justify maintaining their own laboratories. Roentgenograms (X-rays), laboratory tests, and other specialized services typically are obtained from outside sources.

Health and safety team

For companies large enough to maintain full occupational health and safety staffs, Hamilton and Hardy[4] have defined the team requirements to include:

Industrial or occupational hygiene chemist. Tasks include sampling workroom air, assaying biological samples from exposed workers, identifying materials thought to be harmful. When newly introduced substances are thought to be hazardous, assay methods need to be developed. Chemical engineering or chemistry are the usual fields of training, although an increasing number of institutions offer specific training in industrial hygiene.

Industrial or occupational hygiene engineer. Duties are the identification and measurement of size and amount of dust; design of proper ventilation; control of excessive noise levels; decision as to correct respirator. Chemical engineering may be the prior field of training.

Radiation protection officer, health physicist. This profession was developed in World War II to serve the needs of those involved in the production of nuclear energy. This field, which has become highly specialized, requires a background in physics or electrical engineering. Work involves radiation monitoring, development of instrumentation, assay of biological samples for radioactivity, planning for response to radiation accidents, decontamination.

Industrial or occupational safety engineer. Responsibilities are to prevent accidents, explosions, fires, and to supply such protective devices as safety glasses. In small plants, the person who fills this post may also carry out duties of other team members listed above. Technical or academic training for this job is becoming increasingly available.

Other professionals engaged in the prevention of occupational diseases include physicians, nurses, dentists, dietitians, psychologists, and health educators who may be employed full- or part-time, depending on the size of the company and the concern of its executives.[4]

The environment in which occupational health and safety professionals function has become exceedingly complex when compared with conditions prior to World War II. New products and processes have proliferated, and technology abounds. Today, every new product and process must be evaluated for its potential to cause immediate hazardous effects, contaminate the work environment, cause long-term disabilities among workers, or damage the general environment. Occupational health and safety professionals hold the key to providing such evaluations. Working as a team, they must identify hazards that may affect workers as well as the community. They must be alert to changes in the physical condition of the workers and determine the causes (either work- or nonwork-related); and they must strive to provide a safe workplace without health hazards.

That is a massive assignment. No single individual could possibly accomplish all of these important objectives. A knowledge of safety administration and technology, toxicology, carcinogenicity, mutagenicity, teratogenicity, epidemiology, radiation effects, industrial hygiene, and environmental protection—is knowledge that cuts across many disciplines. Even in large organizations that maintain interdisciplinary health and safety teams, not every aspect of the assignment will be covered in house. For smaller companies, the task is formidable. To accomplish it, both management and the occupational health and safety team will need a large fount of information or the knowledge of where to acquire such information from others.

A GROWING PROFESSION

The importance of health and safety cannot be overemphasized. More than two million persons suffer work-related permanent or temporary disability annually in this country, and more than 6,000 deaths result from occupational injuries annually.[5] There are more than five million businesses in the United States employing some 85 million persons. However, 4$^1/_3$ million (87 percent) of these businesses employ 25 or fewer persons. Not surprisingly, it is in the smaller workplaces that many of the more severe occupational health and safety problems exist. The National Institute for Occupational Safety and Health (NIOSH) estimates that only 2 percent of the employees in these businesses have access to industrial hygiene service and workplace monitoring programs.

In small company settings, the team concept of occupational health and safety can be accomplished only by assigning those responsibilities to one individual who will recognize his or her technical limitations and appropriately use consultants and private physicians. The person assigned responsibility for health and safety must be both capable and motivated. The task is no small part of management's responsibility. The health and safety officer in a small company must be willing to accept the recommendations of consultants and private practitioners and to translate those recommendations into corporate policy.

REFERENCES

1. Dyktor, H. G. "Integration of Industrial Hygiene with Industrial Medicine." *Industrial Med.* 9(4):193, 1940.
2. Howe, H. F. "Organization and Operation of an Occupational Health Program—Part I," *J. Occ. Med.* 17:360-400, June 1975; "Part II," 17:433-456, July 1975; and "Part III," 17:528-540, Aug. 1975.
3. "Industrial Health Conference." *Industrial Med.* 9(2):588,1940.
4. Hamilton, Alice, and Harriet Hardy (revised by Asher J. Finkel). *Industrial Toxicology.* Boston: John Wright Publishing Sciences Group, 1983.
5. *Preventing Illness and Injury in the Workplace.* Washington, DC: U.S. Congress, Office of Technology Assessment, OTA-H-256, April 1985.

Chapter 2

The Occupational Health And Safety Team

by Julian B. Olishifski, Jane E. Parker, Ralph J. Vernon, Carl Zenz

As late as 1977, the National Institute for Occupational Safety and Health (NIOSH) called *conservative* the estimate of an additional 4,700 safety professionals and 4,000 certified industrial hygienists required by American industry. Even more staggering were estimates of a deficit of 25,000 occupational health nurses and 1,000 certified occupational physician specialists. Approximately 30,000 physicians with short-course training in occupational health were needed to meet the minimum professional manpower requirements perceived by NIOSH.[1]

Occupational health and safety programs involving the major disciplines—safety, industrial hygiene, health physics, occupational medicine, and occupational health nursing—are expanding rapidly in America and a large number of other countries. A nationwide survey of the occupational health and safety work force conducted by NIOSH estimated the demand through the year 1990 for occupational health and safety personnel.[2] Results were based on status quo forecasts that the 1977 distribution of the occupational health and safety work force would remain unchanged through 1990, and second, that future changes in the requirements for occupational health and safety employees would add employment opportunities in both the public and private sectors.

Educational institutions are gearing up to produce more than 2,100 occupational health and safety graduates per year by 1990, which means that annual awarded occupational health and safety degrees will have increased 50 percent from 1980 to 1990, and if we use the year of the OSHAct as a base, a staggering increase of 1,400 percent will have occurred between 1970 and 1990.[2]

To help meet these educational goals, NIOSH has funded centers for training in occupational health and safety throughout the United States. Known as Educational Resource Centers (ERCs), these academic centers provide graduate-level instruction and training in occupational and environmental health sciences. They are found currently at the universities of California (northern and southern centers), Cincinnati, Illinois, Michigan, Minnesota, North Carolina, Utah, and Washington, as well as at Mount Sinai and affiliated New York and New Jersey institutions, Harvard at Boston, and Johns Hopkins in Baltimore. Additional funding is provided to the universities of Iowa and Wisconsin, and Meharry, although these are not ERCs. All incorporate varying amounts of training for physicians, industrial hygienists, occupational health nurses, safety professionals, health physicists, and other specialists. Many ERCs also provide direct occupational health and safety training for management and worker groups.

TRAINING REQUIREMENTS FOR SPECIFIC PROFESSIONS

Safety

One of the difficulties encountered in categorizing occupational health and safety personnel relates to coordinating requirements with job titles. In safety, for example, there are more than 100 different titles in use today. But basically, safety professionals are responsible for preventing accidents and injuries in the workplace.

To become a safety professional, formal training is required, typically consisting of an appropriate bachelor's degree and often a master's degree or equivalent experience. Although the trend is toward more advanced training, a master's degree level of academic background is less commonly attained in safety than it is in industrial hygiene and health physics. However, the eventual level of responsibility assumed by the safety professional in the industrial setting and his or her ability to establish a safety program and to communicate with top management, usually meets or exceeds those of the other specialties.

A directory of colleges and universities offering programs of training in safety has been compiled by both NIOSH and the National Safety Council,[3] and further information on specific programs can be obtained from these sources. Close to 400 academic institutions offer accredited training programs in occupational safety in the United States.

Safety training has few restrictive criteria, both at the academic level and in industrial job placement. The more than 22,000 members of the American Society of Safety Engineers (ASSE) do not reflect the total number of persons assigned to safety roles in American companies. No study clearly defines the background of this diverse group of specialists. It is safe to say that a variety of backgrounds—from chemical engineering to health education—serves as a springboard for successful careers in occupational safety. The important ingredients for effective safety management are concern for health and safety in the workplace, and an ability to learn and to apply what one learns to develop an effective program.

The Board of Certified Safety Professionals reports that the number of certified persons in this country in 1985 was 5,869. About one quarter of these certified safety professionals (CSPs) are also professional engineers (PEs), which is a state engineering licensure or registration to practice engineering where the safety of the public is concerned. The number of newly certified safety professionals per year is about 300, with the total number of certified safety professionals (reflecting the loss of retiring persons) increasing by less than 100-150 per year.

Eligibility for certification requires a bachelor's degree in engineering or other acceptable curriculum (emphasizing mathematics and chemistry) and one year of professional safety experience. Additional professional experience can take the place of educational requirements.

Certification is a two-stage process. The first stage, a core examination, follows graduation with a bachelor's degree in safety meeting BCSP requirements. The second stage is an examination following four years of professional safety experience or comparable experience rendered through a study program or registration/certification in related areas. The number of persons taking and completing the core examination is increasing, with 320 associate safety professionals passing the core examination in 1985. This will result in an increasing number of certified safety professionals in the near future.

Although untrained employees may be trained at the work site to be safety technicians, an appropriate one- or two-year associate's degree or certificate of program completion is usually required. The wide variety of training opportunities in safety in other than academic institutions is reviewed in a document from the Center for Occupational Research and Development.[4] OSHA maintains a full-time training institute in Des Plaines, Ill., for educating federal and state compliance officers, and provides a limited training opportunity for private industry safety professionals.[5] NIOSH also sponsors training programs throughout the United States.[6] Its Educational Resource Centers (ERCs), already mentioned, consist of a number of universities that have received grants to underwrite development of new professional courses in the fields of occupational health and safety.

The National Safety Council (NSC) is a private, nonprofit, public-service organization. The Council develops and evaluates methods and procedures that prevent accidents and illnesses, and that mitigate injury and economic loss. At its Chicago offices, the Council sponsors the

Safety Training Institute, which conducts public and private safety and health courses. The Institute trains more than 1,200 persons a year. The contents of the courses range from fundamental concepts to special areas of current importance to occupational health and safety. Courses offered by the Safety Training Institute include:

- Principals of Occupational Safety and Health
- Safety Training Methods
- Safety Management Techniques
- Practical Aspects of Industrial Hygiene
- Fundamentals of Hospital Safety
- Product Safety Management
- Safety in Chemical Operations
- Laboratory Safety
- Motor Fleet Accident Investigation Workshop

Many of these courses are also offered regionally by state or local safety councils. Some are offered as home study courses. The NSC also offers on-site telephone consultation with loss-control specialists in safety and health, as well as literature searches and other library services.

Other training organizations in occupational safety include the National Mine Health and Safety Academy in Beckley, W. Va.; the National Fire Protection Association (NFPA) in Quincy, Mass.; and a wide variety of state safety associations. Many companies sponsor their own training programs, and some industry groups meet jointly to share the effort of safety training. Information on many of these meetings can be found in the Council's *National Safety and Health News,* the ASSE journal, *Professional Safety,* and other publications listed in Chapter 15.

Although no data are published to document the practice, the majority of safety training is done at the employment site through on-the-job training, classes, reading, and exchange of knowledge with other persons assigned to safety within the company or with similar companies. There is little reason to conclude that this means of safety training will not continue to be important in the future. The emergence of more training opportunities in formal academic settings and in larger industry groups is viewed as a positive addition to in-house training, and the certification process for safety professionals gives the field a stature befitting its importance to workers and industry.

Health physics

The field of health physics developed during and after World War II

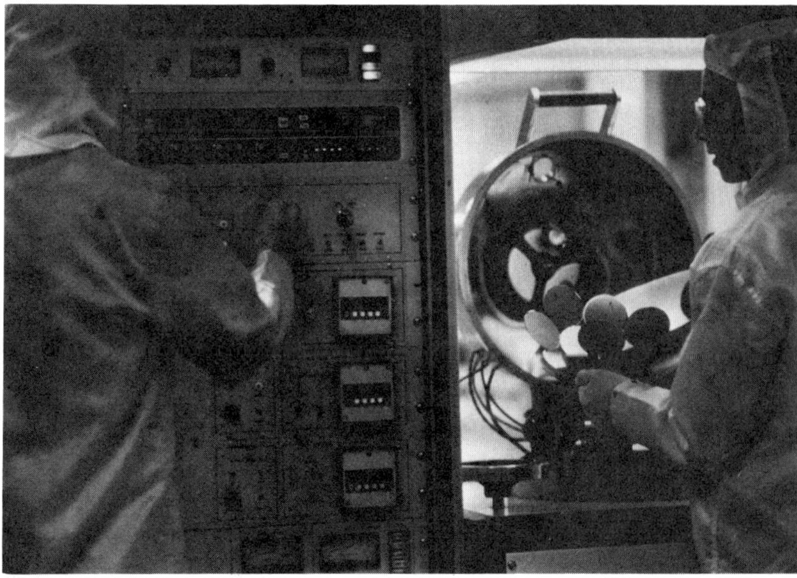

Figure 2-1. The high-technology microelectronics industry requires that considerable attention be paid to occupational health and safety by means of a team effort to provide workplace monitoring, personal protective equipment, hazardous materials disposal, and the assurance of continued health of the workers and the community.

Courtesy of Intel Corporation.

and is concerned with radiation safety in the workplace. The health physicist is responsible for monitoring radiation within the plant and the environment, developing instrumentation needed for this purpose, evaluating biological samples for radioactivity, and developing plans for coping with radiation accidents and decontamination. This highly specialized field requires a diverse background in science, and graduate level training in health physics.

In 1971, the Health Physics Society (HPS) estimated that there were about 5,000 health physicists in the United States. Of the total, approximately 600 had been examined and certified by the American Board of Health Physics (ABHP). The membership in the Health Physics Society in 1971 was 3,400, but rose to 6,348 in 1985. The ABHP has certified 870 health physicists through 1984. It is estimated that more than 10,000-11,000 health physicists are currently employed in the United States.

Projections of manpower needs in radiation safety made in 1976 presumed a rapidly expanding nuclear power industry, a phenomenon that has not occurred yet in this country.[7] Even so, the demand for graduates at the M.S. degree level (of primary importance since they serve as the major source of radiation-protection specialists) and other levels of health physicists remains strong. For instance, the American Association of Physicists in Medicine projects a 7 percent annual growth rate in the number of medical physicists serving in the nation's hospitals.

A sampling of HPS members gave the following breakdown on educational background and source of employment:[8]

Academic Background	*Number of Respondents*
Training program	32
AS or equivalent	40
BS/BA	210
MS/MA	314
PhD/equivalent	144
Employer	
Academic	125
Government, nonmilitary	172
Industrial, noncommercial nuclear power	72
National laboratories	161
Commercial nuclear power	88
Military	32
Consultant	35
Self-employed	5
Medical, other	61

ABHP certification shows a clear correlation of higher salaries with higher academic level in this survey. Industrial and consulting positions in health physics appear to carry the highest annual salaries, whereas academic and governmental positions frequently show salaries less than those with private industry.

Industrial hygiene

Industrial hygiene is concerned with the physical conditions of the environment in which employees perform their jobs. The industrial hygienist, therefore, is responsible for proper ventilation so that harmful vapors, fumes, and particulates are kept to a minimum, avoidance of excessive heat and noise, proper lighting to perform the work, and evaluation of new substances for hazards during production processes (see Figures 2-1 and 2-2).

There are approximately 7,600 industrial hygienists in the United States, the majority being professionally trained and certified as opposed to the job of industrial hygiene technician. The American Board of Industrial Hygiene (ABIH) reports that in 1985 there were 2,523 diplomates holding the title Certified Industrial Hygienist (CIH) and an additional 275 seeking certification in 1985. The rate of growth since 1982 is approximately 13 percent a year, far less than the projected manpower needs estimated by NIOSH and projections by the American Industrial Hygiene Association (AIHA) of a doubling in demand for industrial hygienists in the next ten years.

Of the 800 persons currently in training programs in industrial hygiene, 170 are studying to be technicians. This ratio of industrial hygiene professionals to technicians of four to one is thought to be an accurate reflection of the membership of AIHA and of the general distribution of industrial hygienists and technicians in the country.

Certification in industrial hygiene by the ABIH requires a two-part written examination; the core examination is given to persons with bachelor's degree (majoring in chemistry, physics, industrial hygiene, or mechanical, sanitary, or chemical engineering) following one year of professional activity. The second examination is given after five years of professional activity.

A master's degree is generally held by the ABIH to be equivalent to one year of professional activity, and a Ph.D. degree to be equivalent to an additional year of professional activity. The Board reviews applications for certification on a case-by-case basis. There are cases in which a bachelor's degree in other areas, or even the absence of a bachelor's degree, need not prevent certification in this field, which allows considerable freedom in academic training.

Industrial hygiene technicians are required to have five years of activity at generally less than the professional level before certification, with two years of equivalency granted for an associate's degree. Here again, the ABIH reviews applications for certification from industrial hygiene technician candidates on a case-by-case basis.

Occupational medicine

The occupational medicine practitioner is primarily concerned with preventive medicine in the workplace—to keep the working population well. He or she is responsible for evaluating the cause of employee illnesses, both physical and emotional, and for determining whether or not illnesses are work-related, as well as treating those ill or injured from work-related causes. The occupational medicine practitioner is particularly suited to determining work-related causes because he or she knows the work environment well and the circumstances of the employee's job. A personal physician attempting to determine work-relatedness without ever visiting the workplace, on the other hand, would be severely limited by a lack of knowledge about the workplace and the job.

Occupational medicine is a field that spans many areas. Because of the complexities of occupational health endeavors, occupational medicine has become a specialized field of practice. Since 1955, it has shared board certification with aerospace medicine and public health under the auspices of the American Board of Preventive Medicine. The knowledge required to practice occupational medicine and to attain board eligibility encompasses a multitude of occupational and environmental factors in the work setting and, of equal importance, the understanding of the physiologic and psychologic adaptations to work.

Medical schools seldom emphasize the role of occupational medicine during early training in clinical medicine. Consequently, the practicing physician often has difficulty relating occupational factors to his or her specialty practice. Although clinical problems may appear straightforward to the treating physician, workplace conditions include many other factors that can alter the diagnostic approach.

Graduate study in occupational medicine, often leading to the Master of Science or the Master of Public Health degree, is essential to the practice of sound occupational medicine. Most graduate courses require a minimum of two years. Three or more years, depending on the program, are frequent career pathway choices. These programs often require preparation and submission of a thesis. Graduates of these programs have been in high demand by industry, consulting firms, academic centers, and state and federal governmental agencies. Courses in the following subjects are the core of postgraduate education in occupational medicine:

- Biostatistics and occupational epidemiology
- Work physiology, including responses to heat and cold, vibration, physical working capacity and energy expenditure, and elements of exercise testing
- Radiation (ionizing and nonionizing)
- Effects of special environmental conditions, *i.e.,* work at high altitudes and under hyperbaric conditions
- Noise and hearing conservation
- Principles of toxicology
- Occupational safety
- Fundamentals of industrial hygiene
- Biologic monitoring
- Occupational respiratory diseases

- Dermatology—occupational aspects
- Psychiatric and psychologic factors
- Ergonomics
- Environmental health, pollution control measures
- Governmental regulations
- Record and data collection and management.

The program includes visits to different workplaces to observe workers in foundries, mines, chemical plants, electronics, paper making, machine shop and assembly operations (including automotive and farm equipment manufacturing), aerospace industries, pharmaceuticals, and agricultural activities.

Certification. The American Board of Preventive Medicine certifies physicians in the specialty of occupational medicine. Eligibility for examination follows completion of a fully accredited formal training program at one of 29 institutions.[9] The annual number of graduates of these full training programs is only 60-70. The Board does consider the "equivalency route" in some cases; contact the Board for details. However, the equivalency route is not available to those who received their medical degree after January 1984. This will eventually limit board certification entirely to physicians with formal residency training, as outlined next.

The preferred pathway to eligibility for examination is the satisfactory completion of three fully accredited years of formal training, followed by a year of essentially full-time practice of occupational medicine, or of teaching and/or research in the field.[10]

Under this plan, the three years of training needed are as follows:

- One year in clinical and hospital work involving direct patient care in a program accredited by the Residency Review Committee for Preventive Medicine or by another Residency Review Committee for a field involving direct patient care—such as internal medicine, family practice, surgery or pediatrics.

- An academic year at the postgraduate level which leads to the Master of Public Health degree, or to what in the Board's judgment constitutes an equivalent degree requiring successful completion of courses of postgraduate level and rigor in biostatistics, epidemiology, health services administration and environmental health, among others.

- A practicum year of residency providing supervised training and field experience in occupational medicine that has been accredited by the Residency Review Committee for Preventive Medicine.

Figure 2-2. Bioengineering companies are developing at a rapid rate, creating opportunities for health and safety professionals willing to work in areas with few guidelines typical of more mature industries.

Courtesy of Genentech Corporation.

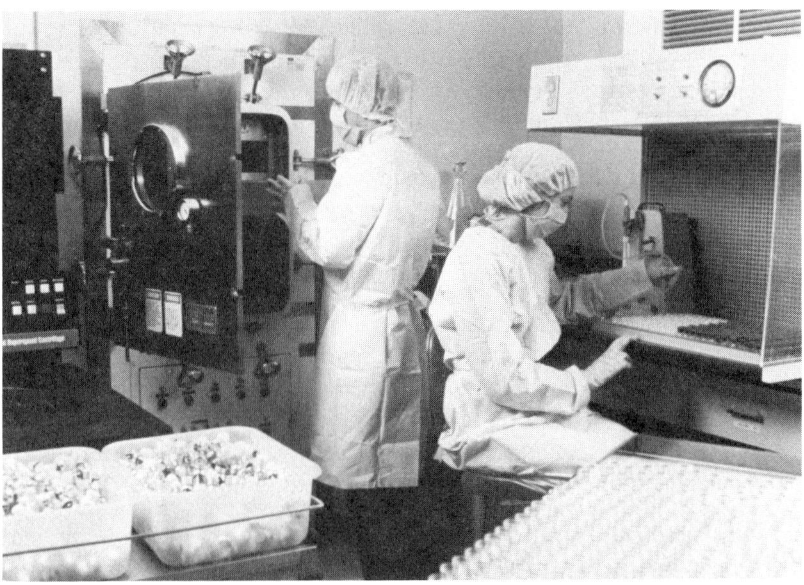

In some programs, the academic and the practicum year of residency training are taken together over a two-year period, with the appropriate proportion of time devoted to each component.

The membership of the American Occupational Medical Association, largely derived from in-plant physicians as opposed to private consultants and practioners, numbers 4,000. The American Board of Preventive Medicine, since 1955, has certified 1,050 physicians in occupational medicine and estimates that about 700 are actively engaged in practice. The number of new certifications annually is about 60-70, which when corrected for the rate of retirement in this group of older physicians, represents a small fraction of actual demand.

Courses and Seminars. A number of short courses are given throughout the United States for physicians in family practice, internal medicine, surgery or other specialties, ranging from hours to several weeks. The Intensive Residency, as programs offered by the University of Cincinnati and the University of California at San Francisco are known, do not determine board eligibility for certification of physicians. They do serve, however, to provide an orientation to physicians making midcareer transitions into the field of occupational medicine from other specialty areas.

During annual meetings of the specialty societies—the American Occupational Medical Association, the American Academy of Occupational Medicine, and the Society for Occupational and Environmental Health—postgraduate seminars and scientific programs are presented, and physicians, engaged part or full time in occupational medicine, are encouraged to attend and have their participation recognized as accredited educational experience by the American Medical Association and by various state medical associations. Private practitioners are finding increasing opportunities to attend scientific programs dealing with occupational medicine at various society meetings, a trend that will likely remain constant during this period of high demand for occupational medical training.

Occupational health nursing

Occupational health nurses are uniquely situated to deliver health care to people at work. Their primary focus should be on prevention of illness and promotion of health, a role contrary to the older practices of crisis care still perceived by much of the industry as the function of the medical department. Although it is true that nurses play an important role in the care of injury and illness occurring in the industrial setting, the best utilization of the occupational health nurse is in the area of preventive health care for both occupational and nonoccupational problems.

In many small plant or other establishment settings, the occupational health nurse is the only health professional employed to coordinate health, and sometimes safety, activities. The importance of interdisciplinary training for the occupational health nurse is readily apparent. The nurse is not only instrumental in promoting the health of the entire population, but may well extend his or her sphere of influence to the safety area as well. This approach requires a willingness to use consultants and to work with persons assigned to health and safety who do not have professional nursing training.

Many people who have witnessed the success of the occupational health nurse in safety functions consider extension of his or her authority and training in an occupational setting a logical move. One of the aims of this book is to provide the necessary introduction of health and safety team members to the special nature of other disciplines so that they can expand their range of activities. In the case of the occupational health nurse, the most opportune area is that of plant safety.[11]

An example of a traditional team approach illustrates the pivotal role of the occupational health nurse. A plant with a noise exposure problem will call upon the industrial hygienist to conduct noise surveys to determine which areas need more stringent engineering controls to reduce noise exposure and also to identify areas still requiring hearing protection. The safety professional will direct the instruction, issuance, and supervision of hearing protective equipment, while the occupational physician would coordinate health surveillance activities. The occupational health nurse initiates audiometric testing and evaluation, counsels the employee about potential hazards and the need to cooperate with the use of personal protective equipment, and answers questions of concern to the worker about noise and hearing loss. The nurse's records are kept in a medical file that assures the employer that the company is complying with regulations and that every effort is being made to prevent hearing loss in exposed workers.

In actual practice, the occupational health nurse may be asked to work with consultants and private practitioners to accomplish all of the above activities. In the smaller establishment or plant setting he or she must often do this alone, which demonstrates the importance of occupational health and safety in industry, despite the relative lack of other resources. The occupational health nurse shares with the safety professional major responsibility for health and safety in industry today.

An occupational health nurse must be a registered nurse (R.N.) who has a license obtained in the state where he or she will be practicing. A licensed practical nurse (L.P.N.) or trained practical nurse (T.P.N.) and other paraprofessionals are not able to act as occupational health nurses without the direct on-site supervision of a registered nurse or physician.

It is not legal for an LPN or others to give medications, treatments, or make any kind of independent nursing diagnosis in an unsupervised setting.[12]

Special training in occupational health nursing is now available in a number of programs throughout the United States. It is possible to recruit an occupational health nurse, some with a master's degree or master of public health. A more practical expectation is that the occupational health nurse with special training in the field will have a bachelor's degree.

Knowledge, skills, and abilities needed and used by the professional occupational health nurse will usually include:

Professional nursing

Community resources

Communicable disease control

Human behavior

Methods of health appraisal

Nonoccupational health problems (adult health)

Nutrition

Immunization

Rehabilitation

Communication skills

Principles of teaching

Interviewing and counseling

Keeping records and reports

Business administration

Legislation affecting nursing practice in occupational health

Principles of industrial hygiene, safety, and toxicology

Epidemiology and biometry or research methods

Occupational diseases and injuries

Labor-management relations.

If the occupational health nurse does not already possess the above training and skills that are essential to the practice of nursing in the occupational health field (and most will not, because nursing curricula until recently have not addressed industry's need for qualified nurses), the nurses hired must be committed to formal or at least continuing education programs to fill some of the above requirements. Furthermore, the company must be ready to provide release time, tuition, and expenses to support this endeavor as they would for any other employee who requires professional development and upgrading of skills. It will be mutually beneficial as both the employer and the occupational health nurse see the growth and application of new practices in their developing occupational health programs.[13]

The American Board of Occupational Health Nurses, Inc., is an independent nursing specialty board established in 1972 to initiate a program to certify qualified occupational health nurses. Certification is a voluntary process designed to provide recognition for nurses who have demonstrated excellence, knowledge, and experience in this field.

Nurses who are certified occupational health nurses (COHN) have five years of experience in the field, are R.N.s, have submitted proof to the certifying Board of 75 course hours of recognized, continuing education programs related to the practice of occupational health, and, finally, have taken a comprehensive examination. More than 3,000 nurses have been certified by the ABOHN. Recertification is required every five years.

The American Association of Occupational Health Nurses (AAOHN) is the professional association concerned with the education and development of the occupational health nurse. In January of 1983, the AAOHN had 11,817 active members. A survey of AAOHN members conducted recently gives a relatively accurate picture of the background and training of occupational health nurses actively working in industry today.

The majority of respondents—63.2 percent—were diploma graduates, but only 2.9 percent were certified occupational health nurses. The entry level education for occupational health nurses was 87 percent diploma nurses (R.N.: three-year education), 10.3 percent associate degree nurses (R.N.: two-year education), 2.9 percent with a bachelor of science in nursing (B.S.N.: usually a four-year college or university course), 4.4 percent with other bachelor's degrees, and 1.4 percent nurse practitioners (usually extra courses and by examination.)[14]

These findings suggest that the entry level of education of occupational health nurses has not been conducive to a greater number of certifications. Age is also a factor in this regard since 45 percent of occupational health nurses are 46 years of age or older. Experience is another factor in explaining the small number of certificates. Two-thirds of all occupational health nurses surveyed report that they have been in the field less than ten years.

As with the other disciplines in this field, occupational health nursing is moving in a positive way toward specialty training, certification, and rigorous programs of continuing education.

The following can provide further information regarding certification:

Board of Certified Safety Professionals
208 Burwash Avenue
Savoy, Ill. 61874
(217) 359-9263

American Board of Health Physics
1340 Old Chain Bridge Road, Suite 300
McLean, Va. 22101
(703) 790-1745

American Board of Industrial Hygiene
302 South Waverly Road
Lansing, Mich. 48917
(517) 321-2638

American Board of Preventive Medicine
P.O. Box 927
Dayton, Ohio 45401
(513) 278-6915

American Board of Occupational Health Nurses
2210 Wilshire Blvd., Suite 771
Santa Monica, Calif. 90403
(213) 450-4468

REFERENCES

1. Hatch, Loren L. "Problems in Occupational Health Programming," in *Occupational Safety and Health Symposia 1977,* Cincinnati, Ohio. U. S. Department of Health, Education, and Welfare, NIOSH Division of Technical Services, June 1978.

2. *A Nationwide Survey of the Occupational Safety and Health Work Force,* DHHS (NIOSH) Pub. No. 78-164; NTIS No. HR-P29-527/A04. Cincinnati, Ohio. U.S. Department of Health, Education, and Welfare; National Institute for Occupational Safety and Health, July 1978.

3. *Directory of Academic Programs in Occupational Safety and Health,* DHHS Publication No. 79-126, NTIS No. PB-80-176-068/A03. U.S. Department of Health, Education, and Welfare, NIOSH, DOTMD, Jan. 1979.

4. *Safety and Health, 1981.* The Center for Occupational Research and Development, 601 Lake Air Drive, Suite C, Waco, Texas 76710.

5. *Training Requirements in OSHA Standards.* U. S. Department of Labor, OSHA, Revised November 1979.

6. *"NIOSH Schedule of Courses,"* U.S. Department of Health and Human Services, NIOSH, DOTMD (published annually).

7. Christensen, Ralph C. "Trends in Health Physics Salaries and Employment: 1980" *Health Phys.* 42(2):107-117, 1982.

8. Moeller, Dade W., Rolf Eliassen, "Assessment of Health Physics Manpower Needs." *Health Phys.* 31(July):1-11, 1976.

9. "Directory of Residency Training Programs," American Medical Association, P.O. Box 821, Monroe, Wisc. 53566, (published annually).

10. Roberts, Norbert J: "Certification in the Specialty of Occupational Medicine," *J. Occup. Med.* 24(7): 546-547, 1982.

11. Brown, Mary L. *Occupational Health Nursing,* New York: Springer Publishing Company, 1981.

12. Lee, Jane A. *The New Nurse in Industry: A Guide for Newly Employed Occupational Health Nurse,* DHHS (NIOSH) Pub. No. 78-143. NIOSH, 1978.

13. Parker, Jane E. "Continuing Education, Who Needs It?," *Occup. Health Nurs.,* 29(4):22-24, 1980.

14. Archer, Helen M. "A Study of the Impact of Continuing Education on Occupational Health Nursing Practice," *Occup. Health Nurs.* 31(2):15-27, 1983.

Chapter 3

History of Occupational Health and Safety

by Jean S. Felton

Occupational safety and health practitioners, especially if they are new to the field, may think that this area of human endeavor started with the Occupational Safety and Health Act of 1970 and that little in the way of historical development preceded it. Although strict laws *are* of more recent origin, the ill effects of the environment on the worker have been known for thousands of years.

But just because hazards have been made known does not imply that they have been corrected; correction results from a deep social concern coupled with effective legal action. Most employers will rarely make changes by themselves because almost every move to create a safer and more healthful worksite costs money and time, and contributes to higher overhead costs. Since higher costs must be avoided insofar as possible, in order to remain competitive, legal action becomes the remedy.

WHAT IS PAST IS PROLOGUE

In an era of rapid change, it is imperative that one retains a memory of what has gone before, not only to appreciate the struggles experienced by the pioneers, but to avoid the errors in theory and practice that might have delayed earlier progress. Elliot Lee Richardson,[1] once secretary of the Department of Health, Education, and Welfare (now Department of Health and Human Services), expressed it well when he observed that without an appreciative sense of the old, we cannot properly measure, we cannot sensibly evaluate, and we cannot discriminately reject or affirm, the new and, further, that we cannot distinguish that which is to be promptly passé from that which is profound, or the farcical from the fundamental, or the ephemeral from the emergent.

Recognizing, then, that a sense of historical proportion must be retained and applied in what one does in occupational health and safety, a look backward should be enlightening. Such a view cannot be isolated from the context of the then-current cultural background. The health of the workforce can only be scrutinized in the social milieu in which it was bettered or, more often, worsened.

In pursuing history, one may be inclined only to see the best of a series of events or to gloss over much of the hideousness and most of the inhumanity that accompanied the growth of industry anywhere in the world. One cannot view everything in occupational history as being glorious and erudite and faultless. There were many periods of inexcusable behavior. In this discussion, these episodes will not be ignored nor an attempt made to hide the ugliness.

Prehistory

Although, by definition, there are no extant writings to offer descriptive discourses on mankind's early efforts to stay alive, we do know from the findings of the archaeologist that there was early cooperation in performing tasks, and that this was the means human intelligence employed to ensure the continuance of the race. Industry was necessary for survival. Shell mounds have been located and flint heaps have been uncovered, giving testimony to man's early manufacture of offensive and defensive weaponry.

The preparation of flint by knapping—the shaping by a series of sharp blows—led to effective, usable points for spears and arrows, and it has been theorized that these primitive knappers probably were the first humans to acquire silicosis from the fragmentation of the hard quartz. Researchers have observed evidence of carbonization and silicosis with pleural adhesions. (Pleural plaque of asbestosis?)

It was further noted that the search for flint—by both paleolithic man and his neolithic successor—became an important industry, for freshly extracted flint was more easily worked than flint that had long been weathered. Remains of the mine sources of this material have been discovered throughout Europe, and the presence of skeletons has indicated death by falls of earth. In these mines were shafts, rather primitive tunnels, and illumination was by crude lamps.

Artisans during the first Sumerian Dynasty—in ancient Babylonia, currently Iraq—produced socketed axes and adzes, and following the discovery of copper and the use of bronze, tools were developed of a more sophisticated nature, circa 3000-2500 B.C. The use of sun-dried bricks for the construction of buildings began about 3400 B.C.

In the Assyro-Babylonian civilization, excellent facilities for sanitation were created—drains, sewers, and stone privies. A further sense of cleanliness was seen in the insistence upon trench burial of warriors.

The Egyptians

That the industrious Egyptians were productive is seen in their monuments—temples, pyramids, tombs, and obelisks. Although the workers were slaves, the concept of productive industry was readily seen for there were discrete tasks, supervised groups of workers, architecture, and methods of material preparation, transport, and erection. The systems of sanitation and irrigation also required great amounts of labor with the attendant injuries depicted in various excavated tombs.

Early documentation of surgical knowledge and medical therapy is encountered in the *Ebers Papyrus*, found in 1862, described as a "miscellany of extracts, recipes and jottings collected from at least forty different sources. . .exactly analogous to the collections of household and medical recipes of Europe in later times." There are treatments given for various traumatic events such as crocodile bite, burns, and foreign bodies (splinters).[2] The *Edwin Smith Papyrus,* acquired in the same year, dating to about 3000 B.C., was more of a textbook of surgery, and it discussed a variety of injuries and treatments involving splints, dressings, and ointments.

Legge[3] cites Diodorus, an historian, who, in about 44 B.C., described the working of the mines in his history of Egypt and neighboring countries. The miners, usually slaves, condemned criminals, prisoners of war, or even exiled innocent families, worked in chains, were practically naked, and were subjected to the cruelty and blows of overseers. The sick, the disabled, the children, the women, all worked until worn out or until they died. The heat, smoke from the oil lamps, perspiration, and the lack of any hygienic measures must have created an unbearable atmosphere. There is no remaining evidence of occupational disease, for it is believed that no one in those conditions lived long enough to develop a chronically disabling lung disorder.

However, silica and carbon particles have been found in the lungs of mummies, indicative of a pneumoconiosis, probably caused by the inhalation of sand during desert dust storms. If there were symptoms, it is not possible to ascertain. Measured mercury levels in the bodies of ancients have been equal to contemporary assays. Of particular interest, though, is the development of early "occupational health services." Rameses II, possibly about 1500 B.C., not only revived a project of constructing a canal from the Mediterranean to the Red Sea, but also ordered a colossal mortuary temple to be built, the Rameuseum. To construct them, many workers were acquired from nearby villages, and the Pharaoh strengthened his industrial medical service by adding physicians, as civil servants, to care for the mine and quarry workers, and other working teams and settlers engaged in the projects.[4] It was not love of his subjects that motivated such action, but the desire to retain a healthy work force.

Each worker was examined regularly and was forced to bathe in the Nile several times daily. During epidemics, the sick laborers were isolated in special camps, and every year the workers had to burn their huts and erect new ones.

They could not relieve themselves at the worksite. By means of "site newspapers"—notations on fragments of limestones—it was learned that one worker "has been *examined* by beating with a stick," an action interpreted by the reviewer as an energetic, and possibly an efficient, procedure for reducing absenteeism. The physician was placed in the work hierarchy between a foreman and a guard, suggesting a salary about equivalent to that of foreman. Again, according to Diodorus, the physician at the time of the Middle and New Empires was still a civil servant of the Egyptian state, and his services were provided free, as they were during military expeditions and voyages.

The Babylonians

Hammurabi, the sixth and most eminent ruler of the first dynasty of Babylon, circa 2000 B.C., systematized the old laws of the land, revised them, and produced a complete codification of all statutes in the form of a body of law of some 280 paragraphs. The Code of Hammurabi was unusual for it covered bodily injuries and fees for physicians, and probably was the first document that bore any resemblance to today's workers' compensation insurance laws.

Some of the pertinent clauses follow:

> §*196.* If a man has caused the loss of a gentleman's eye, his one eye shall cause to be lost.
>
> §*199.* If he has caused the loss of the eye of a gentleman's servant or has shattered the limb of a gentleman's servant, he shall pay half his price.
>
> §*206.* If a man has struck a man in a quarrel, and has caused him a wound, that man shall swear "I do not strike him knowing," and shall answer [be responsible for] the doctor.
>
> §*213.* If he has struck a gentleman's maidservant and caused her to drop that which is in her womb, he shall pay two shekels of silver.
>
> §*214.* If that maidservant has died, he shall pay one-third of a mina of silver.

Hammurabi's laws, although not a code in today's sense, must be considered as representative of his concern in being a just ruler, an ideal which was followed by the Mesopotamian kings at all times. The code is the oldest in existence, and antedates the first books of the Old Testament by about 1000 years.

The Greeks

Although innumerable writers have credited Hippocrates for the first direct clinical account of lead poisoning, none of the writers has provided the exact source. Since none of the authors cited has given reference to the passage in Hippocrates on which they base their conclusions, however; it is difficult to substantiate them.

Nicander, a Greek grammarian, poet, and physician, who lived in the second century B.C., is cited by historians as providing in his poem, *Alexipharmaca,* the earliest description of lead poisoning, with its colic, paralysis, and ocular disturbances. Major[5] offers the following translation from the Latin version of *Euricius Cordus,* 1532 A.D.:

> The harmful cerussa, [white lead] that most noxious thing
> Which foams like the milk in the earliest spring
> With rough force it falls and the pail beneath fills
> This fluid astringes and causes grave ills.
> The mouth it inflames and makes cold from within
> The gums dry and wrinkled, are parch'd like the skin
> The rough tongue feels harsher, the neck muscles grip
> He soon cannot swallow, foam runs from his lip
> A feeble cough tries, it in vain to expel
> He belches so much, and his belly does swell
> His sluggish eyes sway when he totters to bed
> Complains that so dizzy and heavy his head
> Phantastic forms flit now in front of his eyes
> While deep from his breast there soon issue sad cries
> Meanwhile there comes a stuporous chill
> His feeble limbs droop and all motion is still
> His strength is now spent and unless one soon aids
> The sick man descends to the Stygian shades.

Major, in an earlier writing,[6] commented that Nicander might well be called the poet laureate of lead colic.

Although lead water systems created a considerable risk to the health of the early Greeks, a greater hazard resulted from the coating of copper vessels with lead, or lead alloys, to prevent leaching of the copper and spoiling of the taste of the food. As lead inhibited enzyme activity, the Greeks found that it would improve a poor wine and lengthen the time that any wine could be kept.

Tetanus, the dreaded complication of an unimmunized injured worker, was described by Hippocrates, born on Cos circa 460 B.C., and usually termed the father of medicine. He told of a shipmaster whose second digit of his right hand was crushed by an anchor—"Seven days later a somewhat foul discharge appeared; then trouble with his tongue—he complained he could not speak properly. The presence of tetanus was diagnosed, his jaws became pressed together, his teeth were locked, then symptoms appeared in his neck: on the third day opisthotonos [spasm of the back muscles] appeared with sweating. Six days after the diagnosis was made, he died."

Galen, of the second century A.D., considered the most distinguished Greek physician after Hippocrates, practiced in Pergamum, his birthplace in Asia Minor. He served as physician to the gladiators, a position possibly comparable to the physicians who attend today's professional athletes. In that assignment, Galen gained experience not only in medicine, but was able to increase his knowledge of anatomy by having to treat the extensive wounds of his patients. Further, he visited the Phoenician copper mines on the island of Cyprus, and wrote that he smelled the suffocating fumes emitted from the mines and saw the workers naked, because the acid mist and vapor destroyed their clothing.[3]

The Romans

There were Roman voices, too, that spoke of the health and the practitioners of the day.

Lucretius, a Roman poet and philosopher, 98-55 B.C., produced his *De Rerum Naturum (On the Nature of Things),* a poem in six books, Book VI considering various natural phenomena and the plague of Athens. In this writing, he referred to mining and the ill effects of such environments on the workers.

Lucretius wrote:

> Do you not see that sulphur is generated
> In the earth, like bitumen with its deadly smell?
> And where there is mining for veins of gold and silver
> Which men will dig for deep down in the earth
> What stenches arise, as at Scaptensula!
> How deadly are the exhalations of gold mines!
> You can see the ill effects in the miner's complexions.
> Have you not heard and seen how short is the life
> Of a miner compelled to remain at this terrible task?
> All these exhalations come from the earth
> And are breathed forth into the open light of day.

The presence of metals was acknowledged:

All things of every king: Many they serve for food friendly to life: Many with power and to reflect diseases and so hasten death.

Lucretius also spoke of entry of toxic substances into the body:

Many harmful elements pass through the ears, and through the nostrils, and many make their way that are noxious and rough; and not a few should be avoided by the touch.

Martial (Marcus Valerius Martialis), a well-known epigrammatist, who wrote of the Roman manners and customs of his day with a sharp wit, moved readily from formal orations to ridicule, even in relation to his benefactors.[7] He discussed the hazard of handling sulfur:

. . . and Bellona's raving throng
does not rest. . . nor the Jew
taught by his mother to beg, nor the blear-eyed
huckster of sulphur wares.

And for those physicians in academe who, with their entourages comprising house officers and medical students, visit patients, Martial made this pertinent comment to a practitioner who visited him:

I was indisposed; and straightway you came to see me, Symmachus, accompanied by a hundred of your pupils. A hundred hands frozen by the northern blast, felt my pulse. I had not then an ague, Symmachus, but I have now.

Lastly, among observers of the Roman scene, was Juvenal, a satirist living from 60-140 A.D., who dealt with the fatal degeneracy of the Romans. A few comments were made regarding the ills attending certain occupations. He asked, "what will the soothsayer get but varicose veins for his patience?" The reference is to the priests who were obliged to stand for long periods of time reading the auguries,"[8] specifically *varicosus haruspex,* the broken-veined soothsayer.

Rome's contribution to health lay in the construction of aqueducts, sewerage systems, drains, public baths, and well-ventilated houses. Large blocks of cut stone were used in the sewers and the manholes that provided access to them; the latrines were made of slabs of white limestone and faced with polished marble; and the aqueducts required endless blocks of cut and fitted stone. The preparation of these building materials unquestionably created respirable dust, and disabling pneumoconioses must have resulted. But of such diseases, secondary to the work with stone, there is no record.

However, Vitruvius, the eminent Roman architect and engineer, known only by his writings, publicly condemned the use of lead for piping water to be used domestically. Pipes of beaten lead were introduced to convey water from the masonry aqueducts to the homes, and several outbreaks of lead poisoning resulted because, as water conditions can change, so can its degree of lead solvency; thus recognition was finally given the source of this illness. In fact, as early as the first century A.D., Pliny the Younger mentioned lead poisoning as a disease present among the slaves who worked the mines.[9]

Before leaving the Romans, mention must be made of Pliny the Elder. The only writings extant are his *Natural History,* in 37 books. Apart from discussing copper, the purification of precious metals, the discovery of glass, he did describe the use of a primitive respirator to obviate the inhalation of dust. Also, he wrote of the use of ox bladders by workers engaged in the production of vermilion, the hazard being mercury. Translations differ, as to whether the user saw *through* or *over* the worn bladder. In an effort many years ago to simulate the use of such a device, bladders were obtained and dried. They were translucent but not transparent, implicit in which finding is the idea that they must have been worn below the eyes.

The only other comment that can be made relevant to the Greco-Roman period is one regarding Alexander the Great. He may be thought of as the first general to devise a type of medical service for the army and to employ surgeons on the field of battle.[10] While today's medical support to the armed services is in great part occupational medicine, those moves of Alexander may well have been the first occupational health activity for soldiers.

The Middle Ages

From the sixth to the sixteenth centuries, there was little documentation relating to the health of workers. Unquestionably, construction continued particularly in the great cathedrals, world exploration and trade advanced, textile production began, and some power-driven factories appeared. There must have been ill effects from contact with a broad variety of pigments, the grinding of metalware, and the silvering of mirrors, among other work processes. But epidemiology was awaiting discovery several centuries later, so there were few, if any, cause-and-effect relationships found in the study of the artisans of the day. Further, living conditions shortened the lives of workers so that tuberculosis or the various plagues took their toll before occupational diseases gained their foothold and resulted in death.

One document of significance to the diseases of workers appeared in the 7th century in ancient Lombardy—the code of King Rothari, promulgated on November 22, 643. At age 38, this ruler codified the laws in 388 chapters, all of which had existed previously only in the memories of unlettered judges.[11]

In this body of law is probably the origin of the basic principles of compensation for injury.

The edict applied to personal injuries received in brawls, fights, and feuds, and there was a sliding scale in that injured slaves and servants received less than freemen. There were payments for disability and death. There was a dual purpose in establishing such a system—to fine the offender, and thus create revenue for the court, for only part of the payment went to the victim; and to denounce the barbaric system of resorting to combat to settle disputes. Compromises were encouraged.

King Canute, who was a Danish king of England from 1016 to 1035, was also King of Denmark and Norway. Not only did he establish names for the digits of the hand, but he also stated the principles of compensation for particular injuries, especially in cases of trauma involving the hand. The importance of the loss of a thumb was recognized even then, for the compensation was twice that for the loss of the second digit, and two and one-half times that given for the loss of the third digit.[12]

In 1524 A.D., a small booklet appeared, written in 1473, by Ulrich Ellenbog, an Austrian physician.[13] The tract was directed toward goldsmiths and other handlers of metal, offering counsel regarding the hazards of lead, antimony, silver and mercury. Ellenbog warned against the burning of coal in confined spaces, and the inhalation of the vapor—more dangerous than the parent substance—arising from the heating of the metals. As he wrote, "Workers should not hold their heads above this vapor, but should wear a mask. The vapors of quicksilver, silver, and lead are a cooling poison, for they make a heaviness and tightness in the chest, burden the limbs and often lame them, as one can often see in the foundries where men work with large quantities of these metals." In essence, this eight-page publication was the first writing devoted exclusively to industrial metal poisoning.

The 16th Century

With the beginning of the Renaissance, work began to appear indicating a sensitivity to diseases resulting from contact with toxic work substances and/or environments. The turn of the 15th century saw the appearance of a highly unusual and forceful personage in medical history. Philippus Aureolus, *Theophrastus Bombastus von Hohenheim,* later calling himself Paracelsus (greater than Celsus), born in 1493, grew up in Switzerland, and later studied medicine in Italy. He learned chemistry by laboring in the mines and smelters, and because of a strong spirit of rebellion, he sought out other teachers by traveling throughout Europe. While so doing, he practiced medicine as an itinerant, and visited many mines and workshops.

Although his theories of disease were unacceptable to the faculties, he wrote furiously, but eventually quarreled with everyone. However, he produced a treatise, *On the Miners' Sickness and Other Miners' Diseases,* which was translated by Dr. George Rosen for a publication issued in commemoration of the quadricentennial of his death.[14] The treatise was first published in 1567, and dealt with the pulmonary diseases of miners, the disorders of smelter workers and metallurgists, and diseases caused by mercury.

The work must be interpreted in the light of contemporary knowledge, of course. However, Paracelsus was able to distinguish between acute and chronic poisoning. The significance of the work lies in its being the first monograph dealing with the diseases of a specific occupational group. Outstanding are his direct observations of work in the mines and smelters. Paracelsus made no reference to protective devices, and disregarded dust as an etiologic factor in the miners' diseases. His references to vapors, however, may have been allusions to dust.[14]

About the same time, a Saxon physician, Georgius Agricola (1494-1555), settled among the miners where his interest and observations led to his writing a work of prominence, which was to endure for centuries. In 1561, his *De Re Metallica* was published. Given emphasis was the need for ventilation of mines, and numerous illustrations depicted various devices that would force air below ground.

But Agricola wrote mostly of the afflictions of miners, pointing out that some disorders affect the joints, others the lungs, some the eyes, and, finally, some are fatal. He underscored the hazards of the dust produced in dry mines,"...for the dust which is stirred and beaten up by digging penetrates into the windpipe and lungs, and produces difficulty in breathing...." He relates that, in the Carpathian mines, women were found who had married seven husbands, "all of whom this terrible consumption has carried off to a premature death."[15] More recent inquiry has led to the belief that these deaths resulted from lung cancer, sustained by working with rich radium-containing ores. Agricola was aware of the need for personal protective devices, and shown in several illustrations are gloves, leggings, and masks.

The 18th Century

Many observers in the 17th century commented on the diseases of certain workers, but it remained for the opening year of the next century to bring about the publication of the first comprehensive treatise on the diseases of workers in a multitude of trades. Bernardino Ramazzini (1633-1714), identified with both the Universities of Modena and Padua, visited innumerable shops to study the working conditions and the clinical afflictions of artisans in a variety of trades.

His advice in the classic *Discourse on the Diseases of Workers* still applies to physicians today. Ramazzini pointed out that, in addition to the questions asked of a patient, as suggested by Hippocrates, one more should be added, simply, "What is your occupation?" He wrote, "In medical practice, however, I find that attention is hardly ever paid to this matter, or if the doctor in attendance knows it without asking, he gives little heed to it, though for effective treatment evidence of this sort has the utmost weight." Ramazzini viewed medicine in a social light. In his discussion of the disease of brickmakers, he derides fellow physicians for treating all patients with the same remedies of purging and blood-letting, irrespective of the occupational origin of the illness. He writes, ". . . for the doctors know nothing of the mode of life of these workers who are exhausted and prostrated by unceasing toil."[16]

Ramazzini has been dubbed the father of occupational medicine, and rightfully so, for his work was translated into at least six languages. Ramazzini summarized well the two causes, he believed, of disease occupationally incurred by the workers of his day: "The first and most potent is the harmful character of the materials that they handle, for these emit noxious vapors and very fine particles inimical to human beings and induce particular diseases; the second cause I ascribe to certain violent and irregular motions and unnatural postures of the body, by reason of which the natural structure of the vital machine is so impaired that serious diseases gradually develop therefrom."[16]

Physicians employed by industry were rarities until the 1900s. However, a Mr. Ambrose Crowley had established in London the sales outlet of his iron works. Irons, screws, chains, files, hammers, hoes, and locks were the types of products offered. The business was established circa 1682, and available to employees were medical care, sick benefits, and provision for dependents of workers who were disabled or who had died.

The work was carried out under a set of extremely stringent laws that governed not only the site of manufacturing but also what was the equivalent of a company town. Extravagance was discouraged and punishment was severe for morning drinking, even with customers, and "for whoreing, . . . high eating, gameing, and sotting and trading to the hurt of the business."[17] But more particularly, the last of 113 "Orders and Instructions for Carrying on the Works at Winlaton—1692-1815" called for the appointment of a "Doctor and Chirurgeon."

John Crowley, in consideration of the deplorable state of his "honest and laborious workmen and their families," resolved to appoint "one well skilled in Physick . . . one that's of a sober life and conversation and not addicted so much to pleasures as to be withdrawn from a due attendance on his business." His patients, in keeping with accepted company practice, were encouraged to inform against him if he were suspected of "making use of my [the owner's] Medicines or Druggs in his own private practice." He was to maintain a "Journal of Cases," and attend daily in "my Factory." Presumably, the physician was engaged both at the plant and in town, and was ordered appointed in 1724.

The appearance of mass manufacturing

The economic changes of the 1700s were of great significance. Europe had been primarily engaged in agriculture, and most workers were serfs on manorial estates or, rarely, were free men working small holdings. In England, many of those peasants, freed from serfdom, had been driven into the towns because landlords preferred to enclose their estates and raise sheep for the production of wool. English towns were filled with unemployed vagrants, who had little experience in farming and could find no employment.

In England, there was also a growing demand for goods for export. The production methods were labor-intensive, with cotton and wool still being carded, spun, and woven by hand in the scattered cottages of the rural North and West. Further, transport was slow. Then, James Hargreaves developed the spinning jenny in 1764, which soon was to be improved by Richard Arkwright in 1769, and Samuel Crompton in 1779. From 1784 on, Rev. Edmund Cartwright brought to perfection the power loom, for which Parliament voted him a £10,000 gift, and in 1786 he was knighted for his service to national industry.[18] To be placed in chronological sequence, is the invention in America of the cotton gin by Eli Whitney in 1792.

Other modes of production were also required to make the developing machinery available. Needed also were numbers of laborers who would work regular hours, under the control of their employers, and in structures where the machinery and power had been provided. This change was the beginning of what initially was termed the "factory system," but, later, on a larger and more complex scale, to be designated by A. Toynbee[19] as the Industrial Revolution. As described by his nephew, Arnold J. Toynbee,[20] he was "the first economic historian to think of, and to set out to describe, the Industrial Revolution as a single great historical event, in which all the details come together to make an intelligent and significant picture. In doing this, he created the frame within which all subsequent work on the Industrial Revolution has been carried out."

The new system involved first the manufacture of cotton goods, but the arrangement was extended soon to the production of woolens, and eventually into the fabrication of metal, wooden, and leather items, and, ultimately, into all forms of production. This change in production methods with the concomitant need for masses of workers affected the history of occupational health and safety because it brought with it hazards never before encountered. A new stage had just been set.

The Industrial Revolution

But what, exactly, are the characteristics of the system now affecting the lives of millions of workers, and currently being developed in the Third World? The Industrial Revolution implies "those changes in the processes and organization of production that mark the passage from an agrarian, handicraft economy to one dominated by industry and machine manufacture, from what is sometimes called a premodern or traditional economy to a modern one." The Industrial Revolution—when capitalized—refers specifically, to the "first historical instance of this transformation, which began in Britain in the 18th Century, spread from there to the Continent and to offshoots of Europe overseas (the United States in particular) and transformed in the span of a century the life of western man, the nature of his society, and his relationship to the other peoples of the world."

Specifically, the innovations encountered in the processes and organization of production included:

- The substitution of inanimate for animal sources of power (particularly steam power through the combustion of coal).
- The substitution of machines for human skills and strength.
- The invention of new methods for transforming raw materials (particularly in the making of iron, steel, and industrial chemicals).
- The organization of work in large units, such as factories or forges or mills, making possible the direct supervision of the process and an efficient division of labor.

Paralleling these production changes were the altered technologies employed in agriculture and transportation. These same changes in infrastructure are now taking place in the developing countries, where the identical hazards to the workers' health and safety are being experienced and where the same tensions, pressures, and stresses are being encountered.

GROWTH IN THE UNITED STATES

In the 16th century, there were a number of chartered British companies that provided an extension of overseas trade. The objective of such companies was a monopoly, or collective trading, or regulation of trade, with the Crown holding the right to control the economy.[21]

Hoping to apply the principle of merchant trading organizations to the New World, England was motivated by the desire, through colonization, to break hated Spain's monopoly of newly discovered land.

Sir Walter Raleigh, granted a perpetual patent to discover, occupy, fortify, and enjoy in fee simple all lands "not actually possessed of any Christian prince or people," sent an expedition to America, which arrived at the island of Roanoke in 1584, and which rechristened the Indian kingdom, Virginia. Because of the initial success with the Indians and the glowing story of the rich lands, a fleet with 108 prospective settlers was sent out the following year, only to perish in hostile hands. In the years to come, efforts were made to establish two colonies, new charters were issued, new fiscal support was obtained, but many lives were lost to frontier hazards, and extreme conflicts of will among the designated administrators made governance almost impossible. The Council of Virginia, designated by the Virginia Company, asked in 1610 that "none but honest sufficient Artificiers, as Carpenters, Smiths, Coopers, Fishermen, Brickmen,. . . shall be entertained into this voyage."[21]

Nonetheless, by 1625, the Virginia experience had encouraged additional attempts at colonization of the New World. The Plymouth Company underwent great difficulties in its new colony, as nearly one-half of the 102 adventurers arriving in the *Mayflower* died during the first winter, 1620-1621. But, by 1630, new powers of legislation were granted the colonists, along with the monopoly of the Indian trade in that area. Religious disagreements arose, however, so that of the original colonists, the businessmen who believed colonization to be a profitable undertaking dropped out, leaving the Massachusetts Bay Colony clearly doomed. On the foundations of this colony, though, an important part of the United States was to be constructed.[21]

Although efforts had been made to establish an iron works in Virginia, using iron ore from North Carolina, the first successful works were developed in 1645, in that part of Lynn, Massachusetts, called Saugus, with production lasting until about 1675. There was a demand for iron in both England and the Massachusetts Bay Colony, and ore, timber, and streams were abundant. The Hammersmith works, funded by English capital, was the first large-scale corporate industrial enterprise in America. At peak, a ton of iron was produced per day, but costs could not be lowered, even though a larger work force was brought over, actually indentured servants, Scots who were prisoners of Cromwell after his victory at Dunbar. Eventually, the skilled workers scattered to other parts of New England. Although the enterprise failed, it can be considered as the foundation of the American iron and steel industry.

The New World soon was engaged in typefounding, printing, shipbuilding, quarrying, cabinetmaking, bookbinding, the production of paper, chocolate and cottonseed oil, clockmaking, and even the building of spinets.[22] But it was the growth of the textile industry that saw the true beginning of the factory system in America. Samuel Slater, a former employee of Arkwright in England, was able to recall allegedly the complete design of cottonmaking machinery, and established a mill in Nantucket, Rhode Island. In the next two decades, hundreds of spinning mills shot up in New England.

Even though the Industrial Revolution was continuing its unbounded growth, the toll on the worker began to show early. Lewis Mumford[23] described the health effects in this way:

> Each specialized trade, precisely through its specialization, now acquired its typical 'occupational disabilities'—its lopsided posture, its overdeveloped muscles, its bleached complexion, its myopic eyes, its enlarged heart, or its silicate-clogged [*sic*] lungs, with their associated diseases and permanent anatomical deformities. Too often these ailments were chronic and persistent: a higher death rate betrayed the lower life rate.

At Lowell, Massachusetts, where there was sufficient water power to support a large industrial center, many young women were brought from adjacent farms to work in the mills. Although the owners depicted the advantages and benefits of such employment, and even Charles Dickens wrote glowingly of the working conditions, there appeared in the *Boston Quarterly Review* for July 1840, these comments:

> We pass through our manufacturing villages, most of them appear neat and flourishing. The operatives are well dressed, and we are told, well paid. They are said to be contented, healthy and happy. This is the fair side of the picture; the side exhibited to distinguished visitors. There is a dark side, moral as well as physical. Of the common operatives, few, if any, by their wages, acquire a competence. A few are well paid, and now and then an agent or an overseer rides in his coach. But the great mass wear out their health, spirits and morals, without becoming one whit better off than when they commenced labor. The bills of mortality in these factory villages are not striking, we admit, for the poor girls, when they can toil no longer, go home to die. The average life, working life. . . we have been assured, is only about three years. What becomes of them? Few of them even marry; fewer still ever return to their native places with reputations unimpaired.[24]

Dr. Elisha Bartlett (1804-1855) thought over the statements contained in the newspaper account, questioned the conclusions of the bills of mortality, and pointed out the fact that a hospital had been open for about a year in Lowell, and "with the best medical attendance and nursing," the charge to the patient was but $3 a week. He also indicated that all the girls were vaccinated at the expense of the corporations.[24] In response to the comment regarding morals, Bartlett commented that more than 500 of the girls became "personally interested in practical piety," and more than 600 joined a church. It is interesting that such a lengthy and privately printed response was needed.

A slightly earlier connection between medicine and industry was seen in the relationship established between Dr. Pierre Didier and the Du Pont Company. A native of France, Didier settled in Wilmington, Delaware, in 1795, and through a personal, and professional relationship with the first Eleuthere Irenée du Pont, the company developed an early form of in-plant occupational medicine. A retainer system was developed, families paying $50.10 per year, with additional fees for certain treatments or medications. The company paid a $160 retainer, and the workers' bills were paid by Du Pont, and their accounts were debited. Thus, Didier received immediate payment, the workers could receive costly medical care, and they could repay the company gradually. The physician was called for all emergencies; periodic explosions in the power mills were a "recurrent fact of life" in the community. Charges for treatment of injuries so sustained were borne by the company. Even though the company's responsibility was implicit rather than explicit, it had moved beyond the assurance of availability of medical care to the assumption of moral and financial responsibility and liability for both the working conditions and health of its employees.

OCCUPATIONAL DISEASE—EARLY OBSERVATIONS

Clinical observers of the 18th and 19th centuries, both in England and in America, continued to describe certain morbid states that they considered related to various occupations. Some of the reasons for the increased interest of physicians in the problems of social reform created by the Industrial Revolution follow.

English reformers

The growth of the industrial town pointed to the rising power and wealth of the industrial and the middle class. As towns grew, special literary and philosophical societies supported discussion of problems and opinions regarding resolution. Physicians were prominent in these groups. They were aware of the effects of the ceaseless factory operations on working-class families, and of the high level of employment of women and children. These workers were crowded into unsanitary slums. As young medical practitioners frequently served as "town surgeons" or "club doctors"—workingmen's sick clubs—while their practices were being established, they became aware of the living and working conditions in which they found their patients.

Early on there was an expressed need for the application of statistical methods to social problems. From such demands came a writing like Edwin Chadwick's *Report on the Sanitary Condition of the Laboring Population of 1842,* to which many physicians contributed local data. A description of the period paints the scene to which Chadwick applied numbers.[25]

> It was a time when the conditions of the workmen mattered nothing compared with the mad rush for production. New factories were springing up everywhere. To serve the new machines men, women and children were toiling day and night under inhuman conditions in these factories and in the mills. Accommodations had to be provided for the new workers. New houses, new streets, new towns sprang into existence to house the thousands of workmen pouring up from the South of England to seek employment in the North. Houses were erected with such speed and lack of foresight that they were for the most part scarcely fit to be lived in. Faulty construction and bad drainage were characteristic, and no thought of hygiene or sanitation entered the heads of the builders.

Chadwick's goal was to clean up his country's slums, espousing many social reforms, particularly in public health. He saw the relationship between poverty and chronic disease, and strove to convert the poorhouses into schools and hospitals. He was described as "a stubborn cantankerous man who served Great Britain for more than half a century as a public scold." His efforts led to the passage of the Ten Hours Act and the establishment of the half-time system of education. Working children were required to attend school three hours daily. He fought drunkenness, insisted that railroaders be paid weekly rather than at longer intervals, and although he held factory managers responsible for industrial injuries, a law was not forthcoming until 1898.

In keeping with the spirit of reform in England was Dr. Charles Turner Thackrah, considered Britain's pioneer in occupational medicine. His 1831 volume, entitled *The Effects of the Principal Arts, Trades, and Professions and of Civic States and Habits of Living on Health and Longevity: with a particular reference to The Trades and Manufactures of Leeds: and suggestions for the removal of many of the agents which produce disease, and shorten the duration of life* belies the 124-page length of the work, but not its importance.

Thackrah stressed prevention, "Surely humanity forbids that the health of workmen, and that of the poor at large, should be sacrificed to the saving of half-pence in the price of pots." Meiklejohn,[26] when he published Thackrah's second edition, gathered aphorisms that underscore the beliefs held by Britain's occupational health pioneer:

> Evils are suffered to exist, even where the means of correction are known and easily applied.
>
> A study of medicine. . .which disregards the prevention of diseases, limits its utility and its honors.
>
> The work people are less thought of than the machinery: the latter is frequently examined to ascertain its capabilities—the former is scarcely ever.
>
> Masters, however enlightened and humane, are seldom aware, never fully aware, of the injury to health and life which ills occasion.

The early death of Thackrah was lamented because it retarded the progress of industrial medicine and surgery and amelioration of conditions of employment for half a century.

Lead poisoning

By the end of the American Revolution, medicine was still primitive, with purging, bleeding, emetics, and blisters as standard therapy, and smallpox, dysentery, jaundice, cholera, and typhoid caused more deaths than British bullets.

Even earlier, Thomas Cadwalader of Philadelphia, produced in 1745, over the imprint of Benjamin Franklin, *An Essay on the West-India Dry Gripes,* in which he described the symptoms of lead poisoning,[27] although he did not identify the causes with any accuracy. The essay probably attained recognition more because of the printer's fame than because of any innovative thought of the writer. Franklin himself was concerned with the effects of lead on the human body and responded to a letter from Dr. Cadwalader Evans, a Philadelphia physician, agreeing that the lead used in the coils of stills employed in making rum was the cause of the "dry bellyache." Further, he stated that he had long been of the opinion that the disorder had a metallic cause, "observing that it affects among Tradesmen those that use Lead, however different their Trades, as Glazers, Type-Founders, Plumbers, Potters, white Leadmakers and Painters. . . ."

Although several early 18th-century observers such as Hilary, Musgrave, and Huxham described the abdominal pain of lead colic, it remained for Sir George Baker, physician to Queen Charlotte and the king, to identify the similarity between the endemic Devonshire colic and lead colic. In his classical document, *An Essay Concerning the Cause of the Endemial Colic of Devonshire,* which he read before the College of Physicians in London in 1767, he wrote:[28]

> It is likewise common, in several parts of the country, either to line the cider-presses entirely with lead, in order to prevent their leaking; or to make a border of lead quite round the press, in order to receive the juice of the apples, and to convey it into a vessel, made of wood or stone, placed underneath. And in many other places, where these methods are not used, it is common to nail sheet-lead over any cracks or joints in the presses; and likewise to convey the juice of the apples from the presses in lead pipes.

He demonstrated that the cider of Herefordshire produced no colic, and that the cider from that county was lead-free, having been made in presses constructed without use of the injurious metal.

Other illnesses

Poisoning from mercury had been mentioned by Ellenbog, Paracelsus, and Fallopius, and Mattioli (mid-16th century) observed chronic mercurialism among the miners at Idria in Yugoslavia, where mercury is still being mined today.[14] Some historians[29] have speculated that the acute symptoms of byssinosis might have been prevalent among colonial Americans who processed flax in their barns, placing children and other farm workers also at risk. The researchers were able to conduct pulmonary function evaluations on employees of Philipsburg Manor, a restoration of a 17th-century barn, where processing of flax was conducted. The subjects experienced some chest tightness and demonstrated airway constriction.

Scurvy, the disease that decimated both naval and merchant crews, was finally counteracted by 1796, when lemon juice was added to the rations of the British Royal Navy.

While other writers were bringing to light the cause-and-effect relationship of the work environment to the development of disease, certain efforts were being made to improve air quality in workplaces and to prevent occupational illness and death. One innovation was the invention by Sir Humphry Davy of the safety lamp for coal miners, based on the principle that a naked flame could be exposed in an explosive mixture of gas and air, provided it was surrounded by wire gauze of a particular mesh. The new wire-gauze lamp was adopted almost universally, and saved the lives of thousands of miners throughout the world. Davy published his first edition, *On the Safety Lamp for Coal Miners: with Some Researchers on Flame,* in 1818.

DUST DISEASES OF THE LUNG

Diseases of miners and others coming in contact with respirable particulates have been known clinically for centuries. The mechanisms of how they got there, however, were yet to be clarified. Initially, from the 10th century on, it was believed that the small demons or gnomes presumably inhabiting the mines were responsible for a generic disease entity called *Bergsucht* (miners phthisis), which included pathogenic change—from a neoplasm to metal disposition—in the lung.[15] Agricola, who was mentioned earlier, indicated that, in spite of the currency of the belief in goblin-like creatures, dust was harmful, and that, "If the dust has corrosive qualities, it eats away the lungs, and implants consumption in the body; hence in the mines of the Carpathian Mountains women are found who have married seven husbands, all of whom this terrible consumption has carried off to a premature death, as noted earlier."[15]

When steam power became possible during the industrialization of England, coal was needed in large amounts. With increased production, the problem of "the black spit" appeared early in the 1800s. Working conditions were dangerous, hours were long, and people were kept in debt to the company store. Even women and children worked underground, pulling loaded carts or carrying coal in creels on their backs, 100 to 300 pounds of coal. It was in keeping with the ethos of Victorian England that, "as the evils of this child and female labor racket were gradually exposed, the subsequent popular outcry arose not so much because of revulsion from the abhorrent material conditions, but because the system prevented the children's receiving religious instruction, and because the fact that nearly naked men and women worked together underground was thought to be conducive to immorality."

For many years, the important question was whether the coal miners' "black spit" and pigmentation of the lungs came from coal dust that was inhaled or were produced by the pulmonary tissue itself. Pathologists were concerned solely with tissue change, while practitioners who lived in the same communities with their patients saw the problem as one with social, community, economic, and engineering aspects. It was Zenker, in 1866, who first used the term which, with slight modification, is used today. He proposed[30] that, instead of the lengthy designation used in the title of his article, *Staubinhalationskrankheiten,* one should employ *Pneumonokoniosis,* from the Greek *dust*, and that variations with modifiers could be used, such as anthracosis, siderosis, and the like.

Ventilation in the mines was improved, work hours were shortened, and by 1875, as indicated by Meiklejohn,[31] anthracosis, the black spit, "ceased to exist as a medical problem." Attention subsequently was directed to silicosis, and its complication, tuberculosis. By the 1930s, lung disease among coal miners again appeared in Great Britain, with many claims for compensation. Radiologists reported reticulation or a network, and, by means of a microscope, pathologists saw a dust-laden tissue network, unaccompanied by tuberculosis. Gough, through his studies of large tissue sections, described the early lesion of the coal nodule—collections of coal dust—around which was a peculiar type of focal emphysema, clearly different from silicosis. A second disorder, progressive massive fibrosis (PMF) with localized opacities of larger size, was soon to be described. A worker with simple pneumoconiosis, when removed from dust contact, would not likely develop PMF. If enough simple pneumoconiosis existed, PMF might develop, irrespective of the cessation of dust exposure. Last, it was concluded that once PMF started, it would always be progressive. The term coal workers' pneumoconiosis (CWP) was first used by British investigators in 1942, and the following year the disease became compensable, distinct from silicosis. In 1945, the British Medical Research Council established a Pneumoconiosis Research Unit near Cardiff, Wales.

Strangely, CWP was not accepted as a separate occupational disease in the United States until about 1969. It took the late-1968 coal-mine explosion in Farmington, West Virginia, with the loss of 78 lives, to direct attention to both safety in the mines and the health of miners. A strike of nearly 40,000 miners in that state focused on the issue of compensability of CWP. After a walkout of three weeks, a bill was passed by the West Virginia legislature in keeping with the demands. Later, the U.S. Congress passed the Coal Mine Health and Safety Act of 1969 (CMHSAct). The term "black lung" was given legal standing in the enactment, and the bill probably aided in the passage a year later of the Occupational Safety and Health Act (OSHAct). In this instance, the recognition that black lung was rampant among U.S. coal miners did not evolve of its own accord within the boundaries of medical science. It was forced on the medical community by the intervention of miners themselves. In 1977 and 1978, the CMHSA 1969 was amended to broaden the coverage and to make the benefit program permanent. Periodic radiographic studies of the miners, as called for by law, have followed.

Silicosis, as a clinical entity, underwent historic development. As mentioned previously, the health effects of dust inhalation had been recognized over the centuries, and various terms were used to describe the pulmonary disorders, such as miner's asthma, miner's phthisis, or even spurious melanosis. In 1870, Visconti[32] coined the label "silicosis." Dr. Anthony J. Lanza, later to serve as chief of occupational medicine in the Army's Office of the Surgeon General in World War II, and his co-workers in the U.S. Public Health Service (USPHS) and the Federal Bureau of Mines, initiated studies of workers engaged in dusty occupations as early as 1915. In the tri-state area of Missouri, Arkansas, and Oklahoma, workers in the metal mines had been examined and that industry was to persist with high levels of silicosis, silico-tuberculosis, and tuberculosis for many years to come, almost defying eradication as late as 1980. Investigative work was later to be carried on in the granite-cutting industry and basic research was initiated in the years ahead at the Saranac Laboratory. By far, the greatest research on pneumoconiosis was conducted in South Africa and Australia, resulting in governmental controls of the dust hazard.

Silicosis—Gauley Bridge

In the depth of the Great Depression in the United States, a project was originated that would bring hydroelectric power to an area in West Virginia. The digging of the tunnel through which water was to be brought offered job opportunities to a workforce that had been unemployed, and hundreds of men came with their families from all parts of America. The tunnel was cut, however, through pure silica, and for the first time, notice was given to "acute silicosis," something unheard of until that period. In addition to the innumerable deaths that resulted from the work exposure, the shack housing, cruelty of the foremen, hurried burials of the victims, plus the lack of a system of compensation or benefits had been unmatched in this country.

The tragedy was treated fictionally in telling detail by Hurbert Skidmore in his *Hawk's Nest*. Hearings were held in 1936 by a Congressional committee, and from this episode came the creation of the Air Hygiene Foundation, now the Industrial Health Foundation, under the auspices of the Mellon Institute.

A second result of the Gauley Bridge disaster was a convention in early 1936 of representatives of labor, industry, the public, as well as insurance underwriters, physicians, engineers, and attorneys called by Frances Perkins, then Secretary of Labor, in a meeting designated as the National Silicosis Conference. The conclusion reached was that "silica dust is the important causative factor in silicosis and that other dusts may modify or diminish the toxic effect of the silica."

PROFESSIONAL AWARENESS OF OCCUPATIONAL HEALTH HAZARDS

As industry developed in the United States, there was a slow growth in professional awareness of work-generated disease. The first document to be published in this segment of medicine was a $50-prize-winning dissertation by Dr. Benjamin W. McCready, for 25 years visiting physician at Bellevue Hospital in New York. One of his recommendations involved seamstresses. He believed that they should spend less time in church for, after being confined all week at their sewing tables, they would benefit more were they to remain outdoors on Sunday.

The first American writing on decompression illness appeared in 1843, and in 1860, Freeman of Orange, New Jersey, published a brief report on mercurialism among hat finishers. In the later decades of that century, essays were published regarding "occupational neurosis" from repetitive hand movements, anthrax, arsenic poisoning, plumbism, oyster shuckers' "corneitis," the health of sewing-machine operators, carbon bisulfide inhalation, the health of cigar-makers, and many reports of phosphorus necrosis of the lower jaw.

An unsung leader in the entire move toward control of occupational health hazards was John B. Andrews, Ph.D., long-time secretary of the American Association for Labor Legislation (AALL), who guided that organization from 1906 until its demise in 1942. Through the AALL's *Review,* considerable attention was directed to occupational disease, and it was Andrews' efforts that produced the First National Conference on Industrial Diseases, held in Chicago in 1910. The speakers included not only Andrews, but Dr. Alice Hamilton, Frederick L. Hoffman, statistician and author of many studies on occupationally incurred illness, and Professor Charles R. Henderson, a sociologist at the University of Chicago, at whose instigation the Illinois Governor appointed an Occupational Disease Commission.

From the Conference came the *Memorial on Occupational Diseases* for presentation to the president of the United States, a document that was to be a "carefully prepared memorial of facts and conclusions, emphasizing the urgent necessity and practical expediency of a national expert inquiry into the whole subject of industrial or occupational diseases; their relative degree of frequency in various trades and occupations, the causes responsible for their occurrence, the methods desirable and practicable for their prevention or diminution, and all other matters having a relation thereto including methods of amelioration and relief." The document went to President Howard Taft.

While the history of subsequent action is not known, another effort did produce results through the Executive Office. Andrews was struck by the horrors of phosphorus poisoning in the match industry, manifested by the destructive lesions of the lower jaw and with frequent extension of the necrosis into the upper jaw. He had initiated a study of match workers and, under the auspices of the Federal Bureau of Labor, reviewed 15 plants with 3600 employees. More than 100 affected employees were found, and the classic report was submitted in 1910.

After much testimony in 1911, and after an appeal to Taft, the Esch Act was passed by the Congress, to become effective January 1, 1915, placing a tax of two cents on every 100 white phosphorus matches produced. Obviously, the tax burden precluded profitable production, and phosphorus matches disappeared.[33] Andrews did not receive then, or subsequently, the recognition deserved for removing one of the most hideously disfiguring occupational diseases from the American scene. The AALL went on to fight for workers' compensation and other protective forms of insurance.

The hunt for occupational disease data continued. Alice Hamilton reported on lead poisoning in the sanitary fixture industry; the Bureau of Labor presented its findings on hookworm disease ("cotton-mill anemia") among cotton-mill operatives; and by 1911, California led the list of six states (the others being Connecticut, Illinois, Michigan, New York, and Wisconsin) mandating the reporting by physicians of all cases of certain occupational diseases—poisoning by lead, phosphorus, arsenic or mercury, or their compounds; anthrax; and compressed air illness.

Finally, the first efforts to classify occupational diseases were undertaken by Thompson.[34] By providing the name of the substance, the branches of industry in which the poisoning occurs, the mode of entrance into the body, and symptoms of poisoning, Thompson presented what might be the first textbook in the United States on occupational toxicology.

Before moving to a consideration of regulatory legislation in Great Britain, a summary of the status of health and safety in the United States is illuminating. Hoffman, the statistician from the Prudential Insurance Company cited previously, wrote in 1912:

> The problem [of not conforming to legal and moral requirements] is one of ignorance rather than neglect. Most of the factors which condition health and safety in industry are as yet very imperfectly understood, at least in the United States. It is true that within the last quarter of a century immense progress has been made in Germany, but largely because of the requirements of compulsory accident insurance institutions established to replace an antiquated system of employers' liability law. We have not as yet developed in this country the function of the safety en-

gineer, who has attained to such high professional standing throughout the German Empire because of the practical utility of his services. Manufacturers or employers of labor generally are not blamable for their reluctance to install expensive safety devices, or ventilating devices, or other methods or means by which health, safety, and comfort can be improved, unless the evidence is conclusive that the investment will be productive of the desired results. The function of the ventilating engineer in his relation to industrial requirements is practically new, and almost the same may be said of the function of the illuminating engineer.

Trade union activity

Until about the middle of the 19th century in the United States, any effort on the part of labor to improve working conditions was considered illegal. One worker could bargain legally for higher wages and shorter hours, but a joint effort to "regulate the value of labor" was considered a criminal conspiracy. In 1825, carpenters in Boston struck for higher wages, for it was "impossible for a journeyman housewright and house carpenter to maintain a family at the present time with the wages which are now usually given to the journeymen house carpenters in this city." The resolutions were supported by neither the Master Carpenters nor, as would be expected, by the builders.

In response to such constraints and on seeing the rich rewards of their labor going to a few employers, journeymen of several trades in Philadelphia, in 1827, resolved to form a Union of Trade Associations, in the hopes of "establishing a just balance of power, both mental, moral, political, and scientific, between all the various classes and individuals that constitute society at large." The precedent of handling labor union cases of this type persisted until 1842, when the Massachusetts Supreme Court overturned the doctrine of criminal conspiracy.

Miners in the United States attempted their first organization as the American Miners' Association in January 1861 in an effort to secure better mining laws in the coal-producing states. The action was given added support by the Avondale disaster in the Pennsylvania anthracite fields where, in September of the same year, 109 mine employees were killed. The movement lapsed during the Civil War, and post-conflict strikes brought public indignation. Separate movements appeared, but by 1890, the two largest unions affiliated and, in joint convention, formed the United Mine Workers of America (UMWA) with a membership of about 20,000, out of a potential of 192,000. One of the objectives of the UMWA was to introduce safety appliances by legislation.

In subsequent years, there were strikes—in 1902, the anthracite workers were out for five months; and in 1919, the bituminous coal operators were struck, and while a wage increase was awarded, the six-hour and five-day work week were denied; and in 1946, UMWA bituminous coal miners won a health and welfare fund from the federal government which, in the interim, had seized the mines.

The UMWA is discussed because of the unique program in health care developed through this agreement. An industrywide trust fund, financed by royalty on production, jointly administered by the operators and the union, was based on the principle, expounded by organizer John L. Lewis, that the cost of maintaining the manpower in a workable and healthful condition should be considered as much a part of production costs as any other item entering into the expense of operation.

The UMWA Welfare and Retirement Fund was a child of private enterprise within the perimeters of private industry. A spin-off was "a comprehensive survey and study of the hospital and medical facilities, medical treatment, sanitary and housing conditions in the coal mining areas." The study, under Rear Admiral Joel J. Boone, submitted in 1947, depicted in striking words and pictures, "the enormity of the social blight coincidental to the exploitation of native wealth without obligation or compensation to the resources of the ravaged communities."

An appraisal of medical care resources and the creation of 10 memorial hospitals in rural settings followed. Union members received medical services initially at no cost; later, payment was required. With time, income diminished and, one day, the hospitals were padlocked and sold.

While many trade unions were not greatly concerned with occupational disease or its prevention until the passage of the OSHAct, great concern has developed, particularly in those organizations whose members worked with such hazardous materials as asbestos. Currently, many international unions have specialists in occupational health and safety for purposes of counseling and education.

Other early 20th-century movements

Apart from the trade unionists, others also sought to be heard. In the days before workers' compensation coverage, a study was conducted on work-caused injuries, as part of a larger Pittsburgh Survey, that covered investigations of steel workers, the households at Homestead—a steel community—women and the trades. Crystal Eastman, a member and secretary of the New York State Employer's Liability Commission, reviewed the injuries incurred within three months' time, and the industrial fatalities taking place within a year in Allegheny County, Pennsylvania. The county was essentially the Steel District, with 70,000 workers in the steel mills, 20,000 in the mines, and 50,000 on the railroads. The frontispiece for the monographic report[35] was a "death calendar" for one year, July 1906 through June 1907, with crosses indicating fatalities in industry (see Figure 3-1). The deaths ranged from 35 per month to a peak of 60, for a total of 526 over the 12 months, or nearly one and one-half fatal injuries per day, during a seven-day work week. Many photographs taken by Lewis Hine are included in the book, one of the most striking being that of a former worker, now an above-elbow amputee, with his family, over the caption, "One Arm and Four Children."

Eastman concluded her lengthy study with this condemnation of the law then in force:

> In regard to the existing law of employers' liability and its consequences, our findings, founded on the accident experience of one great industrial district, lead inexorably to certain conclusions. The law in many of its principles is unjust; in operation it uses up time, money, and good will, to little purpose; it furnishes small incentive for the prevention of work-accidents, and leaves well-nigh the whole economic burden of work-accidents to be borne by the injured workman and his dependents, with consequent hardship and privation. It is to be condemned from the standpoints of justice, method, and practical utility.

It was, finally, the period of the muckrakers—a term coined by President Theodore Roosevelt to describe the journalists who shocked the public into awareness by their descriptions of corruption in all of society's institutions—politics, business, the church. These writers, in the period between 1902 and 1912, opened the eyes of the public to these shocking practices. The group included Upton Sinclair, Lincoln Steffens, Mark Sullivan, Ida M. Tarbell, and Edwin Markham. The magazines carrying their articles were *Collier's, McClure's, Cosmopolitan,* and *Everybody's.* It was this kind of expositive journalism that changed the country from its "state of lethargy into the Progressive Movement."

William Hard explored occupational disease and wrote of the bends, lead poisoning, chrome holes, and chlorine; and later he attacked the lack of workers' compensation insurance in an article entitled, "The Law of the Killed and Wounded." The editor of the 1908 monthly *Everybody's,* in which the second piece appeared, introduced the writing with this note:

> On broadly human lines, this article should interest every citizen. On economic lines it should interest every employer of labor. It is not intended to arraign a class, nor to array one set of our people against another. It is rather to demonstrate where a great, heartbreaking waste can be minimized to the physical and financial benefit of all. Congress and the President have failed,

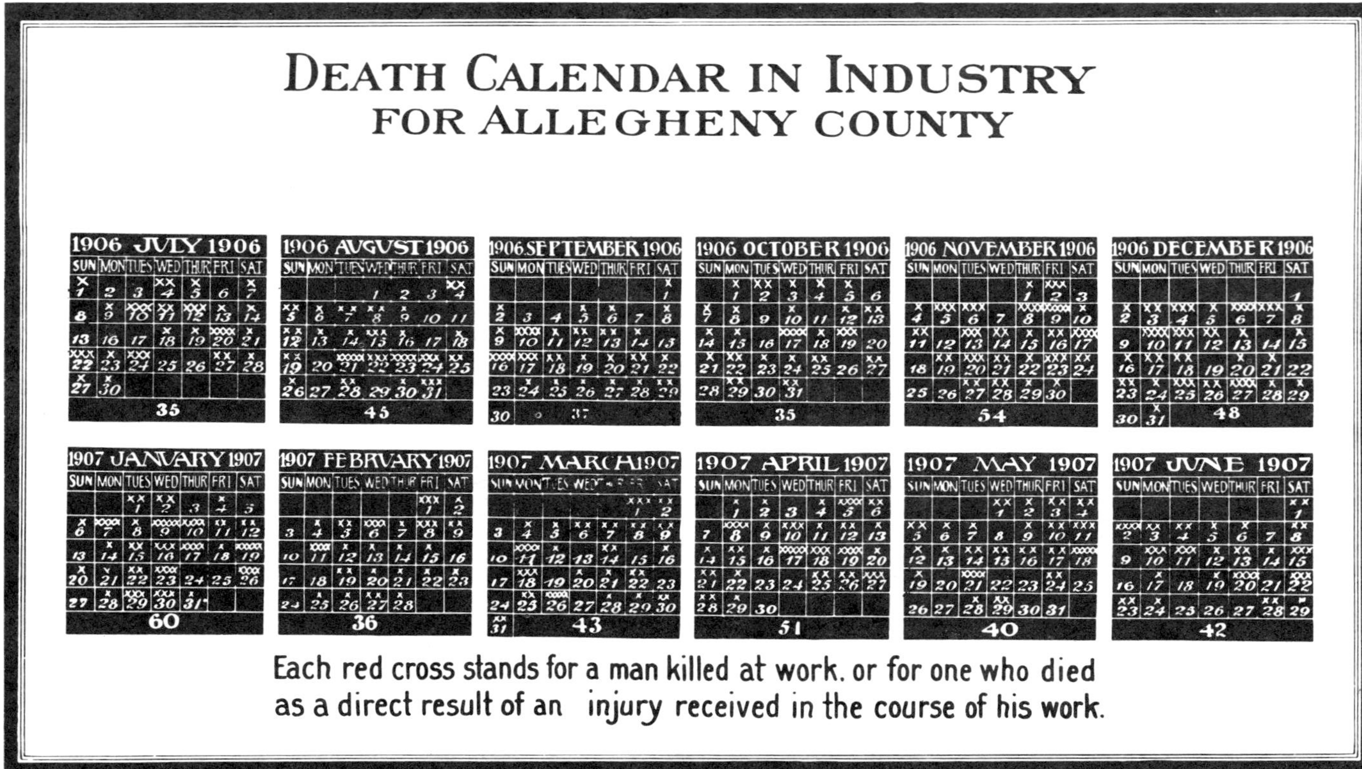

Figure 3-1. Little is known of the exact number of work accidents in the early years of this century. Most communities didn't bother to count them. One exception was Pittsburgh in 1906-1907, where the Russell Sage Foundation of the day made a detailed study. The most memorable feature of this 345-page report has become known as the "Allegheny County Death Calendar," reproduced here.

From *Work Accidents and the Law,* Russell Sage Foundation, 1910.

so far, in their efforts to safeguard the man who works, and the companies for whom men work, and, above all, the women and children who are dependent. But earnest men will continue to fight, with the certainty that when the evil and remedy are both known, all classes will rally in support of the needy. Mr. Hard points the way, clearly, and without prejudice.

Activity was not limited to the muckraking journalists, for there was also an awakening within the medical field and related disciplines. Although, as indicated previously, many articles had appeared in turn-of-the-century periodicals, surveys were also initiated and monographs were appearing.

Even before the public's awareness, certain authors had emphasized the hazards of particular trades. In a two-volume work edited by Buck in 1879, a large section was devoted to the "Hygiene of Occupation," a second one to the "Hygiene of Metal Mines," and a third, to "Hygiene of Coal-Mines."

Dr. George Kober (1850-1931), one-time Army surgeon for the military installation on Alcatraz Island, and later professor of hygiene and subsequently dean at Georgetown University School of Medicine, wrote extensively of health hazards in many different industries, in one of the two 1908 reports of the Committee on Social Betterment of the President's Homes Commission. In his introduction to the report, *Industrial Hygiene and Social Betterment,* Kober wrote that during a visit to Berlin in September 1907, he met Dr. E. J. Neisser who had just completed *An International Review of Industrial Hygiene,* a volume of some 352 pages, and who deplored his inability to present a general review of work accomplished in the United States, since with the exception of an inspector's reports in New Jersey, no recent data concerning "factory sanitation" were available. Kober commented, "Realizing the importance of the subject, not only to wage-earners, but to all interested in the conditions under which our fellow men and women live and work, an effort has been made in the succeeding pages to supply this information." (This episode is in parallel with the often-quoted comment encountered by Dr. Alice Hamilton while attending the International Congress on Occupational Accidents and Disease held in Brussels, during a break in her Illinois lead study. Following her presentation on the white-lead industry in the United States, she stated that it was mortifying being unable to answer questions regarding disease rates, compensation, or regulatory legislation. "Finally Dr. Gilbert [*sic*], of the Belgian Labor Department, dismissed the subject: 'It is well known that there is no industrial hygiene in the United States. *Ca n'existe pas.*' ") Rosen considers Kober's *Report* to be the first American text on occupational health.

Alice Hamilton, as mentioned earlier, reported on the sanitary fixture industry in 1911, and soon was assigned to do a comparable investigation of all the states, under the Bureau of Labor (Department of Commerce).[5] She stated[37] that, as a federal agent, everything was left to her—to find the plants, to gain permission for entry, and to design the survey. She was not to receive a salary; when the report was ready for publication, the government would buy it from her at a price to be determined. Dr. Hamilton later went on to become the first woman faculty member at Harvard, as assistant professor of industrial medicine.

William Gilman Thompson (1856-1927) made a strong contribution in 1914, with the completion of his lengthy text, *The Occupational Diseases*. He was involved in the establishment, in 1910, of the first occupational disease clinic in connection with Cornell University.

The eminent Dr. Richard C. Cabot of Massachusetts General Hospital, Boston, rationalized in 1912 the inability or unwillingness of hospitals to concern themselves with the prevention of occupational disease on the basis of there being too many giants to fight simultaneously—poor housing, alcoholism, poisoned air, and massive immigration, besides work-caused illness. He did say that when physicians realized that the problem could be solved, they would call for trained investigators—some based in the hospital; education provided in a clinic of employers, workers, legislators, and the public; and research conducted in industrial hygiene in a medical school's department of preventive medicine.

Women and children at work

As it was in England, there were also women in the American workforce, and the United States had its own problem of child labor. Before the organization of the National Child Labor Committee in 1904, there were no standards of protection against the exploitation of children (see Figure 3-2). Ten-year-old boys worked in the constant dust of coal breakers, picking out slate, or sweltering all night near the furnaces of a glasshouse. Girls of the same age were employed in the dusty cotton mills of the South. A 12-hour day in a Pennsylvania silk mill met with either approval or indifference. And as Lovejoy expressed it, ". . .tobacco and cigarette factories, canneries, sweatshops, the street trades, and the night messenger service all took unchallenged toll from schoolhouse, playground, or cradle, and despite the evidence gathered from England, Germany, and other European countries attesting the waste and folly of the system, child labor was almost without an enemy in America."

John Spargo, a former worker for the Socialist cause and active in social reform, wrote a stinging, profusely illustrated book in 1906 with the title, *The Bitter Cry of the Children*. He cited some data from the

1900 census relative to the employment of children: 1,750,178 working children between 10 and 15 years; 25,000 boys under 16 years were employed in mines and quarries; 12,000 children under 16 years were employed in the manufacture of tobacco and cigars; 5,000 children were employed in sawmills; 5,000 children were working at or near steam planers and lathes; 7,000 girls worked in laundries; 2,000 girls had employment in bakeries; 138,000 children were employed as servants and waiters in restaurants and hotels; 42,000 boys served as paid messengers; and 20,000 boys and girls were employed in stores. It was Spargo's belief that all of these "official" figures were low. In Syracuse, it was a 1904 lament that "The factories will not take you *unless you are eight years old.*"

Fourteen years of age was the break-point between schooling and work in nearly all states in 1910. The issues were ill health (illness and injury), illiteracy, physical and emotional immaturity, abuse, malnutrition, loss of youth, crime, and violence. Fortunately, with time, protective legislation did come about, and no longer were children employed in physically and mentally destructive jobs at minuscule pay and, as a result, deprived of normal growth and education.

There were child-labor laws in many states by 1920, but they could be evaded easily, and federal legislation was fought strongly by businessmen as being allegedly destructive of individual choice and states' rights. Even with age limits, who was to determine the years of a child with any accuracy? Further, many parents, in collusion with employers, flouted the law for the additional income to be brought in by their children.

It required the passage of the Fair Labor Standards Act in 1938 to clear the work scene of children. The act established 16 years as the minimum age for employment of most young people, with 18 as the lower limit in occupations deemed hazardous by the U.S. Department of Labor. Nevertheless, minors aged 14 and 15 years were permitted work outside school hours in certain nonmanufacturing, nonmining occupations, such as clerical and sales work, for a specified number of hours, but not before 7 AM or after 7 PM. Certain exempt employment included agriculture outside school hours in the school district where the child lived; acting or performing; employment by their parents in a nonhazardous occupation; newspaper sales or delivery; and working at home.

Many of the states passed protective acts for women, in consideration of the hours of work, hazardous employment, and rectification, in most instances, of extraordinary differentials in pay. With the liberation of women in the decades ahead, there would be a movement of women into nontraditional employment, along with the removal of some protective laws. While seemingly successful then—i.e., parity in employment opportunity—many more years were needed to determine the ill effects, if any, of work in the heavier trades. Much of this movement began in

Figure 3-2. In "the good old days," it was more of a problem getting the children to school than worrying about teaching them safety in school.

2a. While these boys sized coal in the mines,

2b. . . . these barefoot girls toiled in the mills.

2c. Here, one false move could mean a mangled toe.

World War I when women entered war production plants. While some left such jobs in 1918, many did not do so at the conclusion of World War II, for their appetites for good wages had been increased. Both wars brought women into the armed services, providing a new experience and, for many, an occupation in the years after 1945.

Workers' compensation

Even though Great Britain had passed its Workmen's Compensation Act in 1897, the United States was still operating under various employers' liability laws. The need for protective insurance had sound historic roots. The growth of capitalism increased the insecurity of the worker, and the situation had grown acute after the onset of the Industrial Revolution in the 18th century. Millions of people were in the labor force, working long hours, receiving wages too low to permit saving. Exposed as they were, initially to physical, and later to chemical hazards, they could be suddenly unemployed by virtue of work accidents, disease, old age, or periodic economic crises. Such persons became public charges, having no reserves upon which to draw, and having needs far beyond the ability of any charity to handle.

The British law passed in 1897 read, "If in any employment to which this Act applies personal injury by accident arising out of and in the course of the employment is caused to a workman, his employer shall... be liable to pay compensation..." The wording, "arising out of and in the course of the employment," which remained in future amendments, was carried over to comparable legislation in the United States and is still applicable in today's adjudications. By 1906, the British Act was repealed, and the principle of compulsory payment extended to nearly all workers, and protection against occupational disease was included for the first time—anthrax, lead poisoning or its sequelae, mercury poisoning or its sequelae, phosphorus poisoning, arsenic poisoning, and hookworm disease. Acts antedating American legislation were passed in Germany (1884), Austria (1887), Norway (1894), Finland (1895), Denmark, France, and Italy (1898), Spain (1900), Sweden (1901), Greece (1901), Belgium and Russia (1903).

There was discontent with the U.S. Employers' Liability Acts. Arguments were voiced about the wrong persons being insured, specifically, employers. Of primary concern was their protection against the consequences of lawsuits. But, as French viewed the situation,[37] "The heavy financial cost for the coverage, the availability of old-time legal defenses that outraged all conceptions of justice, the long delays in the courts, the bitterness that developed, and the resultant unequal burden thrown on injured men to secure redress, formed a formidable combination of reasons why a better method should be introduced." Studies indicated a great need for change. The victims of accidents were usually young men—not oldsters, as believed by many—the majority had families, and their standard of living was greatly lowered by the losses resulting from the injuries.

Finally, after a long study of the German insurance plan, a scheme was submitted to the U.S. Congress by the Commissioner of Labor and, after the passage of legislation protective of federal workers, and after much deliberation by many state commissions, Wisconsin passed on May 4, 1911, the first legislation undertaken by the states. This law, in its opening sections, abrogated the old defenses of assumption of risk, contributory negligence, and the fellow-workman principle. The validity of the act was tested within a few months and was declared constitutional by the Wisconsin Supreme Court. California led a few states in the same year by passage of legislation providing coverage for occupational disease.

While a bill for workers' compensation (the Wainwright Law) was passed in New York in 1910, it was declared unconstitutional by the New York Court of Appeals on the grounds that the law violated both the federal and New York State Constitutions, "because it took property from the employer and gave it to his employee without due process of law." By the mechanism of the adoption of an amendment to the state constitution by the legislature and approval in 1913 at a general election, a compulsory Workmen's Compensation Act finally became effective in mid-1914.

Coincident with the declaration of unconstitutionality of the early act, a fire occurred in a clothing factory in New York City that took the lives of 146 employees. This disaster was called the Triangle Fire; it unified the demand for factory legislation. It gave greater force to reform because the fire took place on the same day (March 25, 1911) that the Court of Appeals decision was publicized. Although recommendations had been made regarding the Triangle Shirtwaist Company's fire escapes, barred windows, and stairways shortly before, no action had been taken. The fire and the public's agitation therefrom resulted in the passage of a new industrial code, which served as a model for several states. Further, the disaster opened the door to progress in job safety. In 1913, New York, California, Massachusetts, Ohio, and Pennsylvania granted rulemaking authority to their labor commissioners. Finally, with the passage of a Workers' Compensation Act in Mississippi in 1948—37 years after Wisconsin's legislation—coverage was introduced into all of the states.

Early organizational growth

The year 1910 was one of significance in the growing concern about the health of workers. In that year, the Bureau of Mines was established, to be headed by Dr. Joseph E. Holmes. Thompson had created the first clinic for occupational disease, the first occupational disease conference was held, Andrews had published his report on phosphorus poisoning, Hamilton completed her study on lead poisoning, and the Bureau of Labor issued a list of industrial poisons, all mentioned previously in this chapter. These events seemed to serve catalytically in the growth of other movements and group efforts in the immediate years ahead, as we see now.

The Conference Board of Physicians in Industry was organized in 1914 with a membership of 29 from nine different states, representing care for about 400,000 workers. By 1922, a definition of the industrial physician had been formulated[38] as "...one who applies the principles of modern medicine and surgery to the industrial worker, sick or well, supplementing the remedial agencies of medicine by the sound application of hygiene, sanitation and accident prevention; and who, in addition, has an adequate and cooperative appreciation of the social economic and administrative problems and responsibilities of industry in its relation to society." This statement was far-sighted and, while not in need of broadening over the years, has been injudiciously narrowed by some practitioners. Subsequent reports from the Conference Board were issued through 1932.

Also, the Public Health Service began to show an interest in the health hazards present in the work environment as early as 1910, and four years later, an Office of Industrial Hygiene and Sanitation was created, which studied New York's garment workers, and found excessive the prevalence of tuberculosis. The findings initiated action toward the eventual elimination of the sweat shop.

During this period, the largest problem of environmental and occupational health yet known was being solved as Army surgeon William C. Gorgas caused the eradication of yellow fever carried by mosquitoes, and this made possible the construction of the canal across the Isthmus of Panama. Life was difficult for Gorgas, because of the appointment of another public servant who was highly critical of the costs of sanitation—$365,000 per year—which actually permitted completion of the project. Gorgas was to join many others in the justification of expenditures for sanitation and occupational health.

The National Association of Railway Surgeons, later the American Association of Railway Surgeons, organized in 1888, continued its meetings, and by 1962, there were 34 railroad hospitals in the United States. The first was established in Sacramento, California, in 1869 by the Central Pacific, later called the Southern Pacific Railroad.

The collecting of statistics in regard to occupational disease during this period was not done regularly, but credit is due the Massachusetts legislatures of 1904-1906 for having taken the first action toward obtaining scientific data concerning the sanitary conditions of factories, workshops, and mercantile establishments. State inspectors of health were subsequently appointed. In 1916, the American Association of Industrial Physicians and Surgeons was organized. Today it is called the American Occupational Medical Association.

A final note about this period of increased awareness concerns Josephine Goldmark, who wrote the book *Industrial Fatigue* in 1912. In 1914, Dr. George Price, author of several works on sanitation and environmental safety, published *The Modern Factory,* a classic, with 275 illustrations.

The safety movement

As nearly all contemporary legislation linked occupational health to safety, it remained difficult to separate illness from injury in the work setting when prevention was considered. October 8, 1911, was the date of organization of the United Association of Casualty Inspectors, composed of insurance company inspectors in New York. Three years later, the name was changed to the American Society of Safety Engineers. After merging with the National Safety Council, which was founded in 1912, it withdrew in 1947, and has been independent since. With no mention of engineering, it is stated in the Society's charter that its purpose is to "expand the use and effectiveness of the application of the arts and sciences to the prevention of accidents and the conservation of health, life, and property by the control of environmental exposures and human behavior."

About the same time, in September 1912, the first National Safety Congress was organized, representing the result of joint action between the Association of Iron and Steel Electrical Engineers and the Cooperative Safety Congress. Two years later, the Congress became the National Safety Council, and currently there are about 80 chapters throughout the United States.

As far back as 1877, Massachusetts passed the first law in the United States mandating safety guards on machinery and regular inspection of factories by state officials. Over the years, much legislation has been activated by the Congress, frequently after the occurrence of one or more disasters, as seen in the following:

1852 Steamboat Inspection Act

1866 Steamboat Inspection Act amended to include passenger ships

1893 Railroad Safety Appliance Act—required air brakes and automatic couplers

1901 National Bureau of Standards created

1908 Federal Workmen's Compensation Act—limited coverage

1916 Federal Highway Aid Act

1926 Federal Workmen's Compensation Act amended to include all workers

1927 Federal Longshoremen's and Harbor Workers' Compensation Act

1936 Walsh-Healey Public Contracts Act

1952 Coal Mine Safety Act

1958 Federal Longshoremen's and Harbor Workers Compensation Act amended to include rigid safety precautions

1959 Radiation Standards Act

1960 Federal Hazardous Substances Labeling Act

1966 National Traffic and Motor Vehicle Safety Act

1966 Child Protection Act—banned hazardous household substances

1967 National Commission on Product Safety created

1968 Natural Gas Pipeline Safety Act

1969 Construction Safety Act

1969 Child Protection Act amended to broaden the coverage

1969 Coal Mine Safety Act

1970 Occupational Safety and Health Act

1970 Poison Prevention Packaging Act.

Much of the legislation resulted from pressure brought by victims or their heirs, and the tragedies involved both "acts of God" and disasters caused by human error. Such events as mine explosions, floods, fires, dam breaks, automobile collisons, plane crashes, ship sinkings, ship collisions, train wrecks, tornadoes, blizzards, earthquakes, and avalanches have been documented as primary causes of multiple deaths over the years.

The fact that, prior to about 1912, safety was given little attention can be traced to an unrealistic reliance on unenforceable laws and by a general industrial apathy probably based on (*a*) the common-law defenses indicated earlier, (*b*) ignorance of the economic losses caused by accidents, and (*c*) "the pernicious idea that accidents were unavoidable, that they were the price which must be paid for industrial progress."

What the future holds for industrial safety, taking into consideration the move toward the deregulation of commerce and industry cannot be evaluated at this time.

One piece of legislation that did not cross over between occupational health and industrial safety was the Walsh-Healey Public Contracts Act, approved in 1936. Although not to be enforced with any strength in the years following, the act represented a second effort to put a floor under wages and a ceiling over hours worked. Most of the activity centered about the law concerned these issues rather than the health and safety requirements that were specified. Apart from limiting workers in plants providing materials for the federal government in any amount exceeding $10,000, to eight hours per day and 40 hours per week, and granting them the minimum wage as established by the Secretary of Labor, these conditions were laid down: No work was to be performed or materials manufactured under contract ". . . in any plants, factories, buildings, or surroundings or under working conditions which are unsanitary or hazardous or dangerous to the health and safety of employees engaged. . ." Further, no work was to be carried out by any male under 16 years of age, or any female under 18 years of age.

By 1942, however, the clause regarding health and safety was given evidence of intent through the publication by the Department of Labor of a booklet, "Basic Safety and Health Requirements for Establishments Subject to Walsh-Healey Public Contracts Act." In addition to such standard subjects in safety as machine guarding and materials storage, there were provisions relating to atmospheric contamination, sanitation, personal service conveniences, illumination, and first aid facilities.

By 1960, the initial major revision was published in the *Federal Register*, and for the first time in the history of Walsh-Healey, the act had the force and effect of law. Reaction was mixed. Public hearings were held, and management, labor, and national safety associations became vociferous. Revisions followed, and in subsequent years, standards were included relating to occupational noise exposure, medical services, and radiation. By 1969, the bombshell created by the revisions seemed to be subsiding, but the OSHAct of 1970 paled the original corporate/professional organizational response.

THE GENESIS OF OCCUPATIONAL HEALTH IN WELFARE

Occupational health as a medical undertaking, as a support segment of a manufacturing, service, financial, or sales enterprise, or as a governmental regulatory service, was not born exclusively from the concern of the medical establishment. In the later decades of the 19th century, millions of people immigrated from all the European countries and came to the United States to settle in its large cities. The steel mills, the machine tool industry, the railroads, the canals, the farms, and the municipal governments hired these people as manufacturing expanded. In the 1840s, 1.5 million persons came to America, and by 1860, about 6 million had arrived. The influx continued up to the years immediately before World War I, and resumed in the 1930s with the rise of Hitler.

Immigrants, in the early years, with few exceptions, entered the work force as unskilled laborers. With comparably low wages in the decades to come, poverty was the characteristic of these urban groups. Demographic knowledge of the population at large was limited. Birth registration was so limited that in 1912, with the creation of the Children's Bureau, attention was directed toward increasing from eight, the number of states registering births in accordance with Census Bureau standards. The attempt to establish a federal Department of Health "called forth horrifying opposition."

That there had to be a reversal of these unanswered needs was behind the genesis of the settlement houses. Jane Addams, a person whose name became synonymous with social work at Hull House in Chicago, gave her impression[39] of the plight of certain city dwellers in one ward of about 50,000 inhabitants:

> The streets are inexpressibly dirty, the number of schools inadequate, the street-lighting bad, the paving miserable and altogether lacking in the alleys and smaller streets, and the stables

Figure 3-3. A Texas Company industrial nurse and her wheels.

defy all laws of sanitation. Hundreds of houses are unconnected with the street sewer.

In partial response to medical needs, a public dispensary was established at Hull House, and a Visiting Nurses Association nurse made his headquarters there.

As the settlement houses developed in the large cities, certain activities began to appear in a slightly more enlightened industrial sector, under the name of welfare work. That such services were provided solely for the advantage of employees is to stray from the truth. In some newspaper articles written in 1907, descriptive of a group of cotton mills in the South, it was stated that a number of these mills "at the expense of the corporation," employed trained nurses to attend persons who were sick in the mill community, and to get them well as quickly as possible (see Figure 3-3). The writer pointed out that the "humanitarian idea of employing trained nurses is not unmixed with commercialism, because the sooner the help is gotten back to work the better for the cotton mill, and from all I can understand this phase of what might be called 'welfare work' is probably the most sensible that is undertaken by any of the cotton mills."

At the New York Telephone Company, there was a "retiring-room" for operators who felt indisposed. If one fell ill, she would be taken to this room, and a physician would be summoned. When the employee was well enough to be moved, she would be sent home in a closed carriage, in the charge of another woman employee.

In a comparable review in 1898-1900, there were listed 15 "measures for the improvement of the condition of workingmen." And of these measures, two are relevant:

> 11. The care for employees' health and comfort by means of bathing facilities, dining and lunch rooms, the furnishing of hot lunches to female employees [male employees either were assumed to bring their own lunches or could subsist on cold lunches] and by improved sanitary construction and appliances and
>
> 12. The care of sick and disabled employees and their families by means or [*sic*] free insurance, free medical attendance or hospital facilities, and by the encouragement of beneficial organizations.

The author stated that all of the plans described were demonstrated and gave "ample proof of their social and economic value."

That there was extreme and continuing interest in welfare benefits by the U.S. Department of Labor is seen in still another study published by the Bureau of Labor Statistics. Fifteen different industries, represented by 51 companies, were investigated. In the conclusion, the investigator wisely observed that, "Indeed, it is safe to predict that the time is not far distant when much of present-day welfare work will be a requirement." So much of this "paternalism" is now considered a birthright of the American worker, but was obtained only through long periods of strikes, negotiations, legislation, illness, and death.

However, it was from these paternalistic roots that occupational health services began to grow into today's in-plant departments.

One author referred to this as "welfare work," but did offer a certain caution. He observed that welfare work was commendable and an essential part of industry, but he thought the term unfortunate, for it implied a "repellent excess of virtue." At times, a too zealous or too precipitate imposition upon an unwilling body of employees has aroused opposition, and actual rejection, leading to abandonment of the effort. The author advised that such welfare work should not be developed too rapidly in a plan, and should avoid arousing suspicion that it is a gift, carrying with it a covert expectation that it is to be paid for by increased production. The activities should be administered as a "bond of mutual helpfulness between employer and employee," for they will "act as lubricants to the human machine and are indispensable to its highest efficiency."

Physical examinations

Eventually, active medical practice began to replace the, perhaps, clumsy initial trials at preventive measures in the workplace. First of such attempts was the conducting of physical examinations. The National Cash Register Company examined applicants for work as early as 1901. Dr. Frank Taylor Fulton, a graduate of Johns Hopkins and a practitioner in Providence, Rhode Island, recorded what probably were the first documented medical evaluations conducted among factory workers.

In a presentation before the Providence Medical Association in 1906, he gave credit to "certain manufacturers" who gave assistance in the examination of a group of employed persons. Dr. Fulton examined, at no charge, a number of employees in one of the big saw factories for the purpose of discovering tuberculosis. Fulton wanted to find patients with active disease so treatment could be initiated. He stated that "there remained nothing excepting patients to treat." At that point, presumably, he examined workers and uncovered many with tuberculosis who were placed on the open-air and rest regimens of the day.

Examinations were introduced at Sears, Roebuck, and Company in Chicago in 1909. And, by 1914, physical examinations of employees were a fixture in many large companies.

It was felt generally that examinations were conducted prior to the passage of workers' compensation legislation. At that point they were introduced much more commonly as "many concerns felt it was necessary to rule out the defectives in order to protect themselves from liabilities." And, as one medical director—or chief surgeon, as he was called—said, it was not until they were convinced that they were getting only the refuse of the railroads and the steel mills, men who could not qualify in competitive physical examinations, that his company began such evaluations in May 1916.

Soon, physical examinations were relatively commonplace. One commentator pointed out the gains through these examinations when he stated that the review of large groups of men at the place of employment provided added insight into the trade causes of disease, an insight which he felt was still rudimentary. Another medical director, in discussing her own examination program in a plant with 3,000 women, tried to put to rest complaints of organized labor by stating that such examinations were not conducted for the purpose of turning down anyone who was fit to work anywhere, but were for the purpose of fitting any applicants, even the physically handicapped, into the work for which they were best adapted physically, where they would injure themselves least, possibly benefit themselves, and continue to work longest. She also expressed the need for an "examiner who talks to the patient—not an "Mm...mm doctor...."

By 1920, following some of the industrial health needs expressed during the period of World War I, several in-plant programs were in operation. The influenza epidemics had come and gone, and although data were available from military camps and from population groups at large—2 percent of all men in the U.S. Army had the disease and one of every 67 who had it, died, no specific figures were derived regarding attack rates among plants or employed groups, although the three waves of the 1918-1919 pandemic caused 500,000 deaths in the United States alone and cost, in 1968 dollars, more than $100 billion. Factory operations must have, indeed, been disrupted.

20TH-CENTURY DEVELOPMENTS

Several breakthroughs made the 1900-1980 era significant in the development of occupational health and safety. Certain occupational diseases had been identified and efforts in control were generated. For example, chronic mercurialism was affecting processors of the source ore and hatmakers alike.

A study of mercury poisoning in the hat-making plants in New York City identified some 80 workers having typical symptoms of toxic mercury salt absorption. In Connecticut, where hatmaking was centered, an accord was reached in December 1941, among the manufacturers, the unions, the U.S. Public Health Service, and the State Health Department, in which the mercury nitrate used in the carotting process would be eliminated, and hydrogen peroxide used in its place.

In the early 1950s, a strange disease of the central nervous system appeared among citizens of Minamata, a Japanese fishing village. By final reckoning, there were 103 deaths, and some 700 were disabled. The disease? Organic (methyl) mercury poisoning, traced to a chemical plant discharge system which dumped the waste into Minamata Bay. The fish became contaminated, the mercury entered the food chain, and the citizens, typically fish-eaters, were poisoned by the chemical. Subsequently, waste treatment systems were installed. But, the impairments of the victims persist in many areas of the world.

Asbestos, once considered a mild fibrogenic substance causing infrequent lung tissue disease, became a subject of extensive study and epidemiologic investigation. It was found to be a carcinogen, acting in company with cigarette smoking to cause death from cancer of the lung, or, without association of smoking, mesothelioma.

Chromium compounds have undergone inquiry since World War II, and, again, while chromic acid mist was known to produce septal perforations, "chrome holes" of the skin, and chronic dermatitis, occupationally caused lung malignancies after contact with certain chromates, represented a new epidemiologic finding.[42]

Beryllium became, in the early post-war years, a part of the nuclear energy industry; later it was used as a heat shield for reentry of the Project Mercury astronauts. It was treated with care because of the earlier findings of both acute and chronic disease resulting from exposure to the berrylium-zinc-silicate of the original fluorescent lamps. Vinyl chloride produced angiosarcoma of the liver in a few patients and Kepone, through total disregard of environmental controls during its production, massively overexposed more than 100 workers, and produced an occupational disease involving several bodily organ systems.

In the area of work-related carcinogenesis, much has developed since cancer of the scrotum in chimney sweeps was described in 1775.

As interest of occupationally associated malignancies increased, a glance at history indicates that such interest usually was squelched. It was Marcel Proust who said that in society, new and original ideas, whether reprehensible or not, cause alarm and apprehension when first announced. They challenge the prevalent spirit of togetherness that so strongly binds society. So it was with one of the early voices proclaiming the relationship between certain industrial substances and cancer in workers exposed to such toxic materials. Early on, in 1925, Schereschewsky, then in Boston, completed a statistical review of the steady rise in cancer mortality over a 20-year period in the death-registration area of the United States. He also organized a conference in 1927 that would advise the Public Health Service on a program of cancer research. Among four recommendations submitted by a committee of eminent investigators was research in occupational cancer.

Activity was increased in Boston in what was then known as the Office of Field Investigations of Cancer, and eventual amalgamation took place with the National Cancer Institute (NCI). In 1949, Dr. Wilhelm C. Hueper joined the staff of the Cancer Control Branch of the NCI. He was strongly critical of industry, a factor that had led to his earlier termination by the DuPont Company and soon complaints came in from other companies that were targets of Hueper's criticisms. The U.S. Surgeon General ordered him not to make any field trips, "in order to preserve Federal programs in industrial hygiene more general than carcinogenic hazards." He became extremely embittered, and retired at age 70, sharing a World Health Organization honor with a Russian scientist for their contributions to environmental carcinogenesis. Hueper had publicized cancer hazards of heavy metals, petroleum derivatives, synthetic hydrogenated coal oils, asbestos, food additives and contaminants, cooking fats, and plastics. He seemed to enjoy shocking and alarming associates and superiors, in addition to industry, and on one occasion had published in the *Police Gazette* an article on industrial and environmental carcinogens. He wanted to reach the workingman who never saw, let alone read, a scholarly journal. In that period, such publication outraged the hierarchy of the National Institutes of Health (NIH).

Industry urged the directorates of both the NCI and NIH to muzzle Dr. Hueper, to prevent his publishing, to restrict his lectures on occupational cancer, and to forbid him to continue his personal inspections of plants where he pointed out carelessness in the handling of potential workplace carcinogens. The repressive measures were implemented, as Hueper expressed it, because of complaints from the chromium industry and continued resistance from the Director of the PHS Division of Industrial Hygiene, "...the Surgeon General of the U.S. Public Health Service abruptly ordered me to stop all further epidemiologic activities in occupational cancers and to discontinue contacts regarding this matter with industries as well as with State health departments." He was asked, also, to remove his name from the last of several papers produced by the studies. Hueper's eventual honors attested to the soundness of his work.[42]

Fortunately, with researchers and activists publishing freely today, both scientific and lay communities are directed not only to significant contributions, but to episodes of occupationally incurred cancer or intoxication by exposure to industrial substances. In an editorial, Shubik and Clayson[43] wrote:

> In the field of occupational carcinogenesis, major recognition of the fine work undertaken by a few dedicated and often harassed workers is finally coming to fruition; the long delayed understanding of these problems by the labor unions is proving helpful. There are more occupational hazards to be discovered; there are actions still required to prevent unnecessary cancers from known occupational carcinogens; there's an overpowering need to train physicians to detect such cancers and to advise industry on potential new hazards. However, society has recognized these needs and finds that it is politic to attack them in a reasoned approach.

Radiant energy

Within months after Roentgen's discovery of the X-ray in 1895, reports began to be published regarding the harmful effects of injudicious exposure to such irradiation. Certain protective devices were introduced shortly after untoward tissue changes were encountered. In 1913, Oliver, the British author of important texts in occupational health, directed attention to the death of Thomas Edison's "private assistant," after experimental work involving X-radiation. H.S. Martland's seminal establishment of the relationship between the absorption of a mixture of crystalline zinc pyrosulfide, radium, mesothorium, and radiothorium, and the development of mandibular necrosis and anemia, and in some of the workers affected, osteogenic sarcoma served, in the words of Dr. Shields Warren in 1951, as the basis for, ". . . a firm scientific foundation and guide for our [Atomic Energy Commission] current work aimed at establishing realistic permissible body burden levels for plutonium and other alpha emitters."[44]

Not given much publicity, however, was the work of Florence Kelley and the National Consumers League in aiding the women workers suffering from, or dying from, radium poisoning. Throughout the period of 1924 to 1928, the United States Radium Corporation fought the issue, claiming that radium was nontoxic, while refusing permission to Dr. Cecil Drinker of Harvard to publish his findings of a survey he made at the company's request. In a second study, sought by the company, Dr. F. B. Flinn of Columbia University, published a paper in which he stated, "I believe we are justified in arriving at the conclusion that an industrial hazard does not exist in the painting of luminous dials." In 1951, *Life Magazine* published a picture-story on the death of what it termed "Radium Victim No. 41."

During the Manhattan District Project, and with the postwar continuation of the succeeding nuclear energy authorities, full-scale occupational health services characterized the preventive medical efforts to preserve the health of the workers in the nuclear energy industry. The story of the increased incidence of lung cancer in uranium miners exposed to high concentrations of radon daughters (radioactive decomposition products) has been told. After the Three Mile Island nuclear accident, a full study of possible effects was initiated.

Occupational health nursing

The early history of nursing in industry presents a fuzzy distinction among public health nurses, visiting nurses in industry, public health nurses in industry, and industrial nursing. Although the first employment of a professional nurse by a company took place in England at the J. and J. Colman plant in 1878 with the addition of Philippa Flowerday, it was 10 years later that a Ms. Betty Moulder, trained at Philadelphia's Blockley Hospital, was credited with being the first U.S. Company-based practitioner. Most frequently cited, though, is the hiring of Ada Mayo Stewart by the Vermont Marble Works of Proctor in 1895. A strong community-minded citizen, Fletcher D. Proctor, one-time governor and president of the company, selected the 24-year old graduate of the Waltham (Massachusetts) Training School to carry out district-nursing activities for his employees. After a year of home visitations, Ms. Stewart became the first superintendent of the newly built company hospital in Proctor.

Other companies, primarily large department stores, added nurses in the early 1900s, and by 1914, 60 firms had nursing services. And, in the years to come, in World War I, the federal government encouraged the employment of such professionals in the shipyards, factories, and mills operating under contract in the production of war material. By 1918, 871 companies had added 1,213 graduate nurses.

The period between the two World Wars witnessed a reduction in professional staffs, partially caused by the Great Depression. The sequence of Lend-Lease, the declaration of war against the Axis, and increased production for the war effort brought nurses back into industry in greater numbers. But organizational efforts had begun earlier when, in 1915, the Boston Industrial Nurses Club was established. In the years ahead, this group eventually became the New England Association of Industrial Nurses. Educational needs were expressed, and the first special educational course for nurses in industry was offered in 1917 by Boston University, and was repeated every five years.

In 1922, the first American Association of Industrial Nurses was formed, but in 1933, was absorbed by the New England Association. With the growth of local and regional groups, it was apparent that an element of some governance was needed, and in April 1942, at the fourth joint conference of the several associations, 300 nurses from 16 states voted for the creation of a national association, designated the American Association of Industrial Nurses (AAIN). The second meeting was held jointly with the American Association of Industrial Physicians and Surgeons, a practice that materialized into the annual American Occupational Health Conference. In keeping with changing concepts, the AAIN became the American Association of Occupational Health Nurses (AAOHN). With the passage of the OSHAct in 1970, educational opportunities were provided the occupational health nurse, and many have partaken of either full-time curricula leading to the master's degree or courses in continuing education.

The future of nursing at various work sites is unquestionably bright. Many nurses have been granted full responsibility for in-plant health programs, and with the rising clamor for greater participation in health planning and health care, and as nursing becomes more specialized, there may be established equity between medicine and nursing in the larger programs.

Industrial hygiene

The greatest growth of professionals engaged in occupational health has been in industrial hygiene. With a shortage of graduates before World War II, the post-war years saw a great number of highly trained specialists emerge from the universities that had established programs in occupational health.

The American Industrial Hygiene Association (AIHA) increased membership several fold after the passage of the OSHAct and a great number of women entered this highly technical field.

Miscellany

No sketch of history can be more than just that, for events, people, legislative changes, ideas, and both natural and human calamities arise in too great numbers to permit a detailed documenting of all episodes. Such topics as the specialized educational programs, the periodical and monographic literature—which has grown incredibly—occupational mental health, and the more recently designated employee assistance programs; the scope of company and national medical conferences; the extent of occupational health and safety activities in the armed services; the function of federal, state, and municipal agencies engaged in the protection of worker health; the upsurge of health education and fitness programs; health at sea; medical care of workers; the effects of industrial chemicals; and of maximal importance today, as vistas open into the understanding of exposure, health-effect relationships and epidemiology have, unfortunately, had to remain untouched.

Occupational health and safety touch all aspects of human endeavor—the group spirit at the worksite, the workers' living community, recreation, spiritual beliefs, family life, personality development, in addition to individual health and physical intactness—that, in many ways, no generic, superficial inquiry can be totally disclosing. Highlights, only, can be touched upon, merely as milestones along a route of human progress.

THE FUTURE

As indicated before, no one can predict with any element of certainty what lies ahead in industry, particularly in the United States, as we enter and become part of the postindustrial society. Computers have entered every human and mechanical enterprise. Will history change drastically? Will legislative mandates be lifted or become more restrictive? Will the ubiquity of high technology require greater or less occupational health and safety support? Will the many new immigrants become a source of pooled labor or will the newly arrived establish their own enterprises? Will heretofore undiscovered, unformulated chemicals come on the market with little, if any, toxicologic testing? Will people be able to cope with all of the changes? The late Director-General of the International Labor Office viewed the technological setting, not too many years ago, in this vein: "Civilization is now at the mercy of the sorcerer's apprentice. The whole future of man is at stake as he conjures with the unknown, with forces he has set loose but neither understands nor controls."[45]

CONCLUSION

But we must learn from history, or else all the errors of our omission and commission will be repeated. One sensitive observer believes that we have "wandered out of history," and that without the perspective of history, he argues, and without a knowledge of the wisdom of the past, we find ourselves unprepared to face the challenging present and the uncertain future.

Henry E. Sigerist, historian of medicine, observed that:

> True history is always contemporary history, as Benedetto Croce once said, because it is a contemporary interest that drives a man to consult the past. Medical history is not only history but *medicine* as well; medical history is not only the concern of history but of medicine as well. And the medical historian, while serving truth as historian, is endeavoring to contribute to the progress of medicine.

It is perhaps this concern with occupational health and safety today that led the editor to include this chapter. We cannot deny the past, for "The denial of history in medicine is thus, as in other human affairs, a denial of the enduring value of the present, as assertion of the futility of human endeavor."

REFERENCES

1. Richardson, E. L. Shattuck Lecture—"The Old Order Changeth, Yielding Place to New: Perspectives on the Health Revolution." *N Engl J Med* 291:283-287, 1974.

2. Bryan, C. P. *The Papyrus Ebers*. London: Geoffrey Bles, 1930, pp. 67-77.

3. Legge, T. R. "The History of Industrial Medicine and Occupational Disease." *Ind Med* 5:300-314, 1936.

4. Soubiran, A. "Medical Services Under the Pharaohs." *Abbottempo* 1:19-23, 1963.

5. Major, R. H. *Classic Descriptions of Disease.* Springfield, Ill.: Charles C Thomas, 1955, p. 312.
6. Major, R. H. "Some Landmarks in the History of Lead Poisoning." *Ann Med Hist* 3:218-227, 1931.
7. Wright, F. A. *Lempriere's Classical Dictionary.* New York: E. P. Dutton, 1949, p. 359.
8. Goldwater, L. J. "From Hippocrates to Rammazzini: Early History of Industrial Medicine." *Ann Med Hist* 8:27-35, 1936.
9. Murray, A. L. "Lead Poisoning in the Mining of Lead in Utah," Tech. Paper 389. Washington, D.C.: U.S. Department of Commerce, Bureau of Mines, 1926, p. 1.
10. Shampo, M. A., and R. A. Kyle. "Alexander the Great." *JAMA* 248:70, 1982.
11. Bartlett, T. N. "Workmen's Compensation Legislation in Relation To Occupational Diseases. 1935 Meeting of the International Association of Industrial Accident Boards and Commissions," Bulletin No. 4. Washington, D.C.: Department of Labor, Division of Labor Standards, 1936, pp. 25-29.
12. Bertelsen, A., and N. Capener, "Finger's Compensation and King Canute." *J Bone Joint Surg* 42B:390-392, 1960.
13. Whittaker, A. H., and D. J. Sobin. "Historical Milestones in Occupational Medicine and Surgery." *Ind Med* 10:203-205, 1941; Ulrich Ellenbog. *Lancet* 1:270-271, 1932. The treatise was reprinted in facsimile in 1927: Ulrich Ellenbog: Von de Gifftigen Besen Tempffen and Reuchen. Eine Gewerbe-Hygienische Schrift des XV. Jahrhunderts. Verlag der Munchner Drucke, Munchen, 1927.
14. Rosen, G. "On the Miners' Sickness and Other Miners' Diseases by Theophrastus von Honenheim, called Paracelsus." Translated from the German, with an Introduction, in *Four Treatises of Theophrastus von Hohenheim called Paracelsus.* Baltimore: The Johns Hopkins Press., 1941, pp. 43-126. See also J. P. Dolan, "Paracelsus and the Origins of the Study of Occupational Medicine." *JSC Med Assoc* 71:175-178, 1975, and W. K. Frankel. "Industrial Medicine's Hall of Fame—Paracelsus." *Ind Med* 22:86-87, 1953.
15. Agricola, G. *De Re Metallica.* Trans. from the first Latin edition of 1556 by H. C. Hoover, and L. H. Hoover. New York: Dover Publications, 1950. Scc also R. T. Legge, "Industrial Medicine's Hall of Fame—Georgius Agricola." *Ind Med* 22:122-123, 1953.
16. Ramazzini, B. *Diseases of Workers,* trans. from the Latin text *De Morbis Artificum* of 1713, by W. C. Wright. New York: Hafner Publishing, 1964.
17. Young, W. A. "Works Organization in the Seventeenth Century. Some account of Ambrose and John Crawley." *Trans Newcomen Soc* Vol. IV, 1925, pp. 73-101.
18. Cheyney , E. P. *An Introduction to the Industrial and Social History of En-*

gland. New York: Macmillan, 1918, pp. 203-212. See also Lincoln, J. T. *The Factory.* Boston: Houghton Mifflin, 1912.

19. Toynbee, A. *The Industrial Revolution.* (First published in 1884.) Boston: Beacon Press, 1956.
20. Toynbee, A. J., in Preface to the Beacon Paperback Ed., Ref. 19, p. ix.
21. Griffiths, P. *A License to Trade—The History of English Chartered Companies.* London: Ernest Benna, 1974, p. 148.
22. McCord, C. P. "Wares Manufactured in the American Colonies" (1600-1775). *Jour Occ Med* 15:975-982, 1973.
23. Mumford, L. *The Myth of the Machine:* New York: Harcourt, Brace & World, 1966, pp. 240-241.
24. Cited in Bartlett, E. *A Vindication of the Character and Condition of the Females Employed in the Lowell Mills, Against the Charges Contained in the Boston Times, and the Boston Quarterly Review.* Lowell, Mass: Leonard Huntress, Printer, 1841, p. 3.
25. Marston, M. *Sir Edwin Chadwick.* London: Leonard Parsons, 1925, pp. 13-14.
26. Meiklejohn, A. *The Life, Work and Times of Charles Turner Thackrah, Surgeon and Apothecary of Leeds (1975-1833).* London: E & S Livingstone, 1957, pp. 45-46.
27. Felton, J. S. "Man, Medicine, and Work in America: An Historical Series. Part IV. Thomas Cadwalader, M.D., Physician, Philadelphian, and Philanthropist." *Jour Occ Med* 11:374-380, 1969.
28. Baker, G. *An Essay Concerning the Cause of Endemial Colic of Devonshire.* London: J. Hughs, 1767, pp. 30-31, reprinted by the Delta Omega Society, [n.p.], 1958.
29. Bouhuys, A., *et al.* "A Physiological Study of Byssinosis in Colonial America." *Trans NY Acad Sci* 35:537-546, 1973.
30. Zenker, F. A. "Diseases of the Lungs Due to Inhalation of Dust (Ueber Staubinhalationskrankheiten der Lungen)." *Dtsch Arch Klin Med* 2:116-171, 1866. See also, D. S. Ross; "The First Case of Pneumoconiosis." *Occup Health* 31:297-299, 1979.
31. Meiklejohn, A. "History of Lung Diseases of Coal Miners in Great Britain: Part I., 1800-1875." *Br J Ind Med* 8:127-138, 1951.
32. Cited in Pendergrass, E. P. *et al.* "Historical Perspectives of Coal Workers' Pneumoconiosis in the United States." *Ann NY Acad Sci* 200:835-854, 1972.
33. Felton, J. S. "Phosphorus Necrosis—a Classical Occupational Disease." *Am J Ind Med* 3:77-120, 1982.
34. Thompson, W. G. "Classification of Occupational Diseases." *Am Labor Leg Rev* 2:185-191, 1912.
35. Eastman, C. *Work-accidents and the Law.* New York: Charities Publication

Committee, 1910.

36. Hamilton, A. *Exploring the Dangerous Trades.* Boston: Little, Brown, 1943.

37. French, J. W. in D. A. Campbell, *Workmen's Compensation Insurance—Principles and Practice.* Parker, Stone & Baird Co., [n.p.], 1935. Vol. I, p. v.

38. Moorhead, J. J. Introduction, in *The Physician in Industry: A Symposium.* New York: National Industrial Conference Board, 1922, p. 103.

39. Addams, J. The Objective Value of a Social Settlement, in C. Lasch, ed. *The Social Thought of Jane Addams.* Indianapolis: Bobbs-Merrill, 1965, p. 46.

40. Mock, H. E. *Industrial Medicine and Surgery.* Philadelphia: W. B. Saunders, 1920, p. 371.

41. Meredith, F. L. "Physical Examinations." *J Ind Hyg* 1:556-563, 1920.

42. Hueper, W. C. "Some Comments on the History and Experimental Explorations of Metal Carcinogens and Cancers." *J Natl Cancer Inst* 62:723-725, 1979.

43. Shubik, P., and D. B. Clayson. "Environmental Cancer and Chemical Agents 1976." *Ann Intern Med* 85:120-122, 1976.

44. Warren S., quoted in Eckert, W. G. "Dr. Harrison Stanford Martland *(1883-1954). Am J Forensic Med Pathol* 2:83-85, 1981.

45. Cited by E. Mastromatteo, in "From Ramazzini to Occupational Health Today From an International Perspective." *J Occ Med* 17:289-294, 1975.

Part II

The Workplace

Chapter 4

Safety and Health Management

by John V. Grimaldi

The complexity of modern production systems, a burgeoning technology of new materials and equipment, and increasing government restrictions and regulations, have resulted in growing concern about the health and safety of workers. The need to provide effective health and safety programs within the fiscal capabilities of companies has led to the concept of health and safety management. Although this concept uses many principles that are well known to corporate management, it does not simply restate ideas about managing. Rather, it emphasizes realistic health and safety goals and employs methods that stimulate others to support the program. The uniqueness and value of this concept is that it uses facts and information and a well-organized system to achieve the goal of a safe and healthful workplace. (The setaside on the next page defines the terms commonly used in health and safety management.)

INADEQUACY OF COMMON SAFETY APPROACHES

Probably the most common approach to plant safety has been to "prevent accidents." Although this approach undoubtedly has resulted in fewer injuries, it has not been completely effective, as indicated by the frequency of injury rates and observed noncompliance with safety measures. The strict accident-prevention approach, when it is not combined with management's responsibility and direction, has two serious limitations. First, it has the impossible goal of preventing all accidents—if an occurrence is accidental, by definition, "it is beyond control." Second, it imposes hazard control responsibilities that employees cannot fulfill effectively, again because of the fact that accidents are simply beyond control.

If accidents are to be prevented, those responsible for preventing

DEFINITION OF TERMS APPLICABLE TO SAFETY MANAGEMENT

Accident
An event occurring by chance or from unknown causes (and therefore unpreventable).

Safe
Protected against hazard.

Hazard
The source of energy and the physiological and behavioral factors which, when uncontrolled, lead to harmful occurrences.

Safety
The elimination of hazards or their control to levels of acceptable tolerance as determined by law, institutional regulations, ethics, personal requirements, scientific and technological capability, experiential knowledge, economics, and the interpretations of cultural and popular practice.

Safety Management
The accomplishment of safety through others.

The Work of Safety Management
Accomplishing safety by applying persuasive methods and information in a closed-loop system.

Persuasive Methods and Information
Discipline, reward, punishment and the power of facts well marshalled when presented in the way necessary to obtain the desired action.

Risk
The assumed effect of an uncontrolled hazard, appraised in terms of the probability it will happen, the maximum severity of any injuries or damages, and the public's sensitivity to the occurrence.

them must have unlimited information not only about unsafe conditions and practices and means of counteracting them, but also about individual, social, economic, and technologic factors that affect the magnitude and occurrence of hazards and their control. Unfortunately, not all of this information is currently available. Consequently, those responsible for accident prevention soon become aware that they cannot achieve this goal and consequently become frustrated. Furthermore, by terming injury occurrences "accidents," it can be argued that since the cause of an accident cannot be controlled, therefore the responsible individual, whether determined to be culpable or not, cannot be guilty. It becomes difficult, if not impossible, therefore, to maintain safety effectiveness, and that is the primary weakness of accident prevention programs, which is a major reason behind the trend to control hazards.[6]

MANAGEMENT RESPONSIBILITY FOR EMPLOYEE SAFETY AND HEALTH

Management, in essence, consists of persuading others to do their jobs properly. Although managers use job standards, performance evaluations, rules and policies, company regulations, and training and other aids to motivate employees to effective work performance, the most important factor is the chain of command that persuades personnel at each level to meet their responsibilities. Some managers, however, are inclined to delegate their responsibilities to those under them—particularly in the area of health and safety programs.

But the health and safety of workers are acknowledged management responsibilities and must *not* be delegated to subordinates or the program's effectiveness will be destroyed. Of course, each worker is responsible for his or her own safety, but in any organization where people work together to produce goods or services, the line authority (management) is obligated to assure, insofar as possible, the health and safety of the employees. The key to an effective program is a commitment to it by line authority. Management must be held responsible and accountable for health and safety in the workplace if the program is to be successful (see Figure 4-1).

The safety management function

Management of safety in the workplace basically amounts to achieving safety through others. But, unlike operational management in which managers have direct control over their immediate subordinates, safety management is conducted generally as a staff assignment. Thus, its "managers" are not managers in the true sense, because the powers of the line of authority cannot be transmitted through an advisory (staff) safety function.

There is evidence that safety is maximized only in well-managed organizations. Thus, the occurrence of serious injuries and damage appear to be directly related to the managerial effectiveness of the line organization. In an 11-year study, it was found that severe injuries and fire losses varied directly with management's fiscal performance even though the safety program did not change.[1] In this case, year-to-year differences in management performance appeared to influence the organization's safety effectiveness, particularly the control of severe occurrences.

Consequently, health and safety staff responsibilities must not be inadvertently intermingled or confused with line management responsibilities. This premise is the most important rule of health and safety

DISTRIBUTION OF RESPONSIBILITY/ACCOUNTABILITY/AUTHORITY

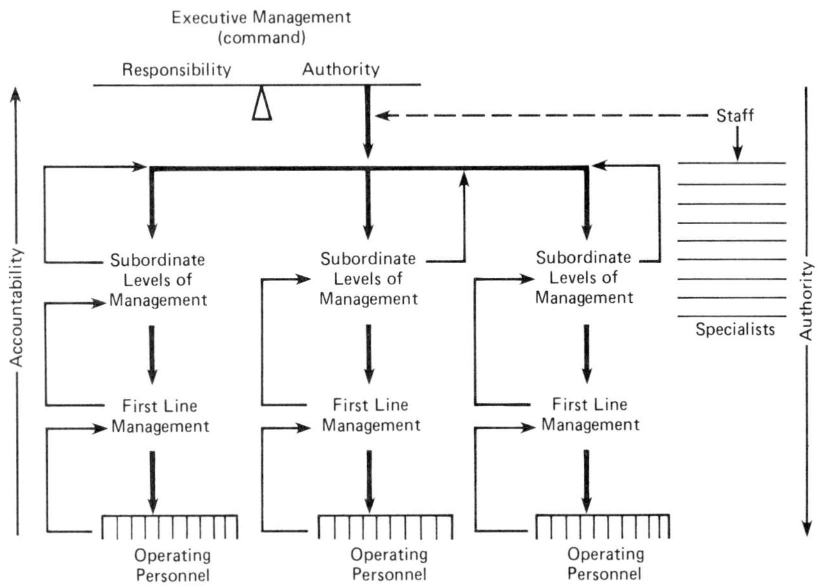

THE DIAGRAM PICTORIALIZES THE FOLLOWING RELATIONSHIPS:

1. Levels of authority must be commensurate (equal) to the responsibility for the mission.
2. Authority may be delegated, mission responsibility may not. Organizational effectiveness, in all its aspects, is the responsibility of the Executive (Command) function. (Subordinate levels of management will have specific task responsibilities of course, that relate to the mission's accomplishment with commensurate authority to fulfill them.)
3. The "power line" of authority follows the heavy black arrows flowing downward.
4. Accountability flows up.
5. The broken line, tying STAFF to the organization, indicates these personnel are outside the "power line."

Figure 4-1. To Succeed, a safety program needs management's commitment of responsibility, accountability, and authority.

management. A common error in this regard is to consider the health and safety professional as a spokesman for management. But the health and safety professional *is* a staff member, and staff's responsibility is to advise and counsel management on technical matters. Although the manager is released from pursuing details, he is still responsible for making sure that his subordinates accomplish the task effectively. If he or she does not use his authority constantly and visibly to stimulate his subordinates, his own job performance will suffer.

Staff support often includes the development of training and other auxiliary programs. But these ancillary activities cannot be allowed to disrupt the health and safety staff from its primary function, which is to alert management to hazardous conditions and practices and to keep current with the most advanced health and safety information. Moreover, managers must not delegate their authority to demand that employees observe health and safety standards; this is especially important because staff safety personnel do not have the power to demand this commitment. Consequently, the safety program will be severely hampered if management considers it to be of marginal importance and delegates its authority over the program.

Health and safety management practice

With the growth of federal and state control of conditions in the workplace, advocates of health and safety management have increased considerably.[2] The practice of health and safety management, however, has not progressed concurrently. Too often it has been a mixture of traditional health and safety methods and some management concepts.

Originally, accident cost information was used to define safety problems in industry, because professionals in this field felt that the dollar amounts of accidents would have a greater impact on management. Although accident cost figures have proved to be fundamentally sound, their significance depends on how much loss an employer is willing to accept. Certainly, economic factors are not the only basis for making risk decisions. Others that are taken into account are (*a*) the probability of the risk, (*b*) the type and magnitude of the consequences, (*c*) how the event could take place, and (*d*) the trade-off considerations in risk versus safe situation. The interrelationship of some of these factors is shown in Figures 4-2, -3, and -4.

Figure 4-2 shows risk acceptance as a function of plant damage or fiscal loss. It suggests that risks would be avoided when there is a high probability of a loss, even if the dollar amounts were minor. Similarly, regardless of the amount of the projected loss, as the probability of loss decreases, risk-taking increases. The acceptance of risks is also higher in some businesses, like minerals, petroleum, and timber, than in companies producing products or services.

Figure 4-3 indicates the degree of risk acceptance associated with human factors—how much risk is acceptable when human injury or fatality is in question? The history of health and safety legislation indicates clearly that the public is more interested in worker health and safety than in property damage and fiscal loss even though the latter can also result in increased product cost to the consumer. Thus, reducing fatalities is the principal motivation for most hazard-control laws.

Safety systems control costs by preventing damaging occurrences or,

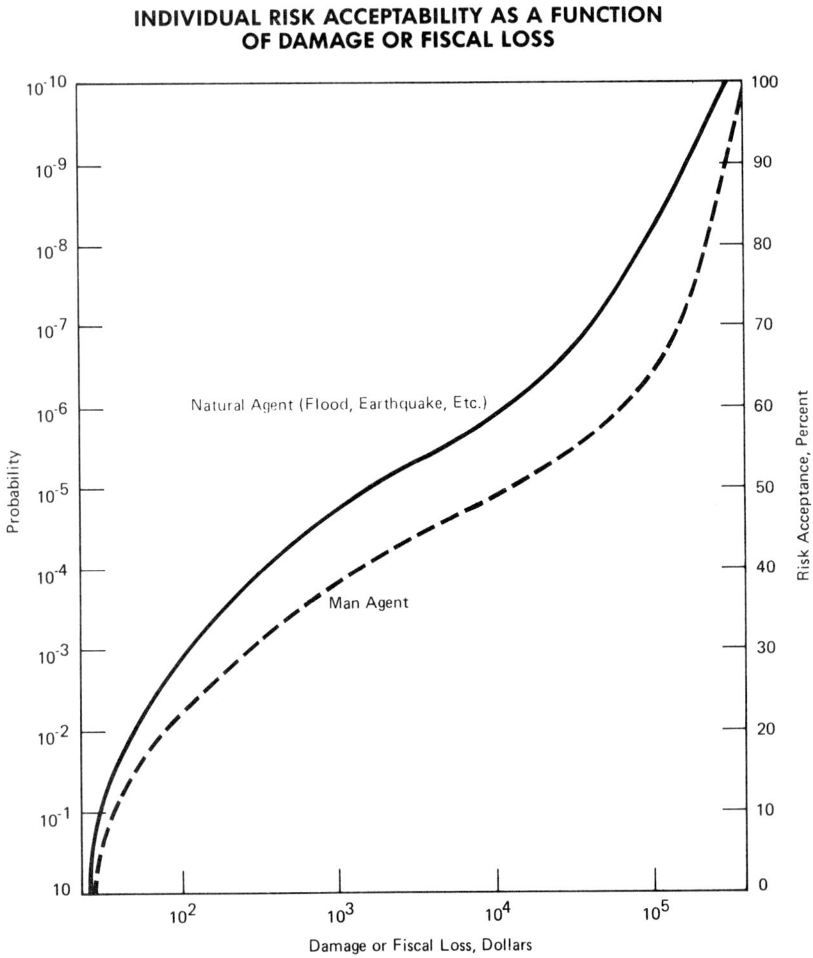

Figure 4-2. Regardless of the amount of the projected loss, as the probability of loss decreases risk taking increases.

at least, by limiting their effects. However, there is a point where the costs of this control exceeds the savings realized. Consequently, cost-effectiveness studies are demanded prior to any major safety action. Figure 4-4 illustrates cost relationships as a function of the level of control. When damage control alone is considered, it should be noted that there is an optimum cost that cannot be exceeded.

The three figures show that safety management decisions are not always made on the basis of safety considerations alone. Other factors that

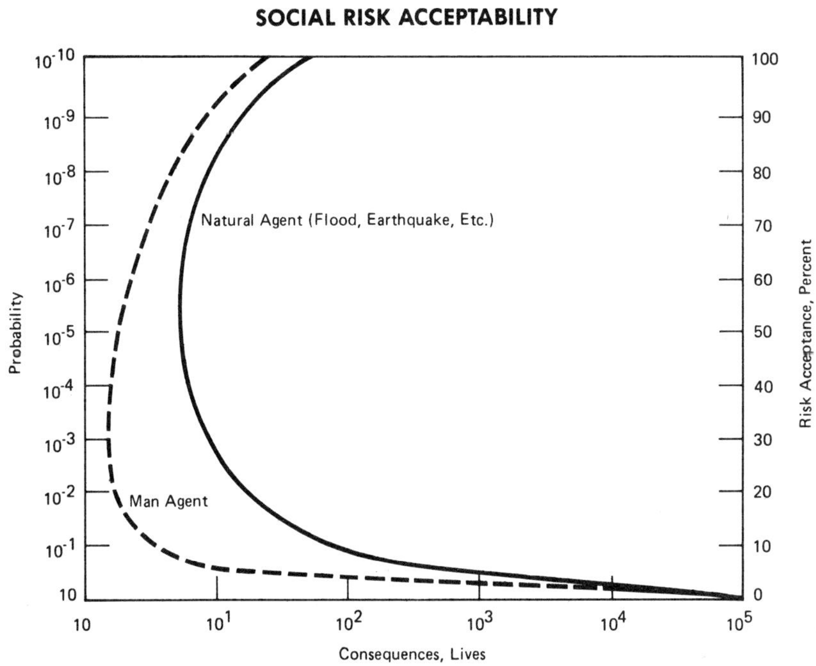

Figure 4-3. The public is more interested in worker health and safety than in property damage and fiscal loss. Risk taking occurs only when the probability of a fatality is removed.

are taken into account include anticipated gain, convenience, time savings, and personal preference. With regard to the last, for example, a worker may resent the safety requirements and may reject or circumvent them—leading to the not uncommon practices of removing machine safety guards, not wearing protective equipment, and taking unnecessary chances. Managers also may choose the wrong option with respect to safety risk because of their own personal values, as well as the desire to save time and money.

As in any decision-making process, many issues may be involved in safety management decisions. In both cases, decisions are usually resolved by determining which factors are most critical and whether safety or risk is to be preferred. All decisions must be based on accurate and inclusive information, and the health and safety staff must provide this information so that managers will have a basis for making appropriate decisions.

Decision making. It may be argued that health and safety decision-

COMMUNITY RISK ACCEPTABILITY

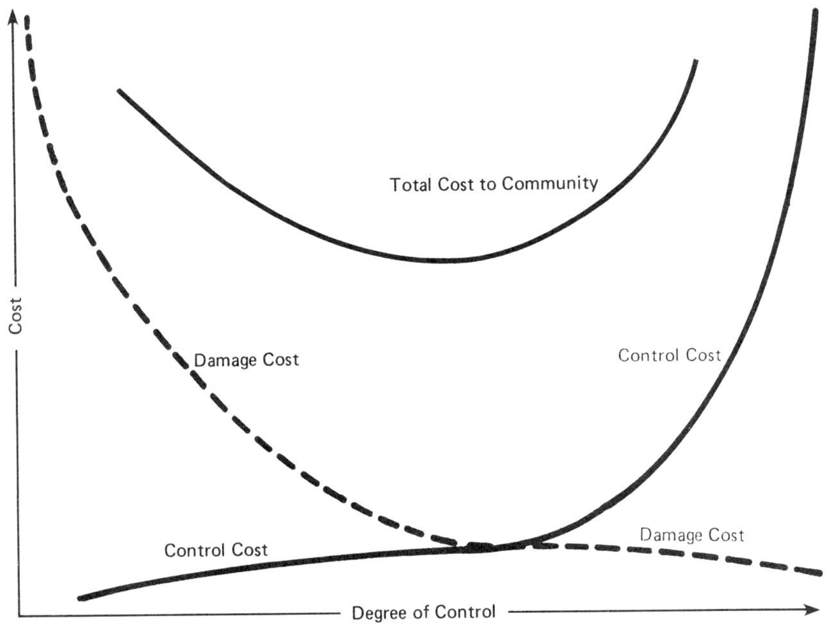

Figure 4-4. Control costs and damage costs intersect at a point of optimal total cost to the community loss control. Cost-effectiveness for studies are demanded prior to any major safety action.

making is a relatively simple matter since laws, regulations, policies, and standards—as well as potential sanctions—delineate the answer. But deviations from safety standards, by all levels of employees, are observed constantly and for the reasons cited above. Obviously, health and safety decision-making is not merely a question of right or wrong.

When occupational safety became a major concern early in this century, conditions were so hazardous that even simple remedies were effective, for example, machine guards reduced or eliminated many hazards. Because such measures could reduce some of the most blatant problems, carriers of workers' compensation insurance encouraged their insureds to install safeguards by cutting the insurance premium for those doing so. The insurance carriers called this approach "schedule rating."

Although simple remedies were adequate for earlier problems, the proliferation of new technologies and new materials has raised new problems for health and safety management that cannot be solved so readily. These advances have resulted in new products, increased productivity, and increased product value, but they have also introduced new hazards into the workplace. (With this change, "schedule ratings" became obsolete and are now used only to cover fire safety.)

Program effectiveness

Health and safety management is not simply a matter of knowing what causes injuries and occupational illnesses and how to prevent them. If safeguards, personal protective equipment, and health and safety regulations are circumvented, the health and safety program becomes meaningless. Furthermore, there are too many simplistic notions about safety. The concept of inducing safety through such slogans as "The safe way is the efficient way!"—does not solve the problem. Other commonly held concepts for safety include the following:

Develop safety awareness.

Develop safety performance criteria and legislate compliance.

Educate for safe living.

Inspect and enforce.

All of the above appear to be reasonable, but they are not sufficient by themselves. This becomes clearly evident when injury trends rise.[3]

If those responsible for health and safety in their workplaces do not throw off the yoke of dependence on incomplete practices and concepts, the program will be ineffective. Penalties and sanctions against ineffective programs can be expected to become increasingly severe. Accordingly, effective health and safety management programs require the following:

- Awareness of the weaknesses in a program; that is, know where the problems are.
- Knowledge of how to improve the weakness.
- Readily available and pertinent information for determining solutions as rapidly as possible.
- Stimulation of management to exert its influence and money to encourage a successful program.

Unique managerial skills are required to implement and sustain an effective safety management program. These skills include the planning, organizing, coordinating, leading, and control stratagems employed by line managers and health and safety managers. However, the health and safety manager must also have skills for convincing and leading managers and for influencing personnel indirectly since he or she will not have the power of line authority. In essence, health and safety management amounts to encouraging, leading, and persuading others to accomplish an effective program.

In addition to maximizing hazard control, which is the basic objective of health and safety management, the management program must, to the greatest degree possible, make use of the influence of line authority and must make sure that line management does not delegate its responsibility and accountability for health and safety to staff-level subordinates. In particular, health and safety managers must not permit line management to delegate responsibility and accountability to them. The functions of line (command) and staff (advisory) must be kept separate and distinct. To accomplish an effective health and safety management program, the following must be done:

- Keep line management responsibilities separate and visible and not mixed with staff functions.
- Hold the line managers clearly accountable for the operation's health and safety effectiveness.
- Hold the health and safety staff accountable for providing effective program concepts and accurate, up-to-date information to line managers and persuading managers to support the program by requiring supervisory and employee adherence to all health and safety regulations.
- Develop operating objectives for health and safety management that not only meet standard health and safety requirements, but also are coordinated with management needs and goals.
- Measure hazard control effectiveness against standard requirements and management's appraisal of its own health and safety performance.

Let's discuss each in turn.

Separation of line and staff health and safety functions. Keeping line and staff functions separate may be easier discussed than accomplished, but the principle is clear. Line has the power of authority and staff the knowledge; consequently the two must be separate, but still interacting. The majority of organizations have assigned health and safety duties to specialists in the field, because they will know more about it and be able to develop and implement the program.

Accountability of line management for health and safety. Management cannot be held accountable for health and safety unless its functions are separated from those of staff. Health and safety management considers the line manager responsible for the health and safety of his operation. But even if management acknowledges this responsibility, some managers may transfer this function to others so that they can reserve their time for operational decisions. When that occurs, literally no one is responsible for health and safety in the workplace. Even if health and safety professionals were given the power to shut down an operation they considered to be hazardous, the accountability issue would still not be settled. Therefore, the line manager must be held accountable for health and safety, and his performance in this regard must be evaluated, including compliance with the laws, standards, guidelines, and other requirements for health and safety.

Accountability of the health and safety staff. The health and safety staff is accountable for providing accurate information on health and safety to line management and for advising and counseling them. Health and safety staff accountability is more difficult to achieve than managerial accountability, because health and safety staff members are often assigned tasks that no one else can do. Furthermore, many health and safety professionals prefer operational work to the encyclopedic task of searching for information developing programs, and the static work of advising and counseling. Finally, when the health and safety professional calls management's attention to health and safety violations or hazardous conditions, he or she is advising management to order employees to comply with safe and healthful practices, although the health and safety professional has no power to enforce such compliance.

The effective health and safety manager has at his or her command only the ability to produce usable and well-prepared information and programs, and to present them persuasively to management. In addition to the standard motivations of salary and advancement, the special motivations are pride in the work, satisfaction of serving clients well, and the effectiveness of the health and safety program.

The most difficult problem facing the health and safety professional is the inability to eliminate entirely or even control some occupational injuries and illnesses. The author believes that this problem arises from not separating the line and health and safety staff functions and from not making managerial and health and safety staff accountability meaningful and operable. This conclusion is based on observing a wide variety of businesses and industries over a 25-year period in the United States and countries overseas. The breakdown in accountability occurs largely when health and safety staff members intrude on managerial accountability, either unwittingly or through a desire to get the job done. Consequently, accountability of the two functions, line and safety and health staff, must be kept separate.

Operating objectives that meet health and safety needs and management goals. Health and safety requirements will not be complied with simply because they have been written down. Therefore, laws and regulations are developed to cause the organizational hierarchy to demand employee compliance. Without the laws and regulations, this would not be so. Lacking effective management and organization of the health and safety program, a company cannot control hazards and compliance with healthful and safe practices. But management is not always interested enough in health and safety and, sometimes, places greater emphasis on operational functions. Consequently, the health and safety program must be made as important as operational matters, but that is not always easy to accomplish.

Two factors can contribute to management's lack of interest in the health and safety program. First, managers may feel that hazards to health and safety are overemphasized, and that a hazard which appears to be harmless cannot be dangerous. Second, the managerial talent for problem solving can limit the effectiveness of the health and safety program. The first factor, of course, stems from the human inclination to emphasize only the dangerous. The second may appear to be a paradox, but is logical upon considering that managers survive according to their ability to solve operating problems—that is "to put out fires." If the fire is serious and the solution effective, the manager's reputation is strengthened. The impact of this on the health and safety program is that some managers can be less inclined to take preventive measures and be more inclined to "fight fires."

If the health and safety program is to be supported actively by management, it must have goals that parallel management goals. In this case, moral issues of preserving the health and safety of workers will not suffice. Proponents of occupational health and safety have argued that effective programs will increase efficiency through reducing injury and illness costs and lost work time. But these assumptions are highly speculative and may be contradicted by readily calculated costs for implementing health and safety programs. These figures may indicate that the dollar savings will not be high (see Figure 4-5).

Experience suggests that the strategy to obtain management support and action should (*a*) use health and safety issues that attract attention and (*b*) demonstrate their significance with illustrative cases. What type of issues would attract attention? In many cases, they would be issues that must be acted upon, such as federal and state (or provincial) laws; organizational policies, rules, and operating procedures; or accepted good practice standards. The fact that they are not being complied with should stimulate corrective action immediately. If the issues do not have an impact and it is determined that requirements have been violated deliberately, the organization must reconsider whether or not the health and safety issues stressed are appropriate. That is, emphasis may be needed in

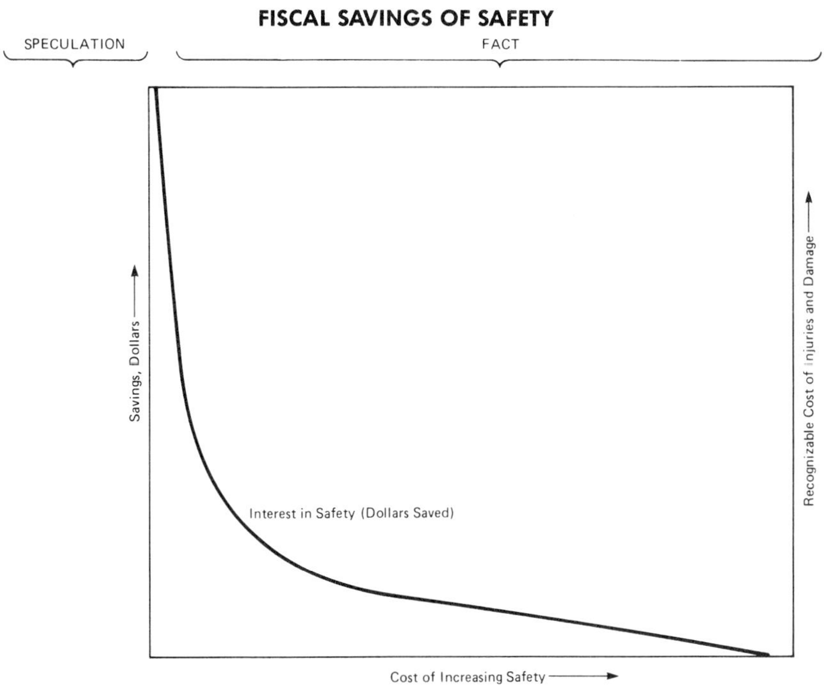

Figure 4-5. The cost of health and safety programs is not always supported by savings in injury and illness costs and lost worktime. The management goal may be the optimal balance of health and safety program costs with savings resulting from it. For many industries, however, this may result in the allocation of inadequate funds to support a health and safety program.

other areas.

Specific hazard control requirements usually do not exist for every serious exposure because of the time lapse between the identification of the hazard and the issuance of regulations covering it. But OSHA imposes the responsibility on employers to foresee such dangers and regulate against them, and the current state of the knowledge of health and safety management makes that possible. Thus, a company must not only emphasize those issues that will stimulate corrective action, but also foresee other potential hazards. The economic costs of failing to do so can be severe in terms of indemnities, fines, direct losses, and adverse public relations. But the costs, in themselves, will not stimulate corrective action unless the consequences of noncompliance are not only severe, but also reasonably certain. To assure compliance with the requirements of health and safety management, it may also be necessary to:

Present the facts, usually as case histories, rather than as a list of requirements.

Use accountability measures and authority (discipline and incentives) to gain the desired goal.

Measurement of health and safety effectiveness

Health and safety effectiveness must be measured in ways that increase managerial response. It is an operational fact that what is measured is what gets done. It is also a fact that managers need to know how well their operations are measuring up to established goals. Peter Drucker has pointed out that managerial measurements need not be rigidly quantitative or exact, but must be sensible, clear, and simple. They also must be relevant so that they "direct attention and efforts where they should go."[4]

Although measures should be as reliable as possible, those that are less than reliable can still be effective *if* their limitations are understood and accepted.

For many years, the severity rate (the seriousness of occurrences) has been thought to be a somewhat accurate measure of the effectiveness of health and safety programs. The reliability of this measure for determining the effectiveness of management control of severe hazards has been discussed in the literature.[1] It is also important for persuading management to take corrective action because it indicates whether or not managers are performing as they should.

Some managers may attempt to justify inadequate health and safety performance on the grounds that they are taking a calculated risk that nothing catastrophic will occur or that they are making a tradeoff between safety/health and costs. The first is unacceptable. The second, also unacceptable, is a fairly common opinion—that is, health and safety programs are merely a kind of "goodness" that cannot be justified by their cost. In this case, not only are dollar costs considered, but also the inconvenience and time required to implement health and safety precepts. Those supporting this argument believe that health and safety programs are often too generalized and nebulous in concept and consequently are not worth the tradeoff. In the minds of many managers, calculated risks and tradeoff health and safety are not a matter of morality, but of sharp business practice. The question underlying such decisions is: How much danger (or common good) is there in going in one direction rather than the other?

Managers are often forced to make risky decisions, and they realize that adverse consequences will be unacceptable, particularly those that result in severe injuries or illnesses, damage, and/or loss. Although such events may be infrequent, they do occur and will continue to occur unless management is alert to the possibility and insists that serious hazards be controlled. The same applies whether the health and safety of workers or the damage to or loss of a project or program is in question. The severity rate helps to measure the effectiveness of control. If managers are consistently attentive to controlling contingencies in general, they will also have better control over hazards that cause severe illnesses/injuries.[5] Conversely, when severe injury/illness or damage occurs, a manager's effectiveness is in question; consequently, managers will feel more responsible for limiting or preventing such occurrences.

Health and safety performance measures that do not indicate whether or not a manager is effective will have little impact because they will not motivate managers to perform their health and safety responsibilities properly. That is why frequency rates have little effect on management health and safety performance.

INTRODUCING HAZARD MANAGEMENT

Health and safety programs can be very complex and, consequently, are difficult to implement. The most difficult problem to solve in implementing such a program is understanding the ramifications involved. Figure 4-6 provides details. The development difficulties include the following:

- Prescribed sociophysiological concepts do not apply to the worker population to be protected.
- Damage-prevention or energy-control concepts do not apply to the organization's operations.
- Standard hazard countermeasures are not always applicable.
- Achievable goals too often permit the reduction of hazards to only a tolerable level.

As Figure 4-6 indicates, the control of hazards is more than simply attacking the agents of injury. Energy sources must be modified or controlled, and, perhaps even more important, the antagonistic logic that often restrains health and safety decisions must be overcome. Powerful arguments must be presented to overcome both the obvious and undetected reasons for rejecting health and safety programs both in management and workers. These arguments must be structured to support the health and safety management issues previously described and should include: risk decisions factors (severity of consequences, probability of legal action, public relations effect); political, technological, and economic outcomes; management responsibility for controlling contingencies; and explicit responsibilities for meeting legal requirements.

Some argue that the difficulties facing health and safety developers and managers can be overcome with technological approaches. They

SUMMARY OF RESPONSIBILITIES OF MANAGERS AND SAFETY AND HEALTH SPECIALISTS

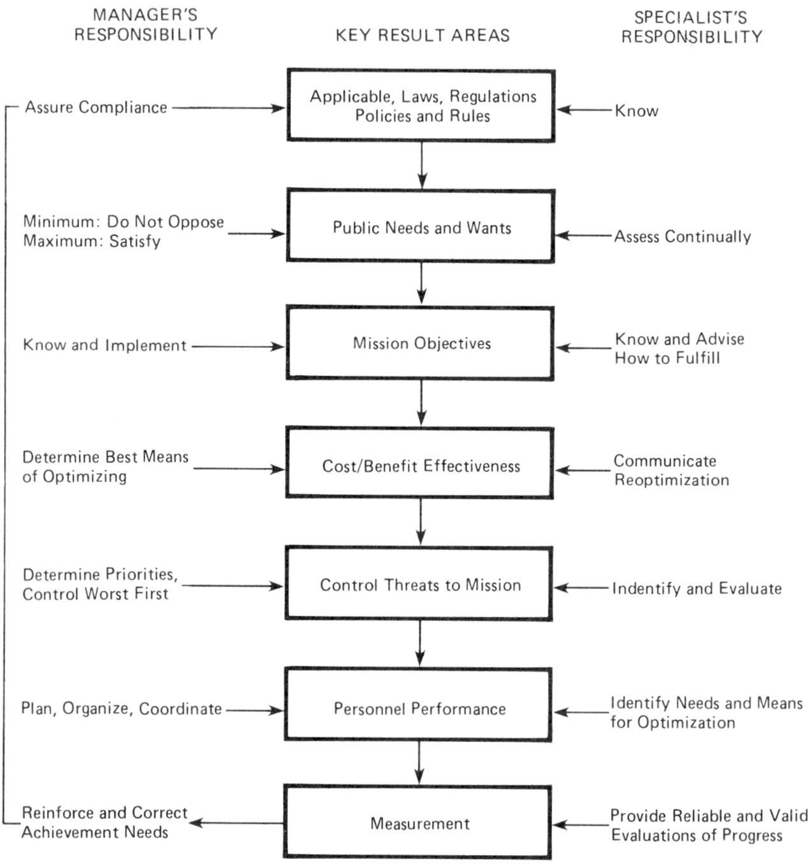

Figure 4-6. Safety and health specialists along with management personnel must accept definite responsibilities in order to overcome resistance to the implementation of effective safety and health programs.

conceive of the problem as simply a matter of controlling energy sources. That is, if safer energy forms are substituted for hazardous ones, or if hazardous energy sources are made less hazardous, the problem of safety development and control would be pretty much resolved.

Haddon has proposed the following strategy options for reducing energy-damage losses:[5]

1. Prevent the initial marshalling [use] of the form of energy.
2. Reduce the amount of energy marshalled [used].

3. Prevent the release of energy.
4. Modify the rate of spatial distribution of release of energy from its source.
5. Separate in space or time the energy being released from the susceptible structure.
6. Separate the energy being released from the susceptible structure by interposition of a material barrier.
7. Modify the contact surface, subsurface, or basic structure which can be impacted.
8. Strengthen the living or nonliving structure which might be damaged by the energy transfer.
9. Move rapidly in detection and evaluation of damage and to counter its continuation and extension.
10. [Use] all those measures that fall between the energy period following the damaging energy exchange and the final stabilization of the process (including intermediate and long-term reparative and rehabilitative measures).

Health and safety strategies with regard to energy sources must not only safeguard workers from exposure, but also prevent them from invalidating the controls.

Protocol for controlling hazards.

The author's experience suggests that a plan for controlling hazards should be based on careful analysis and evaluation of the hazard situation and all factors related to it, including sociophysiological characteristics of the worker population and options available, depending on the significance and complexity of each hazard being considered. The proposed plan within reason is as follows:

Step 1. Analyze the process, operation procedure, or system, and list all latent or kinetic physical hazards. (Omit only the materials, devices, arrangements, designs, motions, routes, equipment, chemical reactions, and so forth, that are known to be safe or harmless. Maintain a questioning attitude. Where any doubt exists, list the item.)

Step 2. Evaluate the significance or severity of each hazardous exposure (magnitude of the most serious possible single damaging event or injury/illness that could occur, number of individuals or valued materials that would be exposed), frequency of exposure, and public sensitivity to the slightest and greatest possible occurrence.

Step 3. Decide which of the following options or combination is most appropriate for the level of control required according to the findings in Step 2:

OPTIONS FOR CONTROLLING HAZARDS		
PREVENTION	**CONTROLS**	
	PHYSICAL	**PERSONAL**
Eliminate the hazard by designing it out of the system or by substituting an acceptable material, arrangement, device, and so forth.	Positive: A. Shield, enclose, contain or restrain the hazardous influence. B. Isolate the hazard spatially (by distance, barrier, and so forth.) C. Dilute the hazardous effect (by ventilation, wet processing, energy absorption: i.e., bumpers, padding, acoustic tile, etc.) shock and vibration mounts, and so forth. Other: D. Provide personal protective devices where positive controls are inadequate.	1. Operator selection 2. Operator training and education 3. General education

Step 4. Reevaluate the process to make sure that no synergistic, hazardous effects have been introduced by any modifications.

Although the four steps described above place strong demands on the health and safety professional, they are the basics that must be adhered to implicitly and explicitly. They are neither less than nor greater than what public sentiment is requiring from health and safety programs. For example, to satisfy OSHA's general duty requirements, Step 1 must be adhered to. To persuade management to formulate appropriate decisions on hazard control, Step 2 is required. The options identified in Step

3 provide the most practical countermeasures to hazards. It should be noted that in Step 3, prevention is distinguished from controls since prevention may be close to impossible because it would mean removing the hazard entirely. Because that is impractical, the hazard is reduced to the lowest extent possible. Physical and personal controls are employed to accomplish this. As shown in the preceding tabulation, the effectiveness of the physical control options is diminished as the choice progresses from Option A through D; consequently, personal controls become more important with each declining option. Finally, Step 4 is needed to reevaluate the entire process so that selected options cannot result in the creation of additional hazards.

CONCLUSIONS

A safe and healthful workplace is not simply a matter of using safeguards and nontoxic chemicals. The achievement is complicated by the need to meet goals through the efforts of others. Consequently, because of the ambivalent attitude of some people toward health and safety, reaching the objective often is a frustrating experience. Supporters of health and safety, therefore, may feel compelled to promote their programs as if health and safety were something unrecognized and unappreciated. But the worth of the health and safety of workers is not in doubt. It is highly regarded for moralistic and humanistic reasons. The limiting factor is that its value is not always clear when risk versus health and safety decisions are made. Unfortunately, depending on the estimated outcome, the risk may appear more attractive at times than health or safety.

Management and workers may be as contradictory about health and safety as they are probably about their jobs. Given a choice, most people probably would not work as hard or as frequently as they do—except that circumstances require it. Similarly, many would very likely circumvent safety and health practices and overlook hazards unless they are made accountable for doing so. Therefore, workers must be made accountable to their managers, and managers must be made accountable not only for their own health and safety behavior, but also for that of their subordinates. They must take the operating steps necessary to achieve the desired health and safety performance of the entire group, themselves included.

Organizations have evolved to accomplish a purpose, whether that purpose is to manufacture a product or to provide a service. They are the best means available at present for optimizing the accomplishment of objectives through the efforts of others. In the manufacture of a product or provision of a service, each member of the group—from manager to worker—must do his job in the manner the organization specifies and each is accountable for doing so. The same is true of safety and health programs—a job must be done, controlling hazards; it must be done according to current health and safety standards; and the individual doing it will be held accountable for doing what is wanted. Thus, there is a hierarchy for health and safety programs just as there is a hierarchy in operations—that is, the worker is accountable to his supervisor, the supervisor is accountable to the manager, and so forth.

Integrating the health and safety program into this hierarchy is the responsibility of safety management. To do that, senior management must be persuaded that health and safety precepts are as important to the organization as operational performance. Management must recognize (*a*) the need for complying with laws, regulations, and standards; (*b*) the importance of identifying hazards and taking action to eliminate or reduce them; (*c*) the significance of keeping managerial and health and safety staff functions separate; and (*d*) the importance of managerial accountability.

REFERENCES

1. Grimaldi, J. V. "The Measurement of Safety Engineering Performance." *J. Safety Res.,* 2(3):137, 1970.
2. Grimaldi, J. V., and R. H. Simonds. *Safety Management.* Homewood, Ill.: Richard D. Irwin, 1984.
3. *Accident Facts.* Chicago: National Safety Council, 1984.
4. Drucker, P. F. *The Practice of Management.* New York: Harper and Brothers, 1954.
5. Haddon, W., Jr. "A Logical Framework for Categorizing Highway Safety Phenomena and Activity," *10th Int. Study Week in Traffic and Safety Engineering.* Rotterdam, The Netherlands, Sept. 9, 1970.
6. McElroy, Frank E., ed. *Accident Prevention Manual for Industrial Operations—Administration and Programs.* Chicago: National Safety Council, 1981.

Chapter 5

Scientific Background Of Occupational Health

by Charles E. Becker

> In biological sciences the role of method is even more important than in other sciences because of the immense complexity of the phenomena and the countless sources of error which complexity brings into experience.
>
> *Claude Bernard*

Science combines techniques of observation, identification, description, and experimental investigation, leading to the theoretical explanations of natural phenomena. Simply put, science is knowledge gained through experience. The word science literally means "to know." In the field of occupational health and safety, knowledge is gained through studying the environment in which men and women work. The scientific goal is to identify the specific causes of safety and health problems in the workplace. This task is difficult in light of the great complexity of modern working environments and the number of factors that interact continuously to affect human health and safety. Scientists are increasingly asked to ascertain if conditions in the workplace can cause or exacerbate a safety or health problem. This chapter touches briefly on the various scientific disciplines that relate to the study of occupational health and safety.

ASSOCIATION/CAUSATION

In the search for knowledge, scientists analyze past experience and attempt to show that a given association of factors is the cause of a specific health problem. Several methods are used to examine the strength of any association in causing certain problems and all require an understanding of the dose response relationship. That is, when an increasing effect can be identified with an increasing dose, the chances are that the association is more than coincidental and, in fact, may very likely be a cause. The fact that an increased dose has an increased effect further suggests that the agent being studied is reaching a specific location. It is critical to select a specific effect (for example, the risk of causing cancer), but that is not always easy. Often, death is chosen as a specific effect. Thus, specialized knowledge of dose response allows the tentative establishment of *thresholds* for given specific effects.

The cause is not always best studied on the basis of life and death. Rather, specialized knowledge of causation can be aimed at more subtle health and safety consequences, such as worker satisfaction or days missed from work.

Obviously, the cause must precede effect, but it is important to remember that a suspected cause is simply a first manifestation and not necessarily a basic cause. Therefore, the true cause cannot be ascertained until it is proven to precede the effect.

The best example of making sure that a cause clearly preceded the event is with a rare tumor known as mesothelioma. Because of the rarity of this tumor, careful studies identified that there must be a prior exposure to asbestos to be the cause. This was first recognized in individuals living near the manufacturers of asbestos. It was because of the rarity of the tumor and the prior exposure, living near the plant, that the strong causal relationship could be established.

Sensitivity and *specificity* are methods of quantitatively estimating errors in predicting association and cause. Sensitivity is the proportion of truly caused events (diseased) that are tested positive; specificity is the proportion of noncaused events (non diseased) that are tested negative, (see Figure 5-1). It is evident that these two quantitative measures are inversely related to one another. Although they are not absolutely reciprocal, if we enhance the sensitivity of a given scientific method, specificity will suffer. Thus, investigators must constantly adjust the sensitivity and specificity of their methods because of the importance of assigning quantitative values to causes. If a health study is designed to identify the true proportion of sick individuals in a population with a given exposure, the testing method would have to be very sensitive so as not to miss identifying any truly sick individual. However, you would also strive to use a specific method to identify correctly all of those workers who are not ill. By altering sensitivity and specificity for a given health problem, we can balance the two extremes and thus get closer to identifying a true cause.

The best example of the importance of sensitivity and specificity for

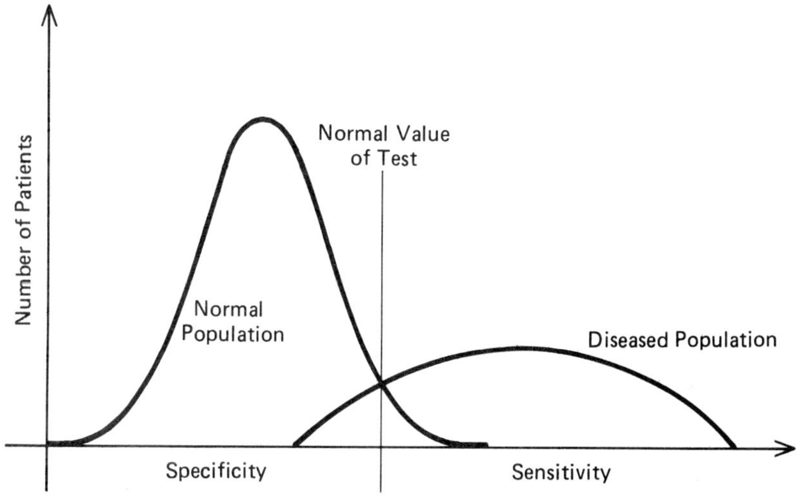

Figure 5-1. Sensitivity is the proportion of truly caused events that are tested; specificity is the proportion of noncaused events that are tested. They are inversely related to each other.

occupational health involves screening of individuals for heart diseases. A routine electrocardiogram, a stress test on a treadmill, or a stress test on a treadmill with a radioisotope known as thallium have all been used to identify cardiovascular diseases in asymptomatic individuals. As screening medical techniques in asymptomatic individuals, the electrocardiogram and the stress electrocardiogram have such a high false positive rate (low sensitivity) that their routine use will cause many individuals to have unnecessary expensive medical evaluation. If, however, the electrocardiogram and the stress test are applied to patients with heart disease, such as chest pain, the test is sufficiently specific and sensitive enough to provide useful information. If the patient is symptomatic and you wish to have the very best sensitivity and specificity to correctly identify heart disease, then the thallium stress test is currently the best method.

Laboratory animals help establish causes of occupational diseases. Animal tests have been used for many years to predict human health factors. Choice of animal species, use of a sufficient number of animals, appropriate dosage, maintenance of animals so that exposure is the only difference between exposure and control groups, careful pathological observations, and administering exposure materials throughout the expected natural life span of the animals are all necessary to establish research validity.

The best example of the assets and liabilities of animal experimentation relating to humans is with methanol poisoning. Many human cases of poisoning with methanol were discovered, but no animal species developed poisoning when challenged with methanol. It was then recognized that only in those species of animals that were deficient in the vitamin folic acid was it possible to produce methanol poisoning. Further research studies showed that most animals cannot be rendered methanol poisoned, unless they are made deficient in the vitamin folic acid.

Thus, the dose response relationships in animals must be related to knowledge of the basic biochemistry of the poison in order to gain certainty as to cause. Likewise, dose response relationships studied in a sensitive and specific manner and verified by laboratory animal studies, all help in gaining certainty as to cause.

Other questions must be answered in establishing the cause of a specific problem. Is the observed association reproducible? Is it sufficiently strong and does it support the findings of other scientific studies? For example, if an investigation is seeking to protect workers from a possible cancer-causing agent, identifying the required strength of the chemical agent that could cause cancer would be the goal.

Do short-term cancer tests provide supportive information for the cancer-causing ability of given chemical substances? In determining cancer risk, a single negative result does not provide sufficient evidence that a chemical cannot cause cancer. Even if a study has established appropriate dose response relationships, and controlled specificity and sensitivity testing have provided sufficient evidence of a cancer-causing effect, the findings must still be confirmed by further independent study.

The most frequent short-term cancer test used in common occupational health practice is the so-called "Ames test." This is a specialized urine test in which bacteria are exposed to agents that alter their basic molecular structure. Common cancer-causing agents alter these urinary tests and become a useful monitor for individuals exposed to carcinogens. It is now recognized that common foods, such as vegetables and cooked meats, cause these tests to become positive in individuals in whom there is no evidence of cancer risk. It is important to realize that these tests are useful for research purposes and should be carefully studied before they are applied to large populations of individuals, lest they cause undue concern before appropriate medical facts are available to interpret their clinical significance.

In the field of occupational health, sound conclusions must be established before suggesting that certain actions are caused by a given agent or process. If the biochemical and physiological background and findings suggest an association, the investigator is, of course, on the right track. Industrial hygiene, toxicology, and epidemiology are used to identify such causes and effects. Industrial hygiene is used to study the environment; toxicology is used to study the effect on the human body; and epidemiology assesses the work force for an outcome.

BASIC PRINCIPLES OF TOXICOLOGY

Toxicology is the study of dangerous or poisonous physical and chemical agents and their injury, or potential injury, to living cells. This injury may be reversible or irreversible and may have an effect on a single cell, an organ system, or the entire animal. Toxicologists gather data on the toxicity of materials to predict their impact on human populations and on the environment. These qualitative and quantitative studies can be related to clinical medicine, occupational medicine, and veterinary medicine.

By attempting to predict injury to living cells, toxicologists may work directly with patients, indirectly in industry, or with the government. Because human experimentation in the field of toxicology is not always feasible, it is often necessary to base health decisions on biochemistry, interpretation of animal studies, and careful assessment of environmental monitoring.

The word *toxic* indicates that an agent can cause reversible or irreversible injury to living cells. *Hazard,* on the other hand, involves the probability that injury will result from the agent under specific conditions of use. A toxicologist may describe a chemical or physical agent as producing a reversible or irreversible injury to a living cell, but the important question is whether or not the living cell will ever be exposed to the toxic effect. Thus, knowing the toxicity of the given physical or chemical agent is not sufficient. It also must be studied during the condition of use. *Risk* is another term sometimes used in toxicology. Risk, like hazard, is the probability that a substance will produce a toxic effect under specific conditions of use. *Safety* is the reciprocal of risk—that is, the probability that no toxic effect will occur under specific conditions of use. These terms must be used precisely by occupational health and safety professionals.

FACTORS INFLUENCING TOXIC EFFECT

Many factors must be considered in determining a toxic effect.

Dose is the single most important factor, but the dose must be delivered to a molecular target or specific cell to have its effect. That is, if a toxic chemical is placed on the skin but does not penetrate the skin to gain access to the circulation, no toxic effect will occur.

The spectrum of undesired effects is so broad that the *nature of the injury* induced by a toxic chemical must be carefully defined. For example, a small quantity of nickel may induce a toxic effect by an allergic reaction resulting from a previous sensitization. This reaction is not a toxic effect from the metal, but rather results from an earlier alteration of the immunological system. The dose response relationship of a chemical causing an allergic effect would look very different if the same chemical caused a direct toxic effect on a given cell or organ. In the basic pathology and mechanism of injury, these reactions are quite different and the distinction between them is crucial in understanding cause and effect.

A toxic agent has *chemical* and *physical characteristics,* as well as *impurities.* Impurities may also account for the observed effect. One of the major problems in toxicology is identifying toxic effects of chemical agents because impurities are not always obvious. For example, it could take as long as 30 years before certain causes of cancer are recognized. Further, the observed toxic effect may not always occur in the adult animal exposed to the toxic agent, but may be transmitted to the next generation. An example of this type of toxicity is that laboratory animals exposed to lead in a sufficient dose to allow procreation cause alteration in the sperm of the next generation of animals.

Toxicity is also dependent on *exposure factors.* The rate of absorption of the toxin and duration of the effect may be critical factors. For example, the body normally is able to metabolize a small quantity of cyanide, but with a large exposure to cyanide, this capacity is easily exceeded. Thus, the dose and rate of absorption become critical factors.

Toxicity is also affected by the presence or absence of *special warning properties.* Certain toxins have pungent odors that provide warning. Some toxins are acids (sulfuric acid) and some are alkaline materials (sodium hydroxide). When either of these is placed on the skin or inhaled into the lungs, it will cause irritation. Local effects, such as skin irritation, provide some degree of warning. However, acid or alkaline substances may injure the protective barrier of the skin and thereby allow the toxic agent to penetrate. These toxins injure living cells directly (like hydrofluoric acid does).

Most toxins, after gaining entrance into the body, do not affect organs uniformly. Usually a specific molecular target or organ receives the primary toxic effect. These targets differ according to the specific agent. For example, the peripheral nervous system is the target organ of arsenic. Thus, scientific study of toxicology requires specialized knowledge of the most likely cells to be injured by a given toxin. Inorganic lead is known to damage the nervous system, bone marrow, testes, and kidney. Thus, if the toxicology of lead were examined in terms of its effect on the liver function, its primary toxic effects would be entirely overlooked.

Another factor in understanding a toxic effect is the degree of protection afforded by a given cell. First, the cell is shielded from the delivery of the toxic chemical by exogenous means. The best example of this would be respiratory protection worn by firefighters. Second, vital tissues may be protected endogenously by certain cofactors that protect cells from injury.

The most important of these endogenous cofactors that protects individuals from injury is the sulfa-containing amino acid, glutathione. This material is derived from dietary sources that are high in sulfa. The body takes the dietary supplement and converts it to glutathione, which protects cells from injury from exogenous chemicals.

Moreover, there are certain inherent factors in cells that cause toxic effect to vary. Species, age, sex, and nutritional factors may also contribute to observable effects. Thus, in assessing factors that influence a toxic effect, it is important to control all variables other than the toxic agent. Finally, environmental factors, such as temperature, humidity, and pressure, are highly important in occupational health and safety because they can influence and alter the potential injury to a living cell from a toxic exposure.

Because of the large numbers of factors that influence a toxic effect, interactions of several toxins should be considered. Toxins may interact by enhancing absorption, altering protein binding, or rendering the molecular targets more or less susceptible to a toxic effect. It is not sufficient to say that two given compounds will produce an individual response and, when given together, will produce a greater response. In biological systems, toxins added together not only may produce an additive effect, but also may create antagonistic effects.

TOXICITY RATINGS

Because of the multiplicity of factors that determine a toxic effect, it is sometimes useful to look first at the inherent toxicity of a given chemical. Table 5-A (see page 108) lists an approximation (mg/kg) basis of the comparison of dose and average response in rats, rabbits, and humans. The system presented here is very similar to many rating systems in textbooks of toxicology. Although this method is useful for predicting toxicity, it is oversimplified and inadequate for determining actual toxicity.

Study of the Lethal Dose 50

In the field of toxicology, the most important dose response curve is usually the lethal response known as an LD_{50}. Since the measurement of death is precise, the LD_{50} provides useful information in a given species as to the potential toxicity of a given agent. It is important to realize, however, that animals may not die but may develop severe toxicity. When testing a new chemical agent, pathological assessment must be performed to determine the cause of the lethal event.

In recent years, LD_{50} testing has become a much more precise art. In most LD_{50} testing, at least 10 animals are used at a given dose with a standard LD_{50} providing more than 3 different doses which gives a broad range so that not all animals die and not all animals survive. A typical LD_{50} curve produces a shape as depicted in Figure 5-2. It is evident that

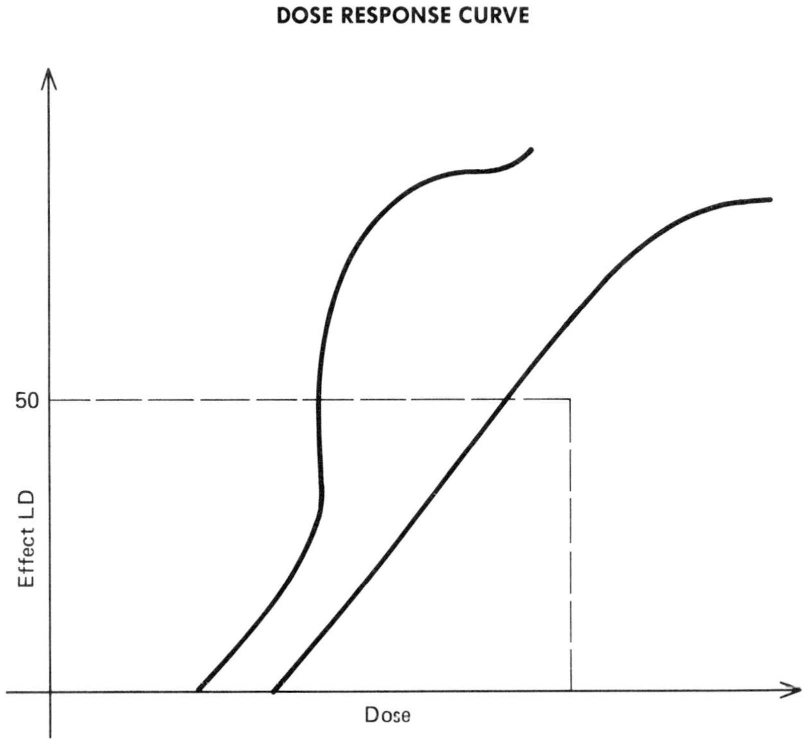

Figure 5-2. LD_{50} curves of two compounds having the same LD_{50} value but widely differing toxicities (shown by the steepness of the curves).

two different compounds could have the same LD_{50} and have widely differing toxicities depending on the steepness of the curve. Furthermore, the lethality of a given agent may have no relationship to the more important levels of clinical toxicity that it may cause. A good example would be those organophosphate pesticides which, upon exposure, inhibit the enzyme acetylcholinesterase, which may have direct lethal results. However, some organophosphates that affect acetylcholinesterase to a lesser degree may inhibit other esterases, accounting for clinical toxicities such as delayed neurotoxicity.

Data at the lowest end of the LD_{50} curve are much less reliable than the data presented at the steepest part of the curve. Extrapolation from a given animal LD_{50} curve to low-dose human exposure is dangerous not only because animal species levels vary in relation to human levels, but also because extrapolation of a *no-dose effect* obtained on a relatively insensitive portion of an LD_{50} curve may provide faulty scientific information.

Toxicokinetics

Given basic toxicology information and knowledge of the importance of dose response relationships, the key variable left is the mechanism of toxin movement throughout the body. The study of this movement is called *toxicokinetics*. To injure a specific molecular target, a given toxin must have gained entrance to that target. The physical and chemical properties of the toxic agent determine its absorption by the gastrointestinal tract, lung, or skin. Once the toxin is absorbed, the body has the ability to metabolize and excrete it via urine, bile, breath, sweat, or stool. It is only when the rate of absorption exceeds the rate of excretion that the toxic compound accumulates in sufficient quantity and for a sufficient duration to injure the living cell.

Thus, the concentration of toxin at the molecular target, being a function of the amount of the toxic agent that is available for absorption, depends on the rate of absorption, the distribution of the agent within the body, the rate of biochemical transformation, and the rate of excretion. A special problem in toxicokinetics occurs when the injury to the living cell occurs at a primary organ of elimination. If a toxin injures the lung, the liver, or the kidney, which are the primary organs of elimination, the body's ability to rid itself of the toxic agent will be retarded. Consequently, higher concentrations of the toxic agent will be delivered to the molecular target, causing enhanced injury.

The best example of a toxin that injures its organ of elimination is mercury. Mercury directly injures the kidney which is its primary route of elimination. In sufficient doses, the mercury which is ingested into the body damages the kidney and causes even higher levels to remain in the body, further damaging the kidney and also other important organs, such as the brain.

RISK ASSESSMENT

Given a basic knowledge of toxic agents, a clear understanding of the factors influencing type and level of toxic effect, and reliable information concerning toxicokinetics, the toxicologist is then able to make a risk assessment. (See Table 5-A.) That is, the toxicologist is often called on to

TABLE 5-A
ANIMAL AND HUMAN ORAL TOXICOLOGY RATINGS

Toxicity Rating	*Commonly Used Term*	*LD_{50} Single Oral Dose Rats, mg/kg*	*LD_{50} Single Oral Dose Rabbits, mg/kg*	*Probable Lethal Oral Dose for Humans*
1	Extremely toxic	1 mg or less	20 mg or less	Taste, 5-7 drops
2	Highly toxic	1-50 mg	20-200	Pinch, 7 drops to 1 tsp.
3	Moderately toxic	50-500 mg	200-1,000	1 tsp. to 2 tablsp. (1 oz)
4	Slightly toxic	500-5,000	1,000-2,000	1 oz to 1 pint (1 pound)
5	Practically nontoxic	5,000-15,000	2,000-20,000	1 pint to 1 quart
6	Relatively harmless	15,000	20,000	1 quart

determine how to extrapolate from known effects of toxic agents at high doses to risks at lower levels, given basic information concerning exposure and dose and a recognition of the danger of toxic effects with long latency. It is clear that new toxicological tools are necessary to refine this process. Since no human action or exposure is without some risk, the investigator must be careful to quantitate scientifically all of the parameters of the risk and then to ask who is at risk. Is a worker at risk? The employer? Society? The process of risk assessment in toxicology is plagued with uncertainty and always is inherently a statistical exercise in which estimates must be made. Since absolute safety is the goal of society in most occupational health and safety endeavors, we must use this imperfect system to our best interests; but we must also strengthen the validity of these estimates by constantly updating our scientific information through experience and research.

Scientific evaluation of occupational health risks may lead, on the one hand, to needlessly prohibiting important action or, on the other, to permitting an unacceptable risk to a working individual. Since risk assessment is always an inherently statistical exercise, the analyst must constantly update animal models, in particular, and their associations to humans. In risk assessment, which is the crucial factor in industrial hygiene studies of exposure levels, investigators must apply epidemiological studies of populations of workers and toxicological information. For instance, predicting a carcinogenic, mutagenic, or teratogenic risk requires all of the following: epidemiological studies of human populations; bioassays of experimental animals; short-term tests of isolated cells or bacteria; and, analysis of the common structural chemistry of the agent. None of these factors alone is sufficient, but, when taken together, they can provide a scientifically sound risk assessment.

The best recent example of risk assessment was the use of malathion spraying in California to control the Mediterranean fruit fly. Scientists reviewed all of the data on the toxicity in humans and animals for malathion. Dose response relationships were clearly established for infants, as well as senior citizens. Based on a synthesis of all of this information, it was concluded by the scientific community that malathion spraying would provide much less human health risk than would the fumigation of the fruit with the very toxic, cancer-causing agent ethylene dibromide, which was considered to be the only acceptable alternative.

EPIDEMIOLOGY

Epidemiology is the scientific study of the how, when, where, and why an illness distributes itself over time in a population. The acute infectious and transmissible diseases were first to be recognized as stemming from environmental agents. Initial epidemiology logic came from the interaction of three factors: the infectious agent, the host, and the environment of both.

The first studies of epidemiology involved the infectious agents of plague and cholera. The most recent example of infectious agents for which there is important epidemiologic data is with AIDS, acquired immune deficiency syndrome. Epidemiological logic clearly identified the infectious nature and that intimate sexual contact and certain social factors were associated with its transmission.

Although epidemiology is a science, it is also a medical philosophy. Epidemiology is the logical process for examining and understanding the interplay between the physical environment, biological effects, and social phenomena that relate to human health and illness in the workplace. Epidemiology is becoming much more sophisticated and has begun to interact with many other scientific disciplines to explain heath problems.

There are basically two types of epidemiological studies—one is descriptive, and the other, experimental. Descriptive epidemiology is an attempt to explain, and thereby prevent, human disease by observing outcomes among study groups not created experimentally. It involves examining dose response relationships in terms of time, place, and the person of exposed individuals. In experimental epidemiology, a hypothetical situation is constructed and controlled by the investigator to gain scientific data that will aid in testing hypotheses relating to nonexperimental situations.

There are several investigative methods of epidemiology. The first is the formulation of hypotheses. This is a never-ending task of epidemiology in which, (*a*) a clinical or pathological observation is made; (*b*) a hypothesis is expressed; and, (*c*) new observations are gathered and tested to refine the hypothesis. A given hypothesis is tested by statistical analysis, dose response relationships, time sequence, and basic knowledge of science. Often this testing is merged with basic experimental laboratory information. This type of testing is often carried out in animal studies in which a controlled environment is deliberately manipulated in an attempt to avoid biases inherent in non-experimental studies.

The three primary methods or types of epidemiology study are (see Table 5-B):

- Prevalence studies, or cross-sectional studies
- Retrospective studies, or case control studies
- Prospective studies, or cohort or incident studies.

Each of these basic studies has a number of advantages and disadvantages that are briefly discussed now.

TABLE 5-B
THREE BASIC EPIDEMIOLOGY STUDY METHODS

Prevalence (Cross-sectional)	*Case Control*	*Incidence/Cohort*
Studies existing living cases	Studies disease/nondisease	Studies exposure/ nonexposure
Present time	Retrospective	Prospective
Quick and inexpensive	Moderate time cost commitment	Expensive and prolonged
Poor for rare disorder	Good for rare disease	Poor for rare disease
Subject to bias in identifying active case	Subject to bias in non-disease group	No bias of exposure
Cannot establish cause	Gives only relative risk but not cause	Gives attributable risk getting at cause

Prevalence or cross-sectional studies attempt to study the relationship between a cause-and-effect at a given point in time. This type of study is based on an existing problem and investigates only present-time cases. As an example, a good cross-sectional investigation might study all the new and existing cases of influenza infection in a given population. It is called a prevalence study because all individuals who are currently affected will be considered.

Prospective or incident cohort studies involve all individuals who are developing a given condition. From this distinction, it is clear that the cross-sectional study cannot establish causation, is most useful for common diseases, and gives very rapid results. Further, this type of study is relatively simple, economical, and rapid because the basic approach is to examine an unselected population, categorize it with regard to presence or absence of disease, and then associate the condition with an exposure.

The most scientifically useful epidemiological study is the *prospective study,* or *cohort* or *incident* study. As was just discussed, the basic approach is to select two similar groups of individuals whose only difference is that one group is exposed to the factor under study and the other is an unexposed, control group. With this selection, the investigator goes forward in time to determine the development of the disease, thus providing incidence rates. The primary bias in this type of study is in quantifying the exposure. Its major advantage is that multiple disease outcomes can be studied; its disadvantages are expense, the long time delay (especially when studying rare diseases), and the large numbers of subjects required with a long follow-up. The attrition rate is high, and changing medical knowledge may change the criteria of study.

Retrospective or case control studies are similar to prevalence studies. In this approach, two groups of people are selected who are similar in all relevant aspects except that one group has the disease and the other does not. The investigator then goes back in time to determine the presence or absence of an exposure. Retrospective studies are relatively simple to perform, relatively inexpensive, and more rapid than prospective studies and are feasible for studying rare diseases. Unfortunately, case control studies suffer from the biases and difficulty in selecting controls and matching a section of controls. Moreover, this type of study cannot be used to establish cause or determine risk directly since it begins with a defined disease. Many sources of bias in this type of study occur with respect to selection of cases and controls and the determination of exposure.

Bias can occur when any epidemiological study is undertaken. Following are some potential biases of scientific studies of occupational health and safety. First, the selection of a study group that does not have the proper characteristics will lead to inadequate or inappropriate sampling for study. Second, in classifying subjects as to exposure of disease, bias may result from poor memory, leading questions, and an inappropriate time frame. Third, use of observers with varying levels of experience, competence, or training may introduce bias when collaborative studies are performed in different locations. Fourth, bias may be introduced by weak classifications of borderline cases caused by poorly designed questionnaires or inconsistent use of medical information by observers. The fifth and final bias in epidemiology occurs when the results of the study may apply only to the study group or another identical group. Extrapolation must be done with extreme caution and is valid only to the extent that the sample study group is adequately demonstrated to be representative of the larger group.

To prevent bias, the sample group must be well defined, precisely described, and strictly followed. The comparability of all study subjects must be established by checking not only for characteristics required in the sampling plan, but also for additional characteristics not included in the plan. The subjects and investigators should be blind as to the outcome of the study, and observer variations must be minimized by recording and adjusting for interinvestigator differences. Despite best efforts, bias is always a danger in epidemiological studies and, when present, it must be quantitated.

In the analysis of epidemiologic studies, all study methods require comparison with a control group. A careful selection of a sample group is the most important means of controlling bias in epidemiology. Here, the adjustment for age comes from recognizing that age is a significant factor in most illnesses. A standard mortality ratio (SMR) can be applied to correct for age in any index of health in the population. Yet, the problem of adjusting epidemiological data for age is not always resolved with these methods. For example, all 50-year-old individuals would not be expected to have the same smoking history or the same exposure to environmental pollutants. In fact, populations with different life expectancies may have identical SMRs. Thus, the adjustment for an SMR is merely accounting for the single variable of age in a situation where there may be other confounding variables. A useful method of dealing with these epidemiological problems especially in incidence/cohort studies is use of so-called life table analysis. A life table is a statistical instrument expressing the mortality of a cohort studied during a definite interval of time. This type of analysis is probably the most precise of occupational health and safety procedures.

Epidemiology attempts to place a disease in proper perspective so that its importance can be assessed in the population and its causes identified and separated from coincidental associations. Logic of study design and the rigorous prevention of bias are essential. Cohort studies combined with life table analysis are the most effective epidemiological methods for occupational safety and health problems. However, epidemiological studies will always involve a compromise with perfection in that anticipated biases can destroy certainty.

CONCLUSION

The ideal goal in occupational health is to *prevent* injury and illness in the workplace. No human experience is risk-free. No absolute scientific principle is any more appropriate than the most recent reproducible data. Because the scientific approach to occupational health and safety is invariably a process of compromise, scientific decisions often occur in a polarized environment requiring rigorous scientific review to avoid bias and conflict. A rigorous scientific approach to occupational safety and health is a constant educational process in which we progressively discover our own ignorance by creating more problems to solve.

SUGGESTED READINGS

Barker, D. J. P., and G. Rose. *Epidemiology in Medical Practice,* 2nd ed. Edinburgh, London, and York: Churchill Livingston Publishing Co., 1979.

An easy-to-read, useful, practical guide to epidemiology.

Baselt, R. C. *Biological Monitoring Methods of Industrial Chemicals.* Davis, Calif.: Biomedical Publications, 1980.

Excellent book on methodologies for identifying industrial chemicals.

Becker, Charles E., ed. *Western Journal of Medicine,* Dec. 1982.

Useful collection of brief topics of scientific interest in occupational medicine.

Fletcher, R. H., S. W. Fletcher, and E. H. Wagner. *Clinical Epidemiology—The Essentials.* Baltimore, Md.: Williams and Williams, 1982.

Friedman, G. D. *Primer of Epidemiology.* New York: McGraw-Hill Book Co., 1974.

An easy-to-read, useful, practical guide.

Hamilton, Alice, and Harriet Hardy (revised by Asher J. Finkel). *Industrial Toxicology.* Boston: John Wright Publishing Sciences Group, 1983.

Good textbook specializing in solutions to problems in its field.

Patty's Industrial Hygiene and Toxicology, vol. I, *General Principles,* vol. II, *Toxicology,* vol. III, *Theory and Rationale.* New York: John Wiley & Sons, 1978 to 1982.

The basic text for industrial hygiene and toxicology—well indexed; a good starting place to look for scientific information.

Proctor, N. H., and J. P. Hughes. *Chemical Hazards of the Workplace.* Philadelphia: J. P. Lippincott, 1978.

A quick reference for common industrial chemical hazards.

Spencer, P. S., and H. H. Schaumburg. *Experimental and Clinical Neurotoxicology.* Baltimore: Williams and Wilkens, 1980.

Everything you want to know about neurotoxicology in one volume.

Part III

Occupational Health and Safety Programs

Chapter 6

Occupational Health

by Sally R. Cowles

In the previous chapter, the scientific basis for the practice of industrial health was discussed. In this chapter, the role of the medical professional is described.

ROLE OF THE OCCUPATIONAL HEALTH PROFESSIONAL

The role of the medical professional in occupational health and safety programs has evolved through several stages since the development of such programs in the early part of this century.

- The first stage, beginning in the 1890s and early 1900s, might be characterized as the surgical or trauma stage, where primary emphasis was on trauma treatment and accident prevention and the physicians involved were referred to as mill surgeons, mine surgeons, railroad surgeons, and so forth.

- The second stage began in the early 1930s with the expansion of workers' compensation laws to include occupational diseases as well as injuries. During this stage, methods of diagnosis (including development of diagnostic tests), environmental sampling, and the clinical description of occupational diseases became prominent.

- The third stage began with the onset of World War II and the manpower shortages that accompanied it. At this, the worker conservation stage, the importance of keeping workers healthy and on the job, was given emphasis, since replacement workers were scarce.

• The fourth, and current stage, the regulatory or management stage, began with the passage of the Occupational Safety and Health Act of 1970. This stage is characterized by a seemingly ever-increasing number of detailed regulations that must be followed. Management skills and the development of programs that meet regulatory requirements are now emphasized along with satisfying the need for trauma care, occupational disease diagnosis, and health conservation.

During this gradual evolution from acute trauma specialist to comprehensive program manager, the occupational health professional's role has become increasingly complex and diversified. Duties and responsibilities have expanded from purely medical concerns relatively isolated from other company policies, to close interactions with a variety of other departments.

The American Occupational Medicine Association (AOMA) has published a committee report[1] that outlines the broad scope of occupational health programs and medical practice. No matter what type of organizational structure is used to deliver these health services, it is important to recognize that today's medical professionals fill several roles, each with a series of complex and demanding duties and responsibilities. Occupational medicine now encompasses far more than the provision of preemployment physical examinations and traumatic injury treatments.

Medical team leader

The major role of the occupational health professional is that of medical team leader. No matter how large or small a company is, if it has a medical staff providing occupational health services, it needs a medical team leader. When a full-time occupational medicine physician is present, that individual will usually be the team leader. When a full-time occupational health nurse is present and physician support is only part-time or contracted, the nurse may well be the more effective leader, relying on the physician only for technical or professional guidance. Marcus Bond[2] has estimated the staffing requirements for a company's medical department based on company size. He has concluded that a full-time department is usually not appropriate for businesses with fewer than 300 employees, and that a part-time nurse may be sufficient for these companies. For such small companies, without an internal department or regular external source of occupational health services, the role of medical team leader must be fulfilled by others. It is usually filled by the external medical professional.

The medical team leader has such important duties as setting the tone and content of medical policy (often with input from other departments); directing other members of the medical staff; determining the emphasis and direction of established programs; initiating, developing, and implementing new medical programs; defending the medical staff from criticism (often an important function for continued productivity and effectiveness); and, when necessary, disciplining other members of the medical staff (another important, and not to be overlooked, duty). The medical team leader determines the effectiveness of the program. Jean Felton[3] has divided this role into the two roles of administrator and supervisor, in an article discussing some of the many roles the occupational health professional fills.

Patient advocate

The next major role of the medical professional is that of patient advocate. To maintain credibility with both the workforce and management, all medical staff members must be extremely sensitive to the needs and rights of their patients. A medical department will be effective only as long as it has the confidence and trust of its patients. To deserve this faith, scrupulous attention must be paid to all aspects of medical care. For example, the key to any successful program is constant vigilance over the confidentiality of medical records. Nothing can ruin a good program faster than unauthorized leaks of medical information. In maintaining confidentiality of medical records, the medical staff will release information only with the employee's consent. The only exception would be giving limited information to authorized individuals to the extent needed to protect the health of the worker, that is, when there is a definite and legitimate need to know to prevent medical injury.

Since every case represents a potential workers' compensation case, the importance of maintaining legible, well-documented, and complete medical records cannot be overemphasized—this is important not only for the patient's sake but for the company's sake. Important, too, is the delivery of high quality medical care and the strict maintenance of medical integrity—the physician or nurse seeing a patient must diagnose and treat the injuries or illnesses, unbiased by management or labor pressures.

When there is honest indecision, it should be acknowledged. The medical part of the program has a responsibility to see that the workplace does, indeed, remain a safe and healthy place to work. If there is any doubt regarding the healthfulness of a work area, the importance of a first-hand inspection cannot be overemphasized. Input from industrial hygiene and safety professionals can also be helpful. An open mind and intellectual curiosity never hurt, either, as new occupational diseases are and will be discovered. Education of the workforce to potential job hazards and what can be done to prevent adverse health effects is a never-ending part of patient advocacy.

Another aspect of this role is the responsibility the medical staff has to actively assist patients who seek additional medical help which, if available, will help them keep their jobs. Knowledge of, and cooperation with, the various social welfare programs in the community, the psychiatric and psychologic resources, and the alcohol and drug abuse programs are all important. Arranging the first appointment for the patient, or helping the patient to do so, can make a difference. Effective treatment should be provided to employees before they lose their productivity and disrupt the department, before, perhaps, ultimately being discharged. Loss of productive employees may be expensive to the company, but it is devastating to the affected employees. Early referral and treatment can make a major difference.

The patient advocacy role also demands that the medical staff provide referral for more intensive or specialized care as soon as the need is recognized. Because most occupational medical programs do not provide total employee health care, it is important to encourage employees to maintain a close liaison with their own private physicians.

Code of ethics

In 1976, AOMA adopted a Code of Ethical Conduct for Physicians Practicing Occupational Medicine[4] (see page 479). The 12 guidelines in the code detail the relationships these physicians should pursue with worker-patients, employers, professional colleagues, and the public. It is obvious that the role of patient advocate is fundamental. Discussion of the difficulties that can be encountered by occupational medicine physicians, especially where it is a question of conflicting loyalties, can be found in the proceedings of a Conference of Ethical Issues in Occupational Medicine by the New York Academy of Medicine.[5] At times, the occupational medicine physician may find that the patient is the corporation rather than the employee.

HEALTH AND SAFETY TEAM EFFORT

The next major role of the medical professional is as member of the occupational health and safety team. The overall functioning of the program depends on active and meaningful participation by all members.

Traditionally, a comprehensive occupational health and safety program has had three major parts: medical, industrial hygiene, and safety. These three areas, when effectively combined, create a strong and cohesive program, far greater than the sum of its respective parts. Just as it takes all three legs for a milking stool to stand, so it takes all three parts for a comprehensive health and safety program. Input from other areas is at times necessary, but without the cohesive combination of medical, industrial hygiene, and safety to provide a solid program base from which to apply toxicology, epidemiology, medical surveillance, and other information, no truly comprehensive program can evolve.

These three major parts, when communication and cooperation are maximal, combine to form an ideal occupational health and safety program—operating as an integrated whole and responsive to both worker and workplace. Too often, however, the interdependence of these three parts go unrecognized, and a health and safety organization is created with three competing departments, sometimes reporting to three separate managers (who not uncommonly report to two or three separate vice presidents). It is not hard to see why a truly comprehensive and effective program is not developed.

The effectiveness of industrial hygiene and safety efforts is ultimately assessed by the health of the individual workers in that workplace. Medical professionals cannot do this effectively without a knowledge of the workplace and its potential hazards, identified by industrial hygiene and safety professionals. On the other hand, the ultimate test of any industrial hygiene or safety control measure in the workplace is the elimination of the medical conditions or accidents caused by the hazard (as determined by medical assessment). All three parts must work intimately together to achieve the common goal of a healthy worker in a healthful workplace. For effective teamwork, the medical staff should not only provide industrial hygiene and safety staffs with relevant medical reports, but also must interact informally with such team members whenever possible.

Industrial hygiene and safety personnel, in turn, are expected to provide the medical staff with both formal and informal reports on potentially dangerous activities. Such interactions will ensure that many potential problems are identified (prior to their full-blown stage) and solved by the team, before they become dangerous. Talking to teammates and keeping up with the activities of the individual members is important for the same reason. It is not uncommon for the medical staff to pass over the industrial hygiene significance of a medical problem, or for the industrial hygienist or safety specialist to completely ignore or not realize the medical significance of an industrial hygiene or safety activity.

The medical professional may or may not be the overall director of the occupational health and safety program. And, it is obvious that the most important attribute of a director is that of being an able manager. If the medical professional is an able manager, that individual may well be the most appropriate one to direct the team. The importance of management skills to the occupational medicine professional has been described in a recent article by Thomas J. McDonagh.[6]

The medical professional

It is no accident that many of the more successful occupational health programs have been directed by medical professionals, and that there are certainly inherent advantages in such an organizational structure. For the medical part to be under control of either hygiene or safety carries a potential danger to a program's integrity, since both industrial hygiene and safety, while focused on a healthy workplace, are judged on their performance on the basis of medical results.

There are other advantages to having a medical professional direct a company's occupational health program. A trained specialist is far more likely to recognize the long-term medical implications of program policy. He or she is in a better situation to assess the impact of recommended industrial hygiene or safety controls on the health of individual workers. The medical professional, especially if well briefed by industrial hygiene and safety associates, commands greater authority and credibility just by virtue of being a medical professional. These individuals can be more effective in selling a comprehensive program to both management and labor. Because their focus is on the individual worker rather than the workplace and by virtue of the independence guaranteed by the patient-doctor relationship, medical professionals are more likely to be able to elicit effective labor cooperation in the program—especially if patient confidentiality is respected.

The ideal situation, with the comprehensive health program managed by a medical professional director, who is both an able occupational medicine physician and an effective manager, is not always possible.

Regardless of who directs the overall occupational health and safety program, the medical professional, whether full time, part time, or contracted, must always decide the medical issues that arise—an industrial hygienist or safety specialist, no matter how well trained or well intentioned, is not competent to make medical decisions.

Another role played by the occupational medicine professional is that of manager of medical information. Recordkeeping requirements have resulted in a deluge of data that requires careful management to meet regulations, to analyze programs, to protect patient confidentiality, and to recognize problems when they occur. Familiarity with computers is becoming a desirable skill for the occupational medicine professional, as is an acquaintance with epidemiologic methods. (See Chapter 11 "Occupational Health and Safety Information Management Systems," for a more detailed discussion of such methods.) Management decisions that must be made regularly include: What information must be kept? How should it be kept? Who has access? How long should the data be kept? How will the information be used? How will it be analyzed? How will results be applied? Is patient confidentiality protected? What is useful, what isn't? And, finally, is there a better, cheaper way to do the same thing?

The final role of the occupational health professional is that of medical policy advisor to management. Because of the high regard and credibility associated with the medical professional, this individual can be a powerful voice. When used selectively in the areas of medical expertise and opinion, such an individual who advises management can obtain far more backing and support than almost any other person. This is especially true if there has been regular planning with industrial hygiene and safety personnel and if adequate industrial hygiene and safety inputs have been available. Because the advisor role is a powerful one, it must be used with discretion, and preferably with a full understanding of the particular problem and with a workable solution in mind. Company management depends on its medical staff for guidance, for feedback on program effectiveness, and to explain the occupational health and safety program to the company.

DUTIES OF THE OCCUPATIONAL HEALTH PROFESSIONAL

Certain duties and responsibilities are inherent in the roles assumed by the occupational health professional. Any medical professional practicing in the occupational health setting has three basic duties: (*a*) fit the worker to the job, (*b*) maintain the health of worker, and (*c*) provide specific medical consultation to management while protecting patient privacy.

Preplacement examinations—fitting workers to jobs

The first duty, fitting the worker to the job, encompasses several medical activities, all aimed at making sure that employees are assigned to jobs that they are able to perform without adverse health consequences. The most common activity is the preplacement examination. Both the unassigned or transfer employee and the prospective job are examined to determine whether or not a successful match can be made. The professional must give an opinion—either the employee is medically able to perform the job or the job is medically unsuitable for the employee. With a successful match, the employee is found to be medically qualified for the job in question. When an unsuccessful match occurs, it is important to specify what aspects of the job are incompatible for the employee and to advise management regarding possible job modifications.

When a successful match cannot be achieved, or when job modifications may not be possible, the medical professional must protect patient confidentiality, while advising management about possible restrictions. Medical diagnoses should not be revealed without employee consent, however, and restrictions should be worded in such a way to emphasize what the employee can and cannot do with regard to the job in question, as well as to other possible jobs within the company.

Once the preemployment/preplacement fit is made, periodic medical follow-up of employees to make sure that both employee and job still match is also important. The early detection and treatment of medical problems with a potential for job impact can prevent many lost productive hours. Elimination of ill employees is not, and should not be, the object of such periodic examinations.

A responsible company will provide job protection for those employees whose skills no longer appropriately match their jobs and who may need either a modification of duties or placement in a more suitable job. Similarly, the early detection and recommendations for correction of adverse health effects on the employee caused by the job not only will help retain a productive workforce, but also will contribute to continued high performance in a healthful workplace. Periodic medical surveillance can identify common health problems arising from a particular workplace; that is, job-related problems that may go unrecognized in the individual, yet when examined as a group, become obvious.

Periodic examinations should, of course, be carefully targeted. Spitzer and associates[7] have emphasized the pitfalls of poorly planned examination programs. Some of the values of a carefully planned and targeted program are detailed by Taylor.[8] When such examinations are designed with the specific medical risks of the job in mind, they can be useful. Even in cases where the employee is found to be in nearly perfect health, such periodic contact with the medical staff presents an ideal opportunity to review with the employee the health risks of his or her job, and how to avoid them. Such enhanced awareness can, in turn, create a safer work environment.

Another aspect of fitting the worker to the job is the return-to-work examination. This examination is most properly used to assess whether or not an employee who has been away from work (because of injury or illness) is medically able to resume fully the duties of the assigned job. The same considerations inherent in the preplacement examination apply. Employees returning to work after serious or prolonged illness or injury may need to be carefully phased-in, and appropriate medical recommendations regarding job modification needed for the employee to return to the job must be carefully developed. Close cooperation between medical staff and job supervisors may be required to ensure productive employment of the worker, while protecting the employee's health. Medically recommended job modifications may be temporary or permanent. If they are permanent, close cooperation with management is even more important in order to develop a responsible solution that both protects the employee's health and provides the company with full productive capacity.

The final test of the success of fitting the worker to the job comes at the time of termination, retirement, or disability examinations. These examinations provide the medical staff with feedback regarding how successful their decisions have been. They provide the final opportunity to identify employee medical problems. Ideally, there should be no surprises at this stage if the program has been functioning effectively.

Health maintenance of employees

The second major duty of the medical staff is health maintenance. This duty is fulfilled in several ways. Periodic medical surveillance of workers is a cornerstone of health maintenance. The early detection and treatment of medical problems—especially such insidious problems as hypertension, diabetes, and obesity—can and does keep employees healthier and on the job longer. The periodic examination offers an excellent opportunity to provide health education to employees—to counsel them regarding smoking, overeating, dealing with high blood pressure, and other problems. Periodic examinations may be confined to high-risk employees, those required by OSHA regulation, or offered to all on a voluntary basis. They may be comprehensive or narrow in content depending on the overall goals of the program and the resources available. Dales and associates,[9] in an 11-year controlled trial of multiphasic screening examinations in a health maintenance organization program, found greater benefits of the program for older males (aged 45-54 at entry) than for other groups screened. Again, this suggests that targeting specific worker groups with examinations based on known risk factors may be a cost-effective approach to health maintenance.

Another way to maintain employee health lies in the treatment of injuries and illnesses—the traditional and most visible activity of occupational medicine, but ultimately the one activity that measures the failure rather then the success of the program. In most situations, treatment provided for nonoccupational injury or illness is confined to emergencies, designed to keep employees at work and on the job for the remainder of the shift. Depending on circumstances, for example, distance to other medical care, other treatments prescribed by the employee's personal physician may be rendered.

Individual companies may vary somewhat in how much medical care is provided for nonoccupational injuries or illnesses. It depends on the company policy, agreements with local physicians, and remoteness of plant location. The occupational medical staff, however, should not become a substitute for personal comprehensive medical care provided by the employee's personal physician. Employees who do not have personal physicians should be encouraged to find them. Close cooperation and interaction with these personal physicians can be extremely effective in helping maintain a healthy and productive workforce. The dividends usually include more rapid return to work after illness or injury and more appropriate medical placement.

Full and complete treatment and medical follow-up to the extent facilities allow of any and all occupationally related injuries or illnesses is necessary in any occupational medicine program. Careful and full evaluation of all on-the-job injuries or illnesses is essential to the prevention of additional cases. When local facilities are inadequate to treat fully a particular occupational injury or illness, referral to appropriate specialty care at the employer's expense is indicated.

Evaluation of work influence. A related duty of the medical staff is that of evaluating the possible work relatedness of an employee's condition. Opinions developed in such situations often become crucial to both the employee and the company when workers' compensation claims are filed. See Chapter 17, "Understanding Workers' Compensation," for a thoughtful and thorough coverage of this topic.

Careful assessment, not only of the work relatedness of a condition, but also of the capability of the ill or injured worker to return to part-time or full-time employment, is an important follow-up. Returning an employee to productive employment as soon as possible after injury or illness (whether occupational or not) not only will help the company, but also will enhance the employee's sense of well-being. It can help to prevent the development of permanent problems such as the compensation cripple or the post-MI invalid syndrome. Regular follow-up of employees with prolonged medical absences is also valuable. What could be more demoralizing for an employee who has been ill or injured and out of work for months than to have no one from the company management or medical staff call?

Rehabilitation programs. Health maintenance is also pursued through active rehabilitation programs—regular follow-up of ill or injured employees absent for prolonged periods obviously plays a part. Rehabilitation programs can range from those that are very simple with no equipment to others that are very elaborate with full physiotherapy departments and rehabilitation counselors. The aim of any rehabilitation program, however, is to return employees to the workforce with increased productivity and job satisfaction. An effective program will also help prevent future injury and foster early return to work. Rehabilitation of the injured or ill worker may require more than physiotherapy and physical conditioning to return an employee to job capacity. In cases where alcohol, drug abuse, emotional, or social problems have contributed to injury or illness, appropriate counseling and treatment for these problems becomes an important part of the rehabilitation process and will help prevent future repetition of the injury or illness and save many lost hours of productive time. Again, where in-house facilities are inadequate for rehabilitative treatment, employees should be referred to appropriate community resources.

Health promotion programs. These have become popular tools these days. They fit neatly under the aegis of health maintenance, yet transfer the burden of responsibility to participating employees. "Health promotion" is a new name for preventive medicine and common sense combined with the realization that all individuals must take responsibility for their own health, in those areas where individual choice may ultimately determine the outcome. Epidemiologic studies have demonstrated that obesity, lack of exercise, increased stress, smoking, and certain dietary habits all contribute to poor health. Chapter 10, "High-Risk Workers," contains much relevant material on this difficult issue. There are positive choices the individual can make to improve his or her individual well-being. Pearson[10] stresses the need to tailor such programs to a particular workforce, however, to yield the greatest benefits for both employee and employer.

Health promotion programs are often designed to identify an individual's major health risks. A recent review of this approach to health promotion by Wagner and associates[11] suggests that these programs are of limited use in health promotion. This should certainly not be the only tool used to effect healthy lifestyle changes in employees. Health promotion programs can provide definite benefits both to participating employees and to employers. A positive medical image is produced in that employee wellness is viewed as something important to the company. Employees benefit both from the positive aspect of the program on their individual health status and from gaining a sense of control over their lives and taking responsibility for their actions which, it is hoped, will carry over into the worksite and home. Employers benefit in terms of fewer lost productive hours and from the positive company image such programs foster.

Although health promotion programs are no substitute for a strong occupational health and safety program, they can be a valuable addition. Health promotion programs can be as simple as periodic blood pressure checks, weigh-ins, and smoking-cessation promotion, or as complex as a full-scale risk-assessment program combined with an integrated and fully equipped exercise facility. Costs can vary widely as can benefits. Health promotion can be incorporated into every employee visit to the medical organization, or it can be structured into a free-standing organization of its own. Whatever the structure, whatever the name, helping employees make positive health choices to improve their chances of remaining healthy on the job is certainly an important assignment for the occupational medical staff member.

Medical consultant to management

The third major job of the medical staff is that of consultant to management. The medical staff becomes an important source of both technical information and advice regarding medical policy and the implications of other management or production policies. The medical staff cannot only explain the medical specifics of programs and policies to all levels of management, but also can provide an important politically neutral perspective on employee issues dealing with health and safety, or medical issues. A less recognized function, but one of potentially increasing importance as product liability litigation increases, is that of providing management with information regarding the potential implications of new product development, production process changes, or new program development. Medical participation in management decision-making can help avoid costly medical problems later.

Medical professionals are once again becoming involved in cost-containment issues. They are an important source of information and expertise to management regarding the content of employee health benefits packages. They are also increasingly involved in audits of health insurance claims against the company and in cost-containment planning. The occupational medical professional must also advise management on workers' compensation cases and at times assist the company in its court defense of such cases if the claim is considered to be medically unsound. Serving as an expert witness occasionally may be demanded.

When the company is large and involves different geographic areas, arranging medical care for both employees and dependents can be a major task as well. Arranging transportation and appropriate medical treatment not otherwise available for ill or injured employees and their dependents in remote or overseas locations may also be necessary. Telephone consultation with these persons at these locations may be required frequently.

RESPONSIBILITIES OF THE OCCUPATIONAL HEALTH PROFESSIONAL

There are several specific responsibilities that the occupational health professional assumes when fulfilling the above roles and duties. The fulfillment of these responsibilities is crucial to the long-term effective functioning and success of any occupational health and safety program and to any successful medical participation in the program (the team concept again).

Know the workplace

The first responsibility of the medical professional is to acquire a sound knowledge of the workplace(s). This familiarity is necessary for a variety of reasons. It is needed by the medical professional to communicate effectively with industrial hygienists and safety professionals, with management, and with patients. But enhancement of communication is not the only advantage a knowledge of the workplace offers. It is essential to the proper diagnosis and treatment of ill or injured employees and their return to work in a timely manner, while avoiding additional injury or exacerbation of conditions. Although a spoken or written description of the workplace from another individual may give the medical professional some idea of what a worksite is like, there is no substitute for periodic first-hand, on-site experience. Without it, important medical decisions are likely to be made in a vacuum. The first and foremost article of faith in the occupational health professional's credo is: know your workplace.

Keep up-to-date

A second responsibility, which in a sense is part of the first, is to have current acquaintance with industrial hygiene and safety reports. Such information is valuable to the medical professional in two ways. It is needed to provide free, open, meaningful communications with the rest of the health and safety team, and it is needed to make sure that the medical role is properly carried out and that patient evaluations are relevant to existing workplace situations.

Follow up on hazards

A third responsibility flows from the first two—the responsibility to follow up on identified job-related hazards. Such monitoring is needed to prevent future injury or illness (either before it occurs, or to avoid recurrences), and to ensure that the occupational health and safety program is medically responsive. If hazard abatement is not possible, then, at least, medical surveillance and monitoring for potential adverse results can be pursued.

Train continuously

A fourth responsibility of the medical profession is a commitment to ongoing training and education. Doctors, nurses, and technicians need continuing education to maintain high quality service and up-to-date care. Stearns and associates[12] have convincingly argued the need for improving management skills as part of this continuing education.

Evaluation

The fifth and final responsibility of the medical professional is to perform periodic, in-depth evaluations of the program itself. Questions that need to be asked and answered regularly include: (*a*) Is the system working the way it is supposed to? (*b*) What should be changed? (*c*) Have priorities changed? (*d*) What needs to be added? (*e*) What needs to be removed? (*f*) Are we keeping workers healthy? (*g*) Does everyone understand the program?

These questions deserve answers and should be asked at least annually, if not more frequently. Regular program evaluation is essential to long-term planning and to setting priorities. It should include a budget review with financial audit and cost-benefit appraisal, a personnel review, and assessment of quality of care. An honest, critical evaluation provides an ideal opportunity for program renewal. Being forced to look at the entire program rather than at its day-to-day problems provides a needed sense of perspective and the opportunity to refocus long-term goals.

RECENT CHANGES IN OCCUPATIONAL HEALTH CARE DELIVERY

Earlier in this chapter, changes occurring in the practice of occupational medicine were mentioned. Since the passage of the Occupational Safety and Health Act (OSHAct) in 1970, there has been a virtual explosion of opportunities in occupational medicine—for all health professionals. In the past, positions in the field were limited for the most part to larger companies or to part-time positions with small companies. Since the OSHAct, however, the demand for occupational medicine specialists has proliferated and the methods for delivering occupational medicine services have markedly expanded. See Chapter 4, "Safety and Health Management," for a lively discussion of various roles and jobs.

Now, in addition to the traditional in-house medical organization to provide a company with its occupational health services, there are private occupational medicine clinics selling services to small- and medium-sized companies either on fee-for-service or contract basis. There are large multidisciplinary group practices that have added occupational medicine staff to meet the needs of corporate clients, and there are hospital-based occupational medicine programs. Furthermore, a number of companies have formed either mobile multiphasic screening units or mobile individual test units to provide clients with screening services and computerized results at the job site. Such services can be of particular value to small companies with a single hazard (for example, noise and the need for audiograms to be made on a regular basis) that requires periodic monitoring.

SELECTING MEDICAL PROFESSIONALS

Such methods of delivery can be effective—both with regard to program quality and cost effectiveness—especially for smaller companies. Company management must carefully study the alternatives and options and select accordingly. Several factors must be examined and considered. A primary factor should be the training and experience of the medical professionals designated to provide care.

Because of the recent expansion in the field, many untrained or poorly trained professionals have entered occupational medicine with little idea of what the field actually is or what a good program encompasses. They are usually physicians with training in family practice or internal medicine, who may have provided a company with preplacement physicals on a part-time basis and who now have decided to relax and do full time occupational medicine. Most have never heard of industrial hygiene, have not seen the workplace, and have little appreciation for existing nontraumatic hazards. They may be able physicians and well meaning, but they have a high likelihood of missing or misdiagnosing occupationally related illness (even sometimes diagnosing occupational illness that is not present) and a high likelihood of misjudging the appropriate time that employees can return to work. The difference in service is similar to that between visiting a general practitioner and a cardiologist for a serious heart problem. Each can provide treatment but the quality of care is likely to be far more specific and appropriate when provided by the cardiologist.

A second group of physicians providing such services are older physicians from a variety of specialities, who, for one reason or another, have become interested in occupational medicine and have taken the time to learn something about the field by taking a variety of short courses and something called a mini-residency—a six- to eight-week intensive course in occupational medicine offered by several medical schools and usually requiring a special project as well. These individuals have a better orientation to the field and at least have some understanding of the issues in the field beyond preplacement examinations and acute treatments. Such additional training demonstrates a conscientious interest and provides a good basis for further supervised experience.

The final group are physicians (usually with prior solid medical training), who have decided early in their training to specialize in occupational medicine and have undertaken a formal residency program. These physicians have intensive training in industrial toxicology, industrial hygiene, epidemiology, general preventive medicine, clinical occupational medicine, biostatistics, management skills, and health care planning, as well as supervised field experience in occupational medicine. They are the people most likely to become board-certified in occupational medicine and become specialists. These specialists are more likely aware of the overall occupational health needs of a company, to be able to relate to the industrial hygiene and safety staffs, and to be able to guide the development of a company's medical policies. Company management should study carefully the qualifications of professionals offering care, since top-quality service is far more likely to come from a board-certified occupational medicine physician than from an obstetrician who has taken a mini residency in occupational medicine, or a family practitioner who expresses an interest. There are notable exceptions, of course, and an interested family practitioner who takes the time to become familiar with the company and the details of the jobs available to employees may well provide better service than an occupational medicine specialist who is not especially interested.

Another factor to consider is continuity of care, because for an occupational health and safety program to be effective, some continuity and consistency is necessary. That means that if medical services are obtained from an outside source, that source should be a single and continuing source. Nothing can obscure the results (successes, or failures) of a program faster than farming out care to several different sources. Such a tactic will guarantee that a comprehensive and cohesive program will not become a reality.

ROLE OF THE OCCUPATIONAL HEALTH NURSE

Many companies that are too small to support a full-time physician maintain a full-time occupational health nurse. As with physicians, nursing training varies. The ideal nurse has had some formal training in occupational health and has been certified by the American Board of Occupational Health Nursing. The next best selection is a young, enthusiastic, well-trained RN who is excited about learning more about occupational health and safety and is willing to work. The National Institute for Occupational Safety and Health publication, *The New Nurse in Industry,*[13] is an excellent discussion of what a well-trained nurse can do for an occupational health and safety program. The occupational health nurse must be able to stand up to a variety of pressures, yet still maintain a sense of professional integrity and program responsibility. Standing orders, to be used as treatment guidelines and signed by the responsible physicians, must be available to provide the nurse with the legal protection required. Pressures from safety staff, personnel departments, supervisors, and management must be resisted if they are in conflict with medical policy or inappropriate invasions of patient privacy.

The nurse is in a situation where at least two masters must be served—the physician who provides technical guidance and standing orders and the company which provides the facilities and the job. This requires a strong person to operate effectively and efficiently; but a good nurse, if given needed support from both physician and company, can contribute significantly to the development of a strong occupational health and safety program and can become a key member of the team. Never underestimate the power of a nurse, or what one can do for a program.[14]

REFERENCES

1. AOMA Committee report: "Scope of Occupational Health Programs and Occupational Medical Practice." Reprinted from *J. Occup. Med.* 21(7), 497-499, 1979.
2. Bond, Marcus B., "Occupational Health Services for Small Businesses and Other Small Employee Groups," in *Occupational Medicine, Principles and Practical Applications,* Carl Zenz, ed. Chicago: Yearbook Medical Publishers, 1986.
3. Felton, Jean S., "The Multiplicity of Roles of the Occupational Physician," *J. Occup. Med.* 18(2), 94-97, 1976.
4. "Code of Ethical Conduct for Physicians Providing Occupational Medical Services, Committee on Ethical Practice in Occupational Medicine," *J. Occup. Med.* 18 (cover), 1976.
5. "Conference on Ethical Issues in Occupational Medicine," *Bull. N. Y. Acad. Med.* 54(8), 705-819, 1978.
6. McDonagh, Thomas J., "The Physician as a Manager," *J. Occup. Med.* 24(2), 99-103, 1982.
7. Spitzer, W. O., *et al.* "Unanswered Questions About the Periodic Health Examination," *Ann. Int. Med.* 83, 257-263, 1975.

8. Taylor, P. J., *et al.* "Preliminary, Periodic and other Routine Medical Examinations," and Wright, H. B., *et al.,* "Screening Well People," in *Occupational Health Practice,* R. S. Schilling ed., 2nd ed. London: Butterworths & Company, 1981.

9. Dales, L.G., *et al.,* "Evaluating Periodic Multiphasic Health Checkups: A Controlled Trial," *J. Chronic Dis.* 32, 385-404, 1979.

10. Pearson, C.E., "The Emerging Role of the Occupational Physician in Preventive Medicine, Health Promotion and Health Education," *J. Occup. Med.,* 22(2), 104-106, 1980.

11. Wagner, E.H., *et al.,* "An Assessment of Health Hazard/Health Risk Appraisal," *Ann. J. Publ. Health* 72(4), 347-353, 1982.

12. Stearns, N.S., *et al.,* "Why Do Occupational Physicians Need To Be Better Managers?," *J. Occup. Med.* 24(3), 219-224, 1982.

13. Lee, J.A., "The New Nurse in Industry: A Guide for the Newly Employed Occupational Health Nurse," Cincinnati: NIOSH DHEW Publication, 78-143, 1978.

14. "Starting an Occupational Health Nursing Service." Chicago: National Safety Council, 1985.

Chapter 7

Occupational Stress

by Leon J. Warshaw

Stress is an inescapable ingredient of life; it affects the lives of all people, everywhere. It modifies perceptions, feelings, attitudes, and behavior, and affects the social, political, and work organizations whose programs and activities its victims direct or carry out.

Stress stems from—and simultaneously affects—our inner sense of well-being, our relationships with family, friends, and neighbors, and our capacity to function in the home, the workplace, the community, and society in general. When excessive, it leads to physical or psychological symptoms; it has been said to cause or exacerbate between 70 and 90 percent of all medical complaints; even the absence of stress has been shown to be harmful. Stress produces stress not only in its victims, but also those who must care for them, and in those who care about them.

In the work setting, stress influences performance and productivity. It may arise from the work itself, the way the work is organized, the physical and social environment, or the interpersonal relationships between superiors, peers, and subordinates. Stress is responsible for accidents in which the victim is frequently an innocent bystander. It is directly related to waste, faulty products and services, imperfect judgments, and inappropriate decisions.

The economic impact of stress is incalculable. The literature is full of estimates, ranging to well over $100 billion per year. Yet, such estimates are subject to challenge on the basis of the values assigned to lost productivity, disability, morbidity, and premature death.

DEFINITION OF TERMS

Stress

Stress is something that everyone experiences, but is difficult to deal with because each person is involved in a highly individualized set of circumstances to which he or she reacts in a different way. Because stress results from attitudes and perceptions and is so bound up in interpersonal relations, the condition is difficult to grasp with the desired degree of scientific precision. We tend, therefore, to describe stress in terms of common denominators that quickly become clichés, stereotypes, and, in the hands of mental health professionals and social scientists, jargon. Stress is equated with tension, anxiety, fear, pressure, conflict, uncertainty, loss of a sense of control over events, or simply with unhappiness and a lack of self-fulfillment.

There is no definition of stress that is universally accepted by all of the disciplines involved. The word, *stress,* is used as a noun, a verb, and an adjective; it denotes a cause or an effect, often both. Consequently, each author or lecturer expounding on stress is obliged to preface the presentation by attempting yet another definition.

Stress is a process—physical, chemical, physiological, mental, emotional, even metaphysical—that involves the entire organism. It is dynamic, ever changing in direction and intensity, and constantly varying with the changing perceptions, capabilities, and resistances of the individual.

Stressors

A stressor is a stimulus, an event or a condition that impinges on the individual. It may be internal, genetically dictated, or physiological, like the biologic rhythms that program the ebb and flow of organ function. The stressor may be a factor in the environment in which we live and work. We are learning more every day about the subtle effects of chemicals and combinations of nutrients on humans. A whole new science of behavioral toxicology has grown out of the way chemicals influence brain and nervous tissue function. Then there are such physical factors as heat, cold, noise, vibration, and radiation, and the effects of the body positions and muscular activity of exercise. The stressor can be our thoughts and feelings as we contend with our psychic and social environment and the events that befall us. The stressor can arise from our own emotions or from the ways in which other people's emotions affect us. The stressor can be real or only perceived; it can be a conscious awareness subject to description and recall or it may be subliminal or entirely unconscious. The stressor can be pleasant or unpleasant, soothing or painful. The stress can be focused or diffuse, brief or prolonged, intermittent or constant. Stress can be any or all of these, simultaneously or sequentially. Although there may be a single isolated phenomenon, more often a barrage of stressors is coming at us from all sides.

In other words, we are continually exposed to constantly changing constellations of stressors. Each stressor produces one or more reactions or responses—the composite of all these reactions is stress. These reactions may be simple, unnoticed minor adjustments, or they may be complex, powerful, and occasionally even overwhelming. Responses can be purposeful or indiscriminate, generic or specific. They can be physical, biochemical, physiological, mental or emotional, singly or in combination, simultaneous or sequential, constant or in changing patterns. Reactions may be brief and evanescent, or persistent and of long duration. Sometimes they form a chain reaction, a sequence of responses that grow in scope and intensity.

The stress reaction, in turn, can act as a stressor. Perhaps the best example is the hyperventilation syndrome in which the physical effort and biochemical effects of rapid shallow breathing lead to a variety of symptoms that may culminate in syncope (passing out).

The stress reaction also can lead to fatigue, depletion, and exhaustion, affecting brain function, hormone balance, and immune mechanisms. Reaction to stress can lead to physical and mental disease or influence its progression as well as the extent and speed of recovery. These reactions certainly can affect our ability to function and perform effectively; they can condition our patterns of behavior and modify our sense of well-being.

FACTORS DETERMINING STRESS

Stress is mediated by a variety of physiologic mechanisms. The immediate response is mediated by the sympathetic nervous system, which modifies heart rate and blood pressure, redistributes circulating blood away from the skin (pallor) and the viscera to the brain and the skeletal muscles, and alters the rate and depth of respiration. These constitute the well-known fight-or-flight response described by W. B. Cannon[1].

Delayed effects are mediated by changes in hormone secretion, notably the adrenal corticosteroids (measurement of urinary corticoid excretion is frequently used as an index of stress), and by emotional reactions such as fear, anxiety, and anger. The long-term effects follow depletion and exhaustion; they may be mediated through impairment of immune mechanisms that sets the stage for what have been labeled the diseases of civilization—heart disease (the now well-recognized Type A personality described by M. Friedman and R. H. Rosenman[2]), atherosclerosis, cancer, and arthritis.

Whether or not a particular pattern of stressors produces an undesirable amount of stress depends on two sets of factors, one internal and the other external. The internal factors determine an individual's vulnerability. They start with the fundamental personality and physiologic responsiveness that is present at birth and include intelligence, developmental experiences, age, education, acquired personality characteristics and conditioning, and dependencies, such as alcohol and drugs. A singularly important factor is the cultural conditioning that determines a person's set of values. These all dictate the way a particular stressor is perceived. For example, a timid, untrained person may be overwhelmed at the thought of addressing an audience while the experienced performer will revel in making speeches, even if they act as a potent stressor to those who must listen.

Even when two individuals are exposed to the same stressors in a particular situation, they may respond differently because of differences in vulnerability. That is why one person may perform well and happily in a particular job, while another with similar skills and experience, but a different personality, may have difficulty. That is why some individuals will break down after a series of unhappy, unpleasant life events while others remain healthy and continue to function without losing their effectiveness.

The external factors constitute the context in which the stressors are operative. They include the social environment, the support (or lack of it) provided by family, friends, coworkers, organizations, trade unions, and others. Ethnic ties, religious affiliations, and cultural mores are also influential. External factors include the political, social, and economic climate of the community and the world at large and the microclimate of the organization or the work group. These factors include morale, the personality of the organization, as well as its management style and the attitudes that it generates. For example, the two people described above who react differently to similar patterns of stressors might exhibit exactly opposite reactions if they moved to another organization or switched bosses.

INDIVIDUAL TOLERANCE FOR STRESS

Each individual has a battery of coping skills and methods of adaptation that represent the confluence of both internal and external factors. Each individual must continually adapt to changes in a stressful and rapidly changing world. Each individual has an optimal zone above which there is too much stress and below which there is the stress of deprivation. Finally, each individual has a limited capacity for stress. When that is exceeded, the excess leads in the short term to fatigue and exhaustion. Then problems arise: performance falters, symptoms appear, and the individual becomes a candidate for accidents and illness. Over time, the excess can lead to disease or aberrant behavior patterns that are almost always self-destructive.

Thus, a relatively small increment of stress might precipitate serious difficulty in individuals who are approaching the limits of their stress tolerance. It matters little how much of that total load comes from stressors in the home, in the community, or on the job. The net effect is the same.

Concurrently, it matters little where the stressors are defused as long as the total stress level is reduced sufficiently to restore the individual's well-being and productivity. Thus, relieving a relatively small increment of work-related stress may be extraordinarily beneficial to an employee burdened by the effects of significant but unapproachable stressors from another source.

Stress is not always bad

Stress is conventionally regarded as harmful and undesirable and associated with unpleasant, unhappy situations. Actually, happy events produce the same physiologic responses in the body as those considered undesirable. This was recognized by T. H. Holmes and R. H. Raye[3] when they developed their rating scale and found that such changes as birth of a child, an outstanding personal achievement, or an enjoyable vacation counted as significant items in computing a life event score that was a predictor of breakdown when it went above a certain level. In general, however, the pleasant stressors are usually not harmful. In fact, people require a certain amount of stress for life to be meaningful. Boredom, monotony, or a purposeless existence are potent stressors. To avoid those, people sometimes seek out sources of stress: participating in challenging or competitive sports, riding roller coasters, viewing horror movies, or similar activities.

Further, up to a point, stress can be energizing. It can stimulate creativity and to some extent improve efficiency and productivity. H. Selye[4], who formulated our modern concepts of stress, has drawn a distinction between distress, which he considers to be the destructive form of stress, and eustress (based on the Greek prefix for good as in euphoria) for stress as a positive force.

Impact of stress on the workplace

Stress is the most pervasive and potent potential toxin encountered in the work environment. Repeated studies have shown that, on the average, about 15 percent of every work force will have significant psychosocial disability in any given year, and another 8 to 10 percent are estimated to have problems relating to alcohol or drug abuse. The influence of stress on the work organization may be acute and disruptive as in the outbreaks of mass psychological illness described by J. J. Smith and others.[5] More often, however, it is subtle and gradual and, therefore, difficult to recognize. It manifests itself in a variety of ways:

1. *Increased rates of illness and disability, in both employees and dependents.* Because stress is frequently manifested by physical complaints and somatic derangements, it leads to greater use of diagnostic and therapeutic services even in the absence of established disease. Increased use of the employee health service by members of a particular work unit is a useful indicator of stress in that section of the organization.
2. *The well-known triad: the 3As.* Absenteeism, alcoholism, and accidents are well-recognized indicators of stress in individuals and in work units. With regard to accident, it should be noted that the accident victims are sometimes the innocent bystanders in mishaps created by stressed individuals, particularly in cases of accidents involving motor vehicles or other complicated equipment.
3. *Reduced productivity.* Decreased productivity may be manifested by reduced output, production delays, or poor quality of goods and services produced. Impaired decision making may be an even more costly manifestation of reduced productivity.
4. *Poor morale.* Morale of employees is an important contributor to company productivity. Poor morale may be manifested by worker unrest, by an increase in complaints and grievances, as well as outright sabotage, or by excessive turnover or a rise in premature retirements.

It should be noted here we are discussing the impacts of employee stress on the organization. (A chapter might well be written on the organization as a functional entity with its own personality and physiology and its need to cope with such stressors as the economic climate, competition, governmental regulations, shortages of critical materials, and the like.) Stress and the steps that organizations may take to deal with it more effectively are dealt with in L. J. Warshaw.[6]

SOURCES OF STRESS AT WORK

Perhaps the most potent work-related stressor is the lack of a job or the imminent threat of its loss. The now classic findings in the prospective study by S. Cobb and S. V. Kasl of workers in two plant closures will be replicated whenever economic conditions become severe and unemployment grows.[7] It is important to note that the severity of this distress is often greater when termination is implied or threatened than after it has become a fact. The demoralizing effects of underemployment or lack of work are seen in the responses of millions of Chinese, who though not actually working, are being paid salaries.[8] The Chinese experience demonstrates that, despite receiving a salary, not having some satisfying work to do can be a potent stressor.

Job change represents another frequently encountered stressor. The new job may be an ardently desired promotion or one that will create problems because the individual is elevated above his or her level of competence. Or, the job change may only be a different supervisor or the addition of women or cultural/ethnic minorities to the work group. In the latter case, workers may fear that the minorities may threaten their position in the group. Transfers of individuals or work units to other areas of the building or plant, or to entirely new locations, also create stress. Relocation of the worker can also affect the entire family if it is to another area of the country. When both spouses have independent careers, the decision to move can threaten the marriage if the couple cannot decide whose career is more important.

Shift work is another form of job change, and when shifts are rotated, repeated adjustments are required. (Shift work is discussed in detail in Chapter 13, "Work Scheduling.") The problems related to rotating shift work were demonstrated by a newspaper printing staff which had a peak incidence of accidents among workers returning after a change of shift.

An unpleasant or dangerous working environment can be highly stressful to many workers. Unsanitary conditions, inadequate lighting, poorly designed furniture, excessive noise, and fear of injury on the job or exposure to toxic substances are sources of stress. Even the location of the workplace and the need to commute to and from work can be stressful.

The job itself provides additional stressors. Inappropriate job content is a particular stressor and includes quantitative and qualitative overload, as well as underload. Quantitative overload occurs when a worker is given more work than can be accomplished in an assigned period of time without adequate rest or relaxation; qualitative overload means that the assigned work is too difficult for the worker. Underload usually means that the work is not challenging enough.

Quantitative overload and restrictive time pressures are common in business and industry today. Unrealistic deadlines, overtime, and inadequate rest periods, as well as too many telephone calls, visitors, meetings, and other interruptions, all contribute to stress in workers. This form of stress can lead to personal habits that are harmful to health, such as smoking, overeating, and alcohol and drug abuse. (See Chapter 10, "High-Risk Workers.") It is often seen in workers who moonlight and especially among those, more often women than men, who have the additional responsibility of housekeeping and family care. Overload may also result from difficulties in long and tedious commuting to and from the workplace.

Qualitative overload is created by work that is too difficult or demanding. It is experienced by those whose jobs require intense concentration and rapid, meaningful decisions. Although much research is still needed on qualitative overload, there is some indication that this stressor can result in increased serum cholesterol and heart rate, job dissatisfaction and tension, lowered self-esteem, feelings of insecurity, and increased smoking.[9]

Underload, or lack of stimulating work, can also be a stressor. Boredom, lack of stimulation, lack of opportunity to use one's ability and acquired skills, and the repetitive performance of seemingly meaningless tasks can all be stultifying. In a study of 1,500 executives of a large financial institution, C. G. Weiman found a high incidence of disease risk factors not only in highly stressed individuals, but also in very low stressed, or understimulated individuals.[10]

Some jobs are so structured that the worker is exposed simultaneously to both overload and underload. The assembly line is a classic example: a boring, monotonous repetition of machine-paced activity that prohibits worker contact and communication.

The individual in the workplace

Increasingly, we are beginning to recognize that the stressor is not the work, but in the way that the work is organized. Inadequate information, too much competition, not enough cooperation, poor supervision, and a loss of personal identity are stressors that stem from this area. These problems often arise from poor interpersonal relationships on the job that have been defined as those that "include low trust, low supportiveness, and low interest in listening to and trying to deal with problems that confront the organizational member."[11]

Frequently, poor communication is both the cause and the effect of unsatisfactory work relationships. Conflicts can occur when inadequate instructions are given, when the job role and responsibilities are not made clear, or when the worker disagrees (perhaps with good logic) with the priorities and methods specified for doing a job. In such conflicts, the stronger and more aggressive individual will take control and the more submissive individual (usually the worker) will experience feelings of low job satisfaction and threatened job security.

The individual's role in the work organization can also be a significant source of stress. In many companies, particularly large, multioperational organizations, little attention is given to clarifying the individual worker's role in the organization, even though role ambiguity and role conflict are recognized causes of occupational stress. Kahn and associates[12] have identified role ambiguity as uncertainty about the scope and responsibilities of the job, about what is supposed to be achieved, and about expectations of co-workers and superiors concerning what is to be accomplished. They define role conflict as a condition of being torn by conflicting demands, by differences of opinion with and between superiors, and by incompatibility between what the job requires and one's personal principles and training.

Individuals affected by role ambiguity respond to the stressor by being dissatisified with their jobs and suffering from job tension and lowered self-confidence. Physiological responses include increased heart rate and blood pressure, which can be accompanied by mental depression. This stressor also produces unmotivated employees and organizational problems of job switching.

Role conflict is believed to affect between 50–70 percent of workers. It appears to occur most frequently in individuals with clerical, managerial, or professional positions. Job dissatisfaction and tension often result, and heart rate and cardiac risk factors (elevated blood pressure, and/or serum cholesterol, obesity, and smoking) increase. Other sources of conflict include territorial disputes over boundaries of responsibility, competition for status, and just plain old office politics.

Finally, lack of personal recognition and absence of opportunities for advancement are well-recognized job-related stressors. When faced with these stressors, employees can become frustrated, dissatisfied, depressed, and even antagonistic. Behavioral and neuropsychiatric problems may well result.

Interpersonal problems cause stress

As in family and community life, problems also arising from interpersonal relationships represent the most frequent source of stress in the workplace. Stress may occur in superior-subordinate relationships, between or among peers, or between workers and persons outside the organization with whom they have to deal—police, correction officers, customer representatives, receptionists, bill collectors, and complaint clerks. Stress may develop out of the need for collaboration or simply from having to get along with one's co-workers. These reflect all of the foibles of human nature including sexual aggressiveness and exploitation. (See Chapter 9, "Women in the Workplace.")

In recent years, much attention has been paid to changes in the meaning of work. For many, work is more than a means of livelihood; it is an important source of satisfaction and a vehicle for self-actualization. We observe the disintegration and demoralization of the chronically unemployed and the unemployable in juxtaposition to those who have opted out of seemingly successful careers because of a need for a different approach to self-fulfillment. We have seen a decay of the work ethic that created our industrial society, and a search for new values and new meaning in work. We are seeing the growing disillusion and dissatisfaction with our economic and political institutions that are driving some people into self-centered isolation, and others into the various forms of communal orthodoxy and fundamentalism. Much of this stems from—and it certainly creates—stress in the workplace.

Rating stress factors

A number of attempts have been made to rate the relative stressfulness of jobs. Some have been based on casual impressions, some on more careful observation, and some on physiologic measurements of people actually working at those jobs. Some have relied on the longevity data or suicide rates among different groups of workers, and at least one was based on compiling the occupations of people hospitalized or seeking treatment for mental illness. In some cases, unions have reached for labels stamping certain jobs as stressful as a basis for negotiating hazard pay differentials for people assigned to them.

Such distinctions are probably valid, but they should be interpreted with caution because they usually represent stereotypes. Most occupational health and safety professionals are well aware that job titles and classifications can be misleading and that even carefully worded job descriptions frequently are only remotely relevant to what the incumbents actually do, much less to the conditions under which they do it.

Even appearances can be misleading. The prototype of the person working under great stress displays frenetic activity, loud, rapid-fire speech, fast and jerky movements, profuse perspiration, and a short temper. Contrasted with this is the quiet worker, moving slowly after deep thought and long contemplation, unhurried, soft-spoken, seemingly calm and unruffled. Those who have watched a diamond-cutter at work will recognize that such outward appearances of equanimity often have little to do with reality.

Much more important and valuable than the stress rating of jobs is an evaluation of the individual doing it to determine the degree of the person–job fit and the level of nonoccupational stress that the worker brings to the job. Then the question must be answered as to why the person is in the particular job. Some, for example, contend that the portrait of the Type A individual at work reflects not so much the nature of the job as what the Type A person makes of it. Such an individual, they say, would convert the most humdrum and pedestrian job into a Type A prototype, if indeed, he or she would remain in it for any length of time. To be sure, generalizations are desirable and often necessary in dealing with organizations and cohorts of their people. But, when confronted by a single individual, these generalizations should be put aside and a personalized assessment made before reaching any conclusions and prescribing any remedies.

One final word: many stress evaluations are based on self-assessment questionnaires and routine surveys. But it must be remembered that for some individuals it is important to be considered as being under stress. This may reflect a kind of macho bravado, or these individuals may perceive that the importance of the job or the judgments about that importance are enhanced by letting it appear that they are under great stress. Consequently, subjective evaluations may provide questionable information.

APPROACHES TO STRESS MANAGEMENT

Many approaches to stress management are being advanced for use in the work setting. Perhaps the best known are those that deal with individuals already suffering from the effects of stress. Based on medical models, these programs seek to identify individuals with signs or symptoms and induce them to come voluntarily or accept referral to professionals who promise to evaluate their problem, diagnose the cause, and offer appropriate treatment. These programs are offered under such labels as employee health programs, counseling services, troubled-employee programs, and employee-assistance programs. Their services range from telephone hot-lines that provide advice in emergency situations, to comprehensive centers offering an array of services administered by multidisciplinary teams of professionals. Some are staffed by qualified persons employed on a full-time or part-time basis, while others are served through contractual or casual referral arrangements with professional individuals or facilities based in the community. Some deal with any and all problems, while others more or less focus on such special stress-related syndromes as hypertension, backache, alcoholism, drug abuse, and marital problems. (See Chapter 10, "High-Risk Workers.")

There is also an ever-increasing group of programs aimed at reducing the vulnerability of the individual. These programs range from prolonged counseling and formal psychotherapy for stressed individuals to workshops and courses that teach people to cope with stress more effectively. Their common denominators are (*a*) training in self-awareness and problem-analysis to detect signs of increasing stress and identify the stressors that are producing it; (*b*) assertiveness training to become more dynamic in controlling stressors; and (*c*) techniques that will reduce stress to more tolerable levels. The simplest involve the distribution of health educational materials that often contain, or are accompanied by, self-assessment tests and the provision of self-help booklets. They may be supplemented by individual or group counseling programs that teach people to recognize their own indicators of a rising stress level.

Meditation

Meditation and relaxation programs have also been popular in many organizations. One such program, Transcendental Meditation (TM), however, has lost much of the popularity it enjoyed in the early 1970s, although it is still practiced by many.

P. Carrington has advanced a variation on TM called "Clinically Standardized Meditation,"[13] which provides a useful approach to meditation without the characteristic mysticism and cultism. She believes that most people can learn the technique of meditation in a very short time if they are taught by an experienced and qualified instructor. If they are not properly trained and motivated to continue using it in their daily lives, many people will stop practicing it.

Carrington warns, however, that individuals taking medication for endocrine or metabolic disorders, for the control of pain, or for psychiatric disorders should be carefully observed by a physician familiar with the effects of meditation. Persons balanced on regular doses of insulin, thyroid preparations, or antihypertensive drugs may require a reduction in dosage while engaging in the practice of meditation regularly. Failure to make such adjustment may lead to drug toxicity. Similarly, individuals undergoing therapy for either physical or emotional difficulties should undertake regular meditation only under the supervision of a competent health professional, preferably one who is familiar with the technique of meditation.

H. Benson[14] of Harvard Medical School has confirmed that meditation, even when shorn of mysticism and ritual, is valuable in reducing tension. His method, called "Relaxation Response," has proven so helpful in defusing the effects of stress that a number of organizations have replaced the traditional coffee break with group relaxation exercises. Physical exercise programs, although they have been promoted generally for their contribution to general well-being and the benefits of improved cardiovascular fitness, have also been singularly useful in reducing personal stress levels. Nutritional counseling has also been found to be helpful in reducing stress. But there seems to be little evidence that the highly advertised megavitamin and stress supplements have any real value.

Supervisor training

One other type of program designed to address stress in the work organization consists of attitude surveys, suggestion programs, and "rap" sessions to probe employees' feelings about their work and the organization. Such programs, however, unless they are supported by a commitment from management to correct any problems that are delineated, may simply add to prevailing dissatisfaction and unrest.

Supervisory and management training programs on employee stress can also be extremely valuable. These programs emphasize the relationship with employees and coordination of the activities. By sensitizing management personnel to people problems and indoctrinating them in their proper management, these programs can make an enormous contribution to controlling job-related stress.

Other ways to reduce stress

Variable work schedules, such as flex time and shortened workweeks, are addressed to resolving stressful conflicts between job and off-the-job. (See Chapter 13, "Work Scheduling," for a more detailed discussion.) In a similar vein, job structuring addresses the same problems. This approach, which includes job enlargement, job rotation, and job enrichment, modifies the content of the job to make it less stressful.

A development that is currently enjoying a wave of popularity is the Quality of Work Life or Quality Circles movement. Also known as "worker democracy," it consists of meetings in which workers analyze the way their work is organized and are given the authority to restructure it in ways that provide greater efficiency, less stress, and more individual satisfaction.

The final category of stress management programs deals with the organization as a living system that needs guidance in deploying and coordinating its resources to cope more effectively with changes in its internal and external environments. These programs deal with faulty internal communications, perhaps the most frequent source of organizational conflict; also with inappropriate or contradictory management styles. Effective planning that establishes realistic attainable objectives should neutralize the effects of managers with personal difficulties whose actions and decisions (or inability to make them) create problems for everyone in the organization.

How to choose

How does one choose among such a panoply of programs? Unfortunately, the choice is often based on the blandishments of those promoting their own wares or the personal prejudice of the individual making the selection. The correct approach involves a study of the organization and its personnel within the context of the industry and the community of which it is a part. It requires an assessment of the current and potential impact of stress and identification of the predominant stressors. This assessment should include a forecast of such major moves as a plant closing, its relocation, or introduction of a new product line. Then, those required to select a program should determine the strength of the organization's commitment to stress management, the magnitude of the resources it is prepared to allocate, and the availability of people in the organization or in the area who are qualified to design, launch, and conduct the program. With this information, effective program elements can be selected, customized to meet the special requirements of the particular circumstances, and appropriately advertised and promoted among the personnel. It is usually helpful in the long run to involve employee representatives from various parts of the organization in the planning effort. When the work force is unionized, however, great care must be taken not to infringe on the prerogatives of the union or the ground rules established by labor-management agreements.

Basics of a stress-management program

In establishing a stress-management program, a number of basic principles must be kept in mind.

First, there are no boundary lines between stresses arising in the workplace, in the home, and in the community. As emphasized earlier, each individual presents a unique composite of all of these elements that is carried with him wherever he or she goes. That means that while the program must focus on those problems that arise in the work-setting, it cannot ignore problems that arise outside the office or the plant. Indeed, it has often been shown that work itself and the social support derived from coworkers and the organization may have a therapeutic value in dealing with personal and family problems.[15] In fact, the loss of this support probably accounts for much of the disability associated with retirement, even when it is voluntary.

Second, stress is highly contagious. It affects not only the individual, but all of those about him with whom he must relate and collaborate. Thus, dealing with stress is simultaneously therapeutic and preventative.[16] This principle has special relevance for those who would focus only on either remedial or preventive programs.

Third, control of stress is inherently an individual responsibility. Troubled individuals can be identified and offered counseling and guidance. They can be given strong support and encouragement and taught to improve their coping capabilities. When needed, they can be referred to mental health professionals in the community for more intensive or prolonged therapy. But each of the foregoing procedures requires the consent and cooperation of the individual, and that depends on the structure of the program, its status in the organization, the capability of the program staff and their reputation, and their accessibility. Perhaps the most important determinant of program success is the establishment of, and adherence to, a strict policy of observing the confidentiality of personal information.

Fourth, control of job stress is fundamentally a managerial responsibility. The program must be based on an explicit organizational policy that places a high value on employee health and well-being. That policy must be reflected in day-to-day operations by the behavior and decisions of supervisors at all levels. To do this, managers and supervisors must be trained in human relations. Disciplinary rules must be fair and well understood, and they must be equitably and consistently administered.

Fifth, a successful occupational stress management program requires an intimate understanding of the job force and the environment. There are few organizations of any size that have a homogeneous work force. Therefore, management must be sensitive to the special problems of the various cohorts and to the ways in which they are changing: ethnic and racial groups, different age groups, new recruits and those facing retirement, women (and also men) with child care and homemaker responsibilities, and workers who care for elderly or chronically disabled family members. All have unique needs that cannot be effectively addressed by a company program that is so generalized and widely encompassing that it precludes attention to such specific problems.

Sixth and finally, in times of economic recession, stress management becomes particularly important. The stresses imposed by unemployment or threat of unemployment, as indicated earlier, are severe. At such times, stress management is critical not only to the employee, but also to the organization, because at no time are employees' peace of mind, collaboration, and optimal productivity more important.

ALCOHOL AND DRUG ABUSE

It would be inappropriate in writing about stress management to omit special reference to the problems of alcohol and drug abuse. These syndromes are widespread in U.S. business and industry. In many instances, they stem from an inability of individuals to cope with occupational, environmental, and/or family stresses. The pervasiveness of these problems among workers has lead many organizations to institute special programs to counteract them.

Alcohol abuse

Alcohol abuse is the more widespread and better recognized of these problems. The National Council on Alcoholism estimates that between 6 and 10 percent of employees are alcoholics. The economic cost in lost production and medical costs for 1975 were estimated to be $20 billion and $13 billion, respectively, with the latter representing more than 12 percent of all health-care expenditures for the nation's adult population. Absenteeism is also a very serious aspect of the problem and probably accounts for additional staggering economic losses. These are compounded by excessive utilization of health care benefits, mostly for diagnoses seemingly unrelated to alcoholism.

There is scarcely an organization that does not have its share of problem drinkers. Further, there is strong evidence that women are becoming increasingly alcohol addicted and suffer health problems after shorter periods of alcohol habituation than do men. M. J. Ashley and associates have reported that women alcoholics showed a greater incidence of anemia and fatty liver than men after fewer years of hazardous drinking (14.1 years for the women; 20.2 years for the men).[17] Women were also found to be greater users of other drugs of dependence, such as tranquilizers. The increasing abuse of alcohol among women employees places additional responsibilities on health and safety personnel to develop programs for identifying problem female drinkers and helping them overcome the problem.

As indicated earlier, many organizations have current programs on employee alcohol abuse, and a company that overlooks the problem and does not institute such programs or upgrade present programs is needlessly dissipating its human and financial resources and failing its responsibility as a corporate citizen. A program for educating and assisting employees with regard to alcoholism should include:

1. Education on the hazards of alcohol and assistance to both men and women in overcoming alcoholism.
2. Cognizance of the social and cultural background of the subjects.
3. Use of all forms of treatment available, and referral of problem drinkers to qualified outside agencies or organizations; this approach should be used even when companies have in-house programs because it preserves continuity of treatment if the employee should terminate.
4. Job jeopardy or the performance approach should be an integral part of the program. When the alcoholic enters the company alcoholism program, he or she should be informed that dismissal will result if the employee discontinues treatment, falls off the wagon too

many times, or fails to maintain satisfactory attendance and performance. And these requirements must be enforced. Organizations using this approach have found that the recovery rate among problem drinkers has been 70 to 90 percent as compared with 20 to 50 percent among individuals treated identically without the threat of job loss.

5. Motivation to abstain must be emphasized.
6. The goal of treatment should be permanent abstinence. Although some have concluded that alcoholics can return to social drinking, the evidence is inconclusive and minimal. Most evidence suggests that total abstinence is the only sure route.

(See also L. J. Warshaw, *Managing Stress*, Reference 6, for additional information on the management of alcoholic employees.)

Drug abuse

The National Institute on Drug Abuse has estimated that in 1978, 22 million people used marijuana, 7 million used prescription drugs without medical supervision, 3 to 4 million used cocaine, and 500,000 used heroin. These numbers have probably grown. It has been well established that drug abuse occurs in every community, in every walk of life, at every age, and in every organization. Like alcohol, drug abuse is costing millions in lost production, absenteeism, and medical care. An additional problem exists in that often both are abused simultaneously.

As L. J. Warshaw[6] has emphasized previously, there is good reason for concern about the abuse of drugs among employees on all levels. The use of pills taken to allay anxiety and feelings of tension—the "downers"—and those taken to enhance mood, counter fatigue, or improve performance—the "uppers"—appears to be on the increase. Involved drugs, prescription drugs obtained illicitly, and the inappropriate use of drugs that were prescribed casually or incorrectly or that are being used improperly. Like alcohol, drugs may impair alertness, perception, and the kinds of psychomotor functions required to drive vehicles or operate machinery with safety. As with alcohol, many individuals become dependent or otherwise habituated to drug use. Then, the same patterns of withdrawn and/or disturbed behavior, impaired performance, and frequent absence are displayed. There is also the same proclivity to denial, concealment, and cover-up that is found in alcoholism, concludes Warshaw.[6]

Consequently, in general, the approach to coping with drug abuse in the organization is identical to that used for detecting and treating alcoholism. It would appear then that the two programs could well be integrated. Moreover, those responsible for identifying abusers of either type should be cautioned against trying to distinguish between them. They are only required to determine whether or not a problem exists and should leave the precise diagnosis to qualified professionals.

Vocational rehabilitation of drug abusers should be a primary concern of business and industrial organizations. Rehabilitated individuals should be hired. This is a responsibility that organizations should not evade, since many rehabilitated addicts remain drug free and need to be allowed back into the economic mainstream. Finally, failure to continue the employment of those who are able to work while undergoing treatment and refusal to rehire those who have been rehabilitated will destroy the credibility of the drug abuse program and drive drug abuse further underground.

CONCLUSION

There is an unfortunate tendency in some quarters to equate stress problems with mental illness. This is entirely inappropriate. In effect, it is an attempt at scapegoating, or blaming the victim. It implies that by gritting one's teeth and applying will power, stress difficulties can be overcome and that failure to do so is an indication of weakness or mental illness or both.

Such self-righteous convictions are doubly harmful for they not only undermine the effort to manage stress, but also stigmatize those who have recovered from a mental illness as unable to cope with the stress of work and daily life. To be sure, excessive levels of stress may precipitate a recurrence of mental illness, but this is also true of heart attacks, bouts of asthma, severe headache, peptic ulcer activity, and even accidents.

Mental illness does occur among workers, as it does among members of the community, but a large proportion can be treated and restored to very satisfactory levels of performance.

The fundamental mission of occupational health and safety professionals is to protect, preserve, and enhance the health, well-being, and productivity of the labor force. It should be remembered that the labor force consists of people, and any program that does not address people problems and the stressors that impinge on work attendance and performance can, at best, achieve only partial success.

ADDITIONAL READINGS

The Addison-Wesley Series on Occupational Stress.

a. McLean, A. *Work Stress*

b. Warshaw, L. J. *Managing Stress*

c. Shostak, A. B. *Blue Collar Stress*

d. House, J. S. *Work Stress and Social Support*
e. Moss, L. *Management Stress*
f. Levi, L. *Preventing Work Stress.*
Reading Mass.: Addison-Wesley, 1979–1981

Agassi, J. B. *Comparing The Work Attitudes of Women and Men.* Lexington, Mass.: D. C. Heath and Co. 1982.

Akabas, S. H., and D. A. Kurzman, eds. *Work, Workers, and Work Organizations.* Englewood Cliffs, N.J.: Prentice Hall, 1982.

Cooper, C. L., and R. Payne. *Stress at Work.* New York: John Wiley & Sons, 1978.

Dorion, R., ed. *Adapting to a Changing World.* Ottawa, Canada: Quality of Working Life Unit, Employment Relations and Conditions of Work, Branch of Labour, 1981.

Egdahl, R. H., and D. C. Walsh, eds. *Mental Wellness Programs for Employees.* New York: Springer-Verlag, 1980.

Keans, J. L. *Stress in Industry.* London: Priory Press Ltd., 1973.

Kiev, A., and V. Kahn. *Executive Stress.* New York: AMACOM, 1979.

Krinsby, L. W., *et al,* eds. *Stress and Productivity,* Vol. IX, "Problems of Industrial Psychiatric Medicine Series." New York: Human Services Press, 1984.

McLean, A., ed. *Reducing Occupational Stress.* Rockville, Md.: National Institute of Occupational Safety and Health, 1977.

Nobel, E. P., ed. *Alcohol and Health.* Rockville, Md.: National Institute on Alcohol Abuse and Alcoholism, 1978.

Wright, H. B. *Executive Ease and DisEase.* London: Pam Books, 1975.

REFERENCES

1. Cannon, W. B. "Stress and the Strains of Homeostasis," *Am. J. Med. Sci.* 189:1, 1935.
2. Friedman, M., and R. H. Rosenman. *Type A Behavior and Your Heart.* New York: Alfred Knopf, 1974.
3. Holmes, T. H., and R. H. Raye. "The Social Readjustment Rating Scale," *J. Psychosomatic Res.* 11:213, 1967.
4. Selye, H. *Stress Without Distress.* New York: J. B. Lippincott, 1974.
5. Smith, J. J., *et al.* "Three Incidents of Industrial Mass Psychogenic Illness," *J. Occup. Med.* 20:399, 1978.
6. Warshaw, L. J. *Managing Stress.* Reading, Mass.: Addison-Wesley, 1979.

7. Cobb, S., and S. V. Kasl. *Termination: the Consequence of Job Loss,* Pub. No. 76-2651. Cincinnati: National Institute of Occupational Safety and Health, 1978.

8. Butterfield, F. *China: Alive in the Bitter Sea.* New York: Times Books, 1982.

9. Cobb, S., and S. M. Rose. "Hypertension, Peptic Ulcer, and Diabetes in Air Traffic Controllers," *J. Am. Med. Assoc.* 224:489, 1973.

10. Weiman, C. G. "A study of Occupational Stressors and the Incidence of Disease/Risk," *J. Occup. Med.* 19:119, 1977.

11. French, J. R. P., and R. D. Caplin. "Organizational Stress and Individual Strain," in A. J. Marrow, ed. *The Failure of Success.* New York: AMACOM, 1973.

12. Kahn, R. L., *et al. Occupational Stress: Studies in Role Conflict and Ambiguity.* New York: Wiley and Sons, 1974.

13. Carrington, P. *Freedom in Meditation.* New York: Anchor/Doubleday, 1977.

14. Benson, H. *The Relaxation Response.* New York: Morrow, 1975.

15. House, J. S. *Work Stress and Social Support.* Reading, Mass.: Addison-Wesley, 1980.

16. McLean, A. *Mental Health and Work Organizations.* Chicago: Rand McNally and Co., 1970.

17. Ashley, M. J., *et al.* "Morbidity in Alcoholics," *Arch. Intern. Med.* 137:833, 1977.

Chapter 8

Safety Programs

by Dan Petersen

The major elements for a successful safety program are discussed in this chapter. Because companies differ from one another, no single program can be recommended for all. However, a minimum of four criteria must be included if any program is to be successful:

- Senior management must lead the way.
- Middle management must be involved.
- The supervisors are the key coordinators.
- Employees (workers) must be involved.

PRINCIPLES OF WORKER SAFETY

Current thinking on safety management and principles of worker safety are based on a number of assumptions. Certainly accidents are symptoms that the safety management system is at fault. Experience has shown that most severe injuries happen as the result of nonroutine or nonproductive activities or those that involve high-energy sources and also certain construction situations. It is axiomatic that safety should be managed like any other company function. Achievable goals should be set, and a program of planning, organization, and control should be established to meet those goals. An important corollary is the assignation of persons who are to be held accountable in the event of problems. Operational errors that result in accidents should be analyzed carefully to determine why they occurred and whether they could have been prevented with more effective controls.

A useful model of accident causation is shown in Figure 8-1. The model indicates that accidents and injuries result from either system fail-

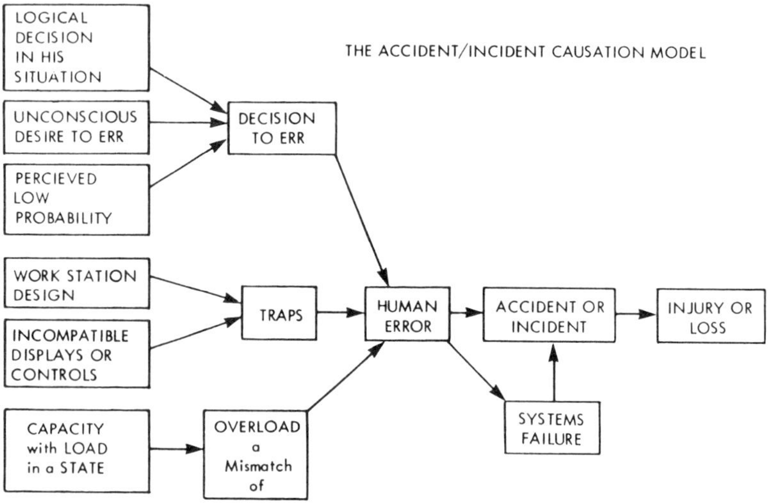

Figure 8-1.

From: D. Petersen, *Human Error Reduction,* New York: Garland Publishing, 1981.

ure or human error. These two factors must be taken into account by those responsible for the safety system.

System failure

Safety management personnel must establish policy for the company, setting rules for authority, responsibility, and accountability. Before an accident or incident occurs, this policy must be clear to everyone who is in charge, just who is responsible and/or accountable, and how those who are responsible and accountable will be dealt with if there is an accident or incident. In making management policy decisions, the following questions must be answered:

- What types of safety inspections should be used?
- How are chemical hazards to be identified?
- What measures should be taken when unsafe conditions are found?
- How are employees who are responsible or accountable for accidents identified and measured for past performance?
- How should new employees be trained with regard to general safety and safety for specific jobs?
- How are employees to be selected for particularly hazardous jobs?

• Is the company's medical program adequate for the most severe problem that can be anticipated?

• What types of records should be kept on accidents or incidents, and how should they be used?

Answers to these questions should provide an outline for developing a company's policy for handling system failures.

Human error

Although an accident or incident may occur because of a failure in the system, it often can occur because of human error. Human error can result from worker overload (too much work, too little time, or both), an individual decision to err, problems in the workplace, or a combination of these.

Overload (also discussed in other chapters) is a significant cause of human error. The cause may be physical (work that is too difficult manually); physiological (work that overstresses the human system); or psychological (too much work, too little work and other stressful conditions). Accidents in which overload may be a factor must be dealt with on an individual basis. Each person involved must be studied with regard to his or her current workload, physiological capacity, and psychological state. Has the individual's workload been increased recently? Has body strength been weakened by recent illness? Does he or she face psychological problems either at work or at home? Similarily, pressure, fatigue, and drug or alcohol abuse must also be taken into account. Practically every aspect of the worker's life, both at work and in the home, can contribute to human error. Thus, the problem of determining whether overload is the cause of an accident is complex and requires considerable time and judgment to resolve.

Human error can also stem from an unconscious or conscious decision of the worker to err. If pressured by supervisors or peers to improve production, the worker may choose an unsafe rather than a safe approach to save time and accomplish more. Some other workers have had accidents before and these must be taken into account, as well as other factors. Finally, some workers are risks to safety because they simply do not believe that anything will happen to them. They are the employees who race fork-lift trucks or take chances when working with cranes—and they can injure or kill themselves or other workers. These problem employees will be the most difficult to handle, but their behavior can usually be managed through the enforcement of strict safety rules and other measures.

Finally, the design, maintenance, and housekeeping of the workplace can cause human errors. The workplace may be incompatible because

OSHA-1B

SAFETY AND HEALTH PROGRAM EVALUATION

	PROGRAMS	YES
5	SAFETY PROGRAM	
	Safety Staff	
	Full Time	
	Part Time	
	Other Duties	
	NUMBER ON STAFF	
6	SAFETY COMMITTEES	YES
	Executive	
	Labor-Management	
	Supervisory	
	Employee	
	Meetings	
	Other (explain):	
7	TRAINING	YES
	Training Staff	
	Type Programming	
	Supervisory	
	Employee	
	On the Job	
	Formal	
	Other (explain):	

	PROGRAMS	YES
8	MEDICAL	
	Pre-Employment Examination	
	Placement	
	Hazardous Occupation	
	Periodic	
	Followup	
	Annual Examination	
	Sight Conservation	
	Hearing Conservation	
	First Aid	
	Doctor	
	(1) Full Time	
	Nurse	
	(2) Full Time	
9	PERSONAL PROTECTIVE EQUIPMENT	YES
	Foot Protection	
	Eye/Face Protection	
	Head Protection	
	Respiratory Protection	
	Hand/Skin Protection	
	Safety Belts	
	Water Hazards	
	(1) Life Vests	
	(2) Life Rings	
	Rescue	
	Ear Protection	

	PROGRAMS	YES
10	ACCIDENT INVESTIGATION (Injury)	
	Supervisor	
	Committee	
	(1) Labor-Management	
	(2) Employee-Management	
	Analysis	
	Corrective Action	
	Followup	
	Enforcement	
	Accident Investigation (No Injury)	
11	INCENTIVE AWARDS	YES
	Supervisory	
	Employee	
	Other explain):	
12	PUBLICITY-PROMOTION	YES
	House Organ	
	Rules and Standards	
	Posters	
	Bulletin Boards	
	Rally-Meetings	
13	MANAGEMENT INTEREST	YES
	Good	
	Lacking	
	Cooperative	
	Injury Rate	
	Costs (Injury)	

Figure 8-2. Safety and health program evaluation form.

From: Occupational Safety and Health Administration.

there is too little light or too much noise. Incompatibility may simply be a matter of equipment or materials not being replaced when or where they should be. There are other problems related to housekeeping. When an accident occurs, safety personnel should reevaluate the condition of the workplace in which the accident occurred to determine whether or not it is properly designed and if housekeeping is adequate.

SAFETY PROGRAM REQUIREMENTS

The requirements of an individual company's safety program will vary according to its type of operation, size, equipment and processes, geographic location, and climate. Hazards vary by type of industry and between companies in the same industry because no two companies have identical design and layout.

If the program is to be successful, all employees of the company, from the president to the shop workers, must be committed to safety. Most important, management cannot view the safety program as separate and distinct from other concerns. Safety must be an integral part of production at all stages if the company is to be efficient and profitable. Although safety must be the highest priority of all employees, the direction of, and driving force behind, the safety program must rest with the highest level of management. Safety should become such an integral part of every management job that abuse of it can affect chances for advancement or salary increases.

EVALUATION OF CURRENT SAFETY PROGRAMS

Numerous tools have been developed for evaluating the effectiveness of a safety program. One of the most useful methods compares the accident rate in a company with its experience in previous years and with accident rates in other companies. To determine the accident/incidence rate, we need to know (*a*) the number of injuries and illnesses recorded during the year; and (*b*) the total number of hours actually worked by all employees during the year. The first can be obtained from the company logs and from the Summary of Occupational Injuries and Illnesses (OSHA Form 200), and the other from payroll or other time records. With these numbers and the following formula, the incidence rate can be obtained.

$$\frac{\text{Number of injuries and illnesses} \times 200{,}000}{\text{Total employee hours worked}} = \text{Incidence rate}$$

The 200,000 figure represents 100 employees working 40 hours per week for 50 weeks per year, the standard base used for determining incidence rates. For example, if a company has 15 recordable injuries and illnesses and total employee working time of 157,000 hours, the formula would be computed as follows:

$$\frac{15 \times 200{,}000}{157{,}000} = 19.1$$

and the company would have experienced an incidence rate of 19.1 recordable injuries and illnesses per 100 full-time employees for the year. The formula can also be used to compute rates for prior years, for portions of years, or for individual departments. If a portion of a year is being computed, the 200,000 figure should be reduced accordingly, so that equivalent units are compared. The incidence rate, while being a helpful tool, will not indicate how effective or ineffective a safety program is in all cases. For larger companies, it is a useful indicator, but for smaller ones, it may only indicate that the company has just been lucky. Moreover, the incidence rate does not indicate the safety performance of individual managers or supervisors. Nonetheless, it does reveal past performance overall and thereby indicates whether a safety program should be reevaluated, reinforced, or altered.

Two other methods may be used to evaluate a safety program—OSHA's Safety and Health Program Evaluation (Figure 8-2) and the American Society of Safety Engineers Safety Audit, pages 160-169. The latter is the more complete and should provide insight into the effectiveness of the program, because it will show more precisely where changes are needed.

SAFETY AUDIT

Activity	Poor	Fair	Good	Excellent
A. ORGANIZATION & ADMINISTRATION				
1. Statement of policy, responsibilities assigned.	No statement of Loss Control policy. Responsibility and accountability not assigned.	A general understanding of Loss Control, responsibilities and accountability, but not written.	Loss Control Policy and responsibilities written and distributed to supervisors.	In addition to "Good," Loss Control policy is reviewed annually and is posted. Responsibility and accountability is emphasized in supervisory performance evaluations.

	Activity	Poor	Fair	Good	Excellent
2.	Safe operating procedures (SOPs).	No written SOPs.	Written SOPs for some, but not for all, hazardous operations.	Written SOPs for all hazardous operations.	All hazardous operations covered by a procedure, posted at the job location, with an annual documented review to determine adequacy.
3.	Employee selection and placement.	Only pre-employment physical examination given.	In addition, an aptitude test is administered to new employees.	In addition to "Fair," new employees' past safety record is considered in their employment.	In addition to "Good," when employees are considered for promotion, their safety attitude and record is considered.
4.	Emergency and disaster control plans.	No plan or procedures.	Verbal understanding on emergency procedures.	Written plan outlining the minimum requirements.	All types of emergencies covered with written procedures. Responsibilities are defined with backup personnel provisions.
5.	Direct management involvement.	No measurable activity.	Follow-up on accident problems.	In addition to "Fair," management reviews all injury and property damage reports and holds supervision accountable for verifying firm corrective measures.	In addition to "Good," reviews all investigation reports. Loss control problems are treated as other operational problems in staff meeting.
6.	Plant safety rules.	No written rules.	Plant safety rules have been developed and posted.	Plant safety rules are incorporated in the plant work rules.	In addition, plant work rules are firmly enforced and updated, at least annually.
B. INDUSTRIAL HAZARD CONTROL					
1.	Housekeeping, storage of materials, etc.	Housekeeping is generally poor. Raw materials, items being processed and finished materials are poorly stored.	Housekeeping is fair. Some attempts to adequately store materials are being made.	Housekeeping and storage of materials are orderly. Heavy and bulky objects well stored out of aisles, etc.	Housekeeping and storage of materials are ideally controlled.

SAFETY AUDIT (Continued)

	Activity	Poor	Fair	Good	Excellent
2.	Machine guarding.	Little attempt is made to control hazardous points on machinery.	Partial, but inadequate or ineffective attempts at control are in evidence.	There is evidence of control which meets applicable Federal and State requirements, but improvements may still be made.	Machine hazards are effectively controlled to the extent that injury is unlikely. Safety of operator is given prime consideration at time of process design.
3.	General area guarding.	Little attempt is made to control such hazards as: unprotected floor openings; slippery or defective floors; stairway surfaces; inadequate illumination, etc.	Partial, but inadequate attempts to control these hazards are evidenced.	There is evidence of control which meets applicable Federal and State requirements--but further improvement may still be made.	These hazards are effectively controlled to the extent that injury is unlikely.
4.	Maintenance of equipment, guards, handtools, etc.	No systematic program of maintaining guards, handtools, controls and other safety features of equipment, etc.	Partial, but inadequate or ineffective maintenance.	Maintenance program for equipment and safety features is adequate. Electrical handtools are tested and inspected and before issuance, and on a routine basis.	In addition to "Good," a preventive maintenance system is programmed for hazardous equipment and devices. Safety reports filed, safety department consulted when abnormal conditions are found.
5.	Material handling—hand and mechanized.	Little attempt is made to minimize possibility of injury from the handling of materials.	Partial, but inadequate or ineffective attempts at control are in evidence.	Loads are limited as to size and shape for handling by hand, and mechanization is provided for heavy or bulky loads.	In addition to controls for both hand and mechanized handling, adequate measures prevail to prevent conflict between other workers and material being moved.

	Activity	Poor	Fair	Good	Excellent
6.	Personal protective equipment—adequacy and use.	Proper equipment not provided or is not adequate for specific hazards.	Partial, but inadequate or ineffective provision, distribution and use of personal protective equipment.	Proper equipment is provided. Equipment identified for special hazards, distribution of equipment is controlled by supervisor. Employee is required to use protective equipment.	Equipment provided complies with standards. Close control maintained by supervision. Use of safety equipment recognized as an employment requirement. Injury record bears this out
C. FIRE CONTROL & INDUSTRIAL HYGIENE					
1.	Chemical hazard control references.	No knowledge or use of reference data.	Data available and used by foremen, when needed.	In addition to "Fair," additional standards have been requested, when necessary.	Data posted and followed, where needed. Additional standards have been promulgated, reviewed with employees involved and posted.
2.	Flammable and explosive materials control.	Storage facilities do not meet fire regulations. Containers do not carry name of contents. Approved dispensing equipment not used. Excessive quantities permitted in manufacturing areas.	Some storage facilities meet minimum fire regulations. Most containers carry name of contents. Some approved dispensing equipment in use.	Storage facilities meet minimum fire regulations. Most containers carry name of contents. Approved equipment generally is used. Supply at work area is limited to one day requirement. Containers are kept in approved storage cabinets.	In addition to "Good," storage facilities exceed the minimum fire regulations and containers are always labeled. A strong policy is in evidence relative to the control or the handling, storage and use of flammable materials.
3.	Ventilation—fumes, smoke and dust control.	Ventilation rates are below industrial hygiene standards in areas where there is an industrial hygiene exposure.	Ventilation rates in exposure areas meet minimum standards.	In addition to "Fair," ventilation rates are periodically measured, recorded and maintained at approved levels.	In addition to "Good," equipment is properly selected and maintained close to maximum efficiency.

SAFETY AUDIT (Continued)

	Activity	Poor	Fair	Good	Excellent
4.	Skin contamination control.	Little attempt at control or elimination of skin irritation exposures.	Partial, but incomplete program for protecting workers. First-aid reports on skin problems are followed up on an individual basis for determination of cause.	The majority of workmen instructed concerning skin-irritating materials. Workmen provided with approved personal protective equipment or devices. Use of this equipment is enforced.	All workmen informed about skin-irritating materials. Workmen in al cases provided with approved personal protective equipment or devices. Use of proper equipment enforced and facilities available for maintenance. Workers are encouraged to wash skin frequently. Injury record indicated good control.
5.	Fire control measures.	Do not meet minimum insurance or municipal requirements.	Meets minimum requirements.	In addition to "Fair," additional fire hoses and/or extinguishers are provided. Welding permits issued. Extinguishers on all welding carts.	In addition to "Good," a fire crew is organized and trained in emergency procedures and in the use of fire fighting equipment.
6.	Waste—trash collection and disposal, air/water pollution.	Control measures are inadequate.	Some controls exists for disposal of harmful wastes or trash. Controls exist, but are ineffective in methods or procedures of collection and disposal. Further study is necessary.	Most waste disposal problems have been identified and control programs instituted. There is no room for further improvement.	Waste disposal hazards are effectively controlled. Air/water pollution potential is minimal.

Activity	Poor	Fair	Good	Excellent
D. SUPERVISORY PARTICIPATION, MOTIVATION AND TRAINING				
1. Line supervisor safety training.	All supervisors have not received basic safety training.	All shop supervisors have received safety training.	All supervisors participate in division safety training session a minimum of twice a year.	In addition, specialized sessions conducted on specific problems.
2. Indoctrination of new employees.	No program covering the health and safety job requirements.	Verbal only.	A written handout to assist in indoctrination.	A formal indoctrination program to orient new employees.
3. Job hazard analysis (JHA).	No written program.	JHA program being implemented on some jobs.	JHA conducted on majority of operations.	In addition, job hazard analyses performed on a regular basis and safety procedures written and posted for all operations.
4. Training for specialized operations (Lift trucks, grinding, press brakes, punch presses, solvent handling, etc.)	Inadequate training given for specialized operations.	An occasional training program given for specialized operations.	Safety training is given for all specialized operations on a regular basis and retraining given periodically.	In addition to "Good," an evaluation is performed annually to determine training needs.
5. Internal self inspection.	No written program to identify and evaluate hazard practices and/or conditions.	Plant relies on outside sources, i.e., Insurance Safety Engineer and assumes each supervisor inspects his area.	A written program outlining inspection guidelines, responsibilities, frequency and follow up is in effect.	Inspection program is measured by results, i.e., reduction in accidents and costs. Inspection results are followed by top management.
6. Safety promotion and publicity.	Bulletin boards and posters are considered the primary means for safety promotion.	Additional safety displays, demonstrations, films, are used infrequently.	Safety displays and demonstrations are used on a regular basis.	Special display cabinets, windows, etc. are provided. Displays are used regularly and are keyed to special themes.
7. Employee/ supervisor safety contact and communication.	Little or no attempt made by supervisor to discuss safety with employees.	Infrequent safety discussions between supervisor and employees.	Supervisors regularly cover safety when reviewing work practices with individual employees.	In addition to items covered under "Good," supervisors make good use of the shop safety plan and regularly review job safety

	Activity	Poor	Fair	Good	Excellent
					requirements with each worker. They contact at least one employee daily to discuss safe job performance.
E. ACCIDENT INVESTIGATION, STATISTICS, AND REPORTING PROCEDURES					
1.	Accident investigation by line personnel.	No accident investigation made by line supervision.	Line supervision makes investigations of only medical injuries.	Line supervision trained and makes complete and effective investigations of all accidents; the cause is determined; corrective measures initiated immediately with a completion date firmly established.	In addition to items covered under "Good," investigation is made of every accident within 24 hours of occurrence. Reports are reviewed by the department manager and plant manager.
2.	Accident cause and injury location analysis and statistics.	No analysis of disabling and medical cases to identify prevalent causes of accidents and location where they occur.	Effective analysis by both cause and location maintained on medical and first aid cases.	In addition to effective accident analysis, results are used to pinpoint accident causes so accident prevention objectives can be established.	Accident causes and injuries are graphically illustrated to develop the trends and evaluate performance. Management is kept informed on status.
3.	Investigation of property damage.	No program.	Verbal requirement or general practice to inquire about property damage accidents.	Written requirement that all property damage accidents of $50 and more will be investigated.	In addition, management requires a vigorous investigation effort on all property damage accidents.
4.	Proper reporting of accidents and contact with carrier.	Accident reporting procedures are inadequate.	Accidents are correctly reported on a timely basis.	In addition to "Fair," accident records are maintained for analysis purposes.	In addition to "Good," there is a close liaison with the insurance carrier.

RATING FORM

	Poor	Fair	Good	Excellent	Comments
A. ORGANIZATION AND ADMINISTRATION					
1. Statement of policy, responsibilities assigned.	0	5	15	20	
2. Safe operating procedures (SOPs).	0	2	15	17	
3. Employee selection and placement.	0	2	10	12	
4. Emergency and disaster control planning.	0	5	15	18	
5. Direct management involvement.	0	10	20	25	
6. Plant safety rules.	0	2	5	8	
Total value of circled numbers	______ +	______ +	______ +	______	× 0.20 Rating ______
B. INDUSTRIAL HAZARD CONTROL					
1. Housekeeping--storage of materials, etc.	0	4	8	10	
2. Machine guarding.	0	5	16	20	
3. General area guarding.	0	5	16	20	
4. Maintenance of equipment guards, hand tools, etc.	0	5	16	20	
5. Material handling--hand and mechanized.	0	3	8	10	
6. Personal protective equipment--adequacy and use.	0	7	20	20	
Total value of circled numbers	______ +	______ +	______ +	______	× 0.20 Rating ______
C. FIRE CONTROL AND INDUSTRIAL HYGIENE					
1. Chemical hazard control references.	0	6	17	20	
2. Flammable and explosive materials control.	0	6	17	20	

RATING FORM (Continued)

	Poor	Fair	Good	Excellent	Comments
3. Ventilation—fumes, smoke and dust control.	0	2	8	10	
4. Skin contamination control.	0	3	10	15	
5. Fire control measures.	0	2	8	10	
6. Waste—trash collection and disposal, air/water pollution.	0	7	20	25	
Total value of circled numbers	_____ +	_____ +	_____ +	_____ × 0.20	Rating _____
D. SUPERVISORY PARTICIPATION, MOTIVATION AND TRAINING					
1. Line supervisor safety training.	0	10	22	25	
2. Indoctrination of new employees	0	1	5	10	
3. Job hazard analysis	0	2	8	10	
4. Training for specialized operations.	0	2	7	10	
5. Internal self inspection.	0	5	14	15	
6. Safety promotion and publicity.	0	1	4	5	
7. Employee/supervisor contact and communication.	0	5	20	25	
Total value of circled numbers	_____ +	_____ +	_____ +	_____ × 0.20	Rating _____
E. ACCIDENT INVESTIGATION, STATISTICS AND REPORTING PROCEDURES					
1. Accident investigation by line supervisor.	0	10	32	40	
2. Accident cause and injury location analysis and statistics.	0	3	8	10	
3. Investigation of property damage.	0	10	32	40	

	Poor	Fair	Good	Excellent	Comments
4. Proper reporting of accidents and contact with carrier.	0	3	8	10	

Total value of circled numbers ______+______+______+______× 0.20 Rating ______

SUMMARY

The numerical values below are the weighted ratings calculated on rating sheets. The total becomes the overall score for the location.

A. Organization & Administration	______
B. Industrial Hazard Control	______
C. Fire Control & Industrial Hygiene	______
D. Supervisory Participation, Motivation & Training	______
E. Accident Investigation, Statistics & Reporting Procedures	______
TOTAL RATING	______

Source: Diekemper, R., and D. Sparz. "A quantitative and qualitative measurement of Industrial Safety Activities." *ASSE Journal,* December 1970. Reprinted in HEW Publication 78-187, NIOSH, Cincinnati, Ohio.

COMPANY SAFETY POLICY

The company's policy on safety should be made known to all levels of the organization. There should be clear-cut restrictions and regulations appropriate to the type of operation and processes involved, the company size, and other factors. To keep employees aware of what is required of them in regard to safety, both management and safety personnel should prepare a document for all outlining company policy. Such a document should indicate that:

- The program requires employee awareness and total involvement.
- Safety is a primary company goal and will take priority over work shortcuts.
- Every effort must be made to reduce the possibility of accidents or any other undesirable occurrence.
- The company intends to comply with all safety laws and to more than meet minimal safety requirements.

The document could also include the following company policies regarding its dedication to safety:

SAFETY AND HEALTH POLICY

People are our most important asset—their safety is our greatest responsibility.

It is the policy of our company that every employee is entitled to a safe and healthful place to work.

When a person enters the employ of our company, he or she has a right to expect to be provided with a proper place in which to work, as well as proper machines and tools with which to do the job, and that the employee will be able to devote his or her energies without undue danger.

Only under such circumstances can the association between employee and employer be mutually profitable and harmonious. It is our desire and intention to provide a safe workplace, safe equipment, proper materials, and to establish and insist on our employees following safe methods and practices at all times.

It is a basic responsibility for all to make the *safety* of human beings a part of their daily, hourly concern. This responsibility must be accepted by each one who conducts the affairs of the company, regardless of the capacity in which he or she functions.

Employees are expected to use the safety equipment provided. Rules of conduct and rules of safety shall be observed. Safety equipment must not be destroyed or abused.

The joint cooperation of employees and management in the observance of this policy will provide safe working conditions and accident-free performances to our mutual advantage.

We consider the safety of our personnel to be of first importance, and we ask your full cooperation in making this policy effective.

Chief Operating Officer   Plant Manager

Figure 8-3. An example of a safety and health policy statement.

Source: American Society for Personnel Administration, 19 Church St., Berea, Ohio 44017.

• Develop and implement safety standards for equipment, working methods, and product processes.

• Conduct safety inspections to identify potential hazards in all areas of the company.

• Investigate accidents and incidents that occur, as a means of developing future preventive actions.

• Analyze accident records and accident causes to determine trends

POLICY STATEMENT

As we are known and recognized for our products and services, so should we be known for our safety performance. No job must ever become so routine or so urgent that every safety precaution is not observed. Prevention of personal injury and damage to the property and equipment of both the Company and its customers must always remain uppermost in the mind of every employee.

It is the policy of XYZ STRUCTURAL STEEL COMPANY, INC. to develop and maintain safe and efficient operations. Our safety program has been designed, in accordance with the Williams-Steiger Occupational Safety and Health Act of 1970, to develop safe working conditions. The success of our program requires the full support of each and every employee.

____

President

Figure 8-4. Shorter policy statement brings in by reference the requirements of the OSHAct, which are the bare minimum to any effective safety program.

and determine targets for protective action.

- Educate and train all employees in safety principles.
- Provide all employees with the most advanced technological protective equipment to prevent injury on the job.
- Promote efforts to encourage employee interest and commitment to the safety program.
- Indoctrinate employees in off-the-job accident prevention.

Figures 8-3 and 8-4 are examples of safety policy statements.

SAFETY RULES AND REGULATION

Management should develop general safety rules for the company that will help avoid accidents (as well as minimize health hazards) on the job. Since most industrial accidents consist of eye injuries, foot injuries from dropping items, and back strains resulting from moving items that are too heavy or that have been moved incorrectly, the most important safety rule will be a mandatory requirement that employees must wear protective equipment. Many incidents or accidents could be eliminated if employees were conscientious about wearing safety glasses and safety shoes for the jobs in which they are required; employees must also be

GENERAL SAFETY RULES

I will:

1. Always ask, before performing any task, the proper methods.
2. Always wear a hardhat in all production areas.
3. Always wear safe clothing and safety shoes.
4. Report any accident or illness immediately to my foreman.
5. Use guards on machinery whenever required.
6. Always wear eye protection.
7. Always lift properly, using my legs and not my back.
8. Always obey all signs, rules, and posted notices.
9. Know that liquor and intoxication are not permitted on company premises.
10. Report unsafe acts of others to my foreman.
11. Never walk or work under products on a hoist, crane, or forklift unless adequate safety supports are used to support the load.
12. Never operate any power machines, vehicles, or equipment unless I have been authorized to do so.
13. Never attempt to repair any defective electrical wiring: I will report such defects immediately to my foreman.
14. Never take part in any horseplay.
15. Never litter; I will keep everything clean.
16. Never ride on vehicles unless I am the authorized operator.
17. Know and obey the safety rules in the department where I work.
18. Never perform any unsafe act that will place myself or anyone else in danger.

Any infraction of the above rules will result in appropriate disciplinary action by the Company. Such disciplinary action will take the form of verbal or written warnings, suspension, or dismissal depending on the seriousness of the infraction.

EMPLOYEE'S STATEMENT

I have read the above safety rules. I understand them and will apply them at all times while I am employed by the Company.

____

Signature of Employee

Figure 8-5. Basic listing of safety rules requires that the employee sign that he read them.

SAFETY RULES AND REGULATIONS

1. Approved safety glasses with side shields and hardhats must be worn in the plant and on job sites at all times. (Any exceptions to this requirement must have prior approval of the Plant Superintendent and Division Safety Supervisor.) Damaged or broken safety equipment will be replaced by the company. It will be the responsibility of the employee to purchase this equipment should it be lost.
2. All employees working in the plant are required to wear safety-toed shoes or boots that will give proper foot protection.
3. Ear protection must be worn by employees working in departments where high noise levels exist. (These areas are posted with signs to indicate ear protection must be worn.)
4. Only authorized operators are allowed on mobile equipment. The operator is responsible for seeing that the equipment is in safe working condition and is operated at a safe speed, and that loads are not moved above workers.
5. Only authorized personnel will do maintenance work on equipment. No work will be done on moving equipment or machinery. Before maintenance work is begun, the equipment must be locked out and tagged.
6. No equipment guards shall be removed except by authorized personnel and/or maintenance employees when it is necessary to work on the equipment. The job is not complete until all guards have been replaced.
7. All injuries happening in the plant, regardless of how slight, must be reported to your foreman at once.
8. Horseplay is positively prohibited.
9. It is the responsibility of the employee to report unsafe conditions or actions so that necessary corrective action can be taken.
10. Under no circumstances is an employee to operate any machine or equipment unless qualified and authorized.
11. Practice good housekeeping. Keep your area clean and return tools, parts, and equipment to their proper places.

IT IS FURTHER EXPECTED THAT ALL PERSONNEL WILL PRACTICE GOOD SAFETY HABITS ON THE JOB TO ENSURE EVERYONE'S SAFETY AND FREEDOM FROM ACCIDENTS.

Date Issued______________ Received by______________

Figure 8-6. Another example of provision for employee to sign that he (at least) received the rules and (hopefully) read them.

trained in proper methods of lifting and moving heavy items. Figures 8-5 and 8-6 provide examples of typical safety rules.

The safety organization

Safety is a line responsibility and must be approached in the same way as other line responsibilities, such as production, cost, and quality control. Consequently, duties should be assigned in the same way—that is, each supervisor should be responsible not only for output, but for the safety of those working under him. In the past, industrial accidents were attributed to unsafe acts, unsafe conditions, or both. But present thinking assumes that management and supervisory staff are partly responsible if they did nothing to eliminate unsafe conditions when, indeed, they may very well have been aware of them.

If a company is large enough, management may wish to employ a safety officer or manager, either part- or full-time, whose sole responsibility is to advise on matters of safety and to assist line managers in controlling accidents and identifying hazards. Or, large companies may opt to employ a full-time safety professional and an industrial hygienist and, perhaps, even a physician and nurse. For smaller companies, this approach is financially not feasible, but they can obtain advice and assistance from federal and state agencies, insurance companies, and safety consultants.

The responsibilities of the safety and health personnel are:

• Assisting line management to fulfill its safety responsibilities through teaching, counseling, and advising.

• Providing for periodic safety appraisals, particularly where new machinery has been installed or existing equipment relocated. Any unsafe or hazardous conditions should be reported in writing to management, with recommendations for improving conditions.

• Promoting and maintaining an on-going safety education program to create at all levels of the organization an awareness of, and commitment to, safety. These efforts should be directed specifically to supervisory staff, training them to maintain safe working conditions, to learn the causes of accidents and incidents, to commit employees to safe practices, and to enforce safety rules.

• Maintaining and interpreting accident and injury statistics and data on the cost of accidents.

• Developing safety standards and measurements for adequately evaluating each responsible supervisor in terms of accident prevention and loss control.

• Preparing instructional materials on safety that will inform supervisors and employees about the program.

• Conducting research into new methods of safety and accident prevention, as well as controlling health hazards in the workplace.

- Maintaining liaison with federal, state, and local governments and safety councils and participating in their activities to learn recent findings in the field.

ASSIGNMENT OF RESPONSIBILITY

All companies, large and small, whether they have full-time safety and health staffs or depend on outside assistance, must clearly define the safety responsibilities of individuals throughout the organization.

The most important areas of responsibility are those of senior and middle management since they must make sure that supervisors perform their safety functions properly. They must visibly demonstrate to all employees that safety is one of the highest priorities of the company and that safety factors must be observed by every worker from the top down. Supervisory staff is the next most important group, because supervisors must coordinate and enforce the safety program. Their tasks in this regard include indoctrinating new workers in safety requirements and developing safe working habits, continuously training and observing both new and old employees, encouraging a commitment to safety among workers, inspecting the workplace for hazards, and investigating and reporting on accidents. If a company has a permanent safety and health staff, they constitute the third important area of responsibility for the program. They are responsible for advising and assisting line managers, as well as senior and middle management.

ACCOUNTABILITY

Who is to be held accountable for accidents and incidents? In other words, who could have prevented the situation? Unless certain employees are aware that they are accountable and that their careers will be affected or endangered by a lack of concern about safety, the program will not succeed. If employees know that accident accountability will appear on their work record, they will become considerably more committed to safety. An endangered paycheck or chance for advancement is a strong argument for safety promotion and commitment. But, all too often, this factor is overlooked when the safety program is designed.

Accountability can be assigned by several means. The simplest is to assign all accidents to the department in which they occur. If a worker suffers a disabling injury, that fact appears on the department manager's or supervisor's work record. Or the costs may be attributed to and charged against the line manager's performance and reflected in his budget. Those responsible for safety should be evaluated annually or semiannually (or whatever period the company specifies) to determine their performance level.

An increase in company insurance premiums is one way to assign accountability to a manager or supervisor. When an accident occurs, the insurance carrier pays the direct costs of the accident, but the company pays for the indirect costs, such as loss of work time. When there are too many accidents, the carrier increases the cost of the premiums. These increases may be charged to various departments or facilities in proportion to their share of accident costs paid. This approach allows an accurate dollar estimate of accountability.

Because line management and supervisors are closest to the worker, they are the people who are most responsible and accountable for safety performance. Consequently, as noted earlier, their safety performance should be evaluated periodically, in terms of the following:

- Number and content of safety meetings held
- Number of toolbox meetings (informal talks with workers on the job)
- Activity reports on safety
- Inspection results
- Accident investigation results
- Accident/incidence reports
- Job hazard analyses.

In evaluating safety as in other job evaluations, the most effective standard for comparison is past performance, because it is better to compare a group or individual against their known past than against the performance of other groups that may not be entirely comparable. Employees are more likely to understand and accept this approach and could resent comparison with other, perhaps competing, groups.

THE WORKER AND SAFETY

Safety of the worker is the basis of any safety program. Efforts at this level must include training and motivation of both new and current employees, safety and health committees in which workers participate, safety inspections of the workplace, accident investigations, and job safety analysis.

Training and motivation

All new employees and employees changing jobs should be instructed in new safety procedures not a week or ten days later—but on their first day. It is important to establish immediately the importance of safety procedures and protective equipment and to motivate employees to observe safety requirements—both for their own well-being and the good of the company. If employees are properly instructed in safety at the beginning, they will be more likely to observe good work habits. If safety orientation is delayed, improper work habits may develop and will be difficult to overcome later. It should be noted that before any employee is assigned to a job, or current workers are reassigned to new jobs or different equipment, they must be determined to be physically and mentally capable of doing the work and be properly trained.

Safety training is based on the assumption that the development of a positive mental attitude predisposes an individual to safe work habits and conduct. Thus, those doing the training and motivating must use a positive approach. First, the instructor must explain why safe work practices are required for each individual job in terms of the worker's safety. Specific examples should be provided to show why safety is necessary—for example, when operating a power press, manning a crane, or testing an interlocking guard. Throughout the session, every effort should be made to motivate employees to do their jobs properly.

During the training session, the instructor should:

- Make the employee feel at ease.
- Explain that he or she should not be afraid to ask questions, especially when new situations arise.
- Explain the job in detail in terms employees can easily understand.
- Indicate why protective equipment, if appropriate, must be worn, and point out consequences of not wearing it.
- Demonstrate at each job step, the precautions and safe work practices that must be followed and explain their importance.
- Question employees to determine whether or not they understand all instructions and, if so, allow employees to demonstrate or practice what they have learned.
- Motivate employees by demonstrating the company's commitment to safety and safe practices.

Training sessions may include first aid courses, cardiopulmonary resuscitation, fire extinguisher use, lifting and moving of heavy objects, and emergency procedures. The instructor's responsibility does not end with the training session. He or she must check the employee regularly to make sure that instructions are being followed and that no unsafe work habits or short-cuts have developed.

Follow-up training should include toolbox sessions—brief, informative get-togethers of supervisors and workers. These informal sessions foster open communication and stimulate safety consciousness. The topic should be planned in advance, and the session should be held at the job site and last no longer than 5 to 15 minutes. Suggested topics include:

- Previous accidents and corrective action
- Job safety in individual tasks and operations
- Housekeeping methods
- Techniques of manual lifting, individually or in groups
- Hearing protection—when to wear protective goggles and types used for different procedures
- Lockout—cleaning, oiling, repairing
- Materials handling safety (cranes, hoists, forklifts)
- Prevention of falls (safety belts, lanyards, tying off).

The National Safety Council publishes a booklet titled "Construction Safety Talks," which gives more ideas on this subject.

Additional training sessions may be indicated if there is a high incidence of injury and illness; an increase in the number of incidents that could have resulted in accidents; excessive waste or scrap after production; a high rate of labor turnover; a change in processes or introduction of a new process; or any sharp, unexplained increase in accidents.

Training provides workers with the knowledge and skill to accomplish the work safely, but unless they are motivated, employees will not normally work that way. Figure 8-7 shows various factors that influence worker motivation, which together indicate whether an employee will perform in a safe or unsafe manner. However, motivation to safety is not that simple. It is also influenced by pressures in daily life, whether or not a person likes or dislikes the job, whether or not the job is meaningful, the personality of the worker, and the job climate. Finally, does the worker get an emotional reward for performing in a safe manner? In other words, does he or she feel satisfied with what was accomplished?

In motivating for safety, the program must take into account the current needs of the workers. Most people want jobs that are meaningful and they also need to feel that they are part of a peer group. If those needs are met, they will very likely become motivated. Thus, the job must offer a chance for achievement and recognition, it must be fun, it must be responsible work, and it must permit personal growth. These factors can be applied to safe working performance: Will the employee be recognized for his or her safety record? Will this person be given addi-

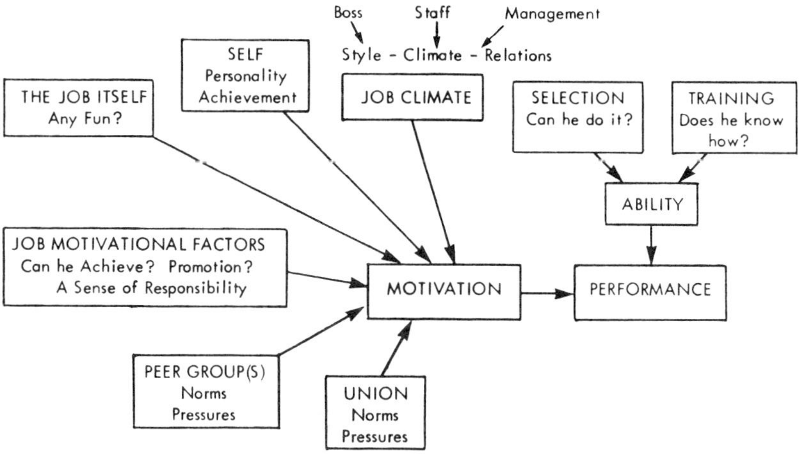

Figure 8-7. Various factors that influence worker motivation and performance.
From: Petersen, D. *Human Error Reduction.* New York: Garland Publishing, 1981.

tional responsibility because of that record? Will the record contribute to career growth? Management and the safety programs that take these factors into account will succeed in motivating employees.

Workers can also be motivated by having satisfactory interpersonal relationships with managers and supervisors. Studies have identified two types of supervisors: the employee-centered supervisor who achieves high production because he is primarily interested in his workers, and the job-centered supervisor who does not achieve high production because he emphasizes *it* more than an interest in the workers. Thus, the most effective manager is one who is interested in both people and production, combining high performance goals with the good of his or her people. The key to employee motivation to safety, then, is a management that cares about and rewards its employees for safe conduct.

Some drawbacks to safety committees

Some companies establish safety committees in which supervisors and workers participate. Such committees are only a small part of the safety program, but if properly organized and used, they augment the safety program. However, they are not essential and can have some inherent limitations. First, only a few workers can participate in a committee's work at any given time. The safety committee concept also has the following additional limitations:

- Busy supervisors can refer safety problems to the committee, problems that they should be handling themselves.

• Only a small percentage of the work force will have a chance to serve on the committee; often the same members are appointed year after year.

• Meetings and attendance of the committee can be irregular, sometimes allowing too much time to elapse between a problem and its solution.

• Committee members often are not qualified to handle complicated safety problems and consequently tend to concentrate on superficial problems.

• Committee recommendations may be bypassed by management, particularly if they require large expenditures.

• Committees may defer action on unsafe conditions that demand prompt action so that they will have something to discuss at the next meeting. Because of these limitations, management and developers of a company safety program should weigh carefully whether or not the committee concept will prove a liability or a benefit.

Safety inspections

Safety inspections are important for identifying potential hazards before accidents occur. (Also see Chapter 14, "Plant Inspection and Worker Protection.") In general, accident causes can be categorized as the results of unsafe acts or unsafe conditions. It should be realized, however, that accidents seldom result from only one cause but are usually a combination of several causes.

Unsafe acts include operating equipment without authority or training; using equipment, tools, materials, or vehicles unsafely; operating at unsafe speeds; operating defective equipment; making safety devices inoperative; failing to use personal protective equipment; unsafe lifting; taking an unsafe position during procedures; or adjusting, cleaning, or oiling machinery that is in motion. Unsafe conditions include unsafe design or construction of equipment or the facility; defective tools or equipment; poor housekeeping in the workplace; inadequately safeguarded or unguarded equipment; or inadequate illumination or ventilation.

Supervisors should regularly conduct safety inspections, and middle and senior managers should periodically conduct unannounced spot-checks for safety abuses or hazards. Prior to inspections, it is helpful to review previous accident reports and job safety analyses to identify problem and high-risk areas. Plant and individual checklists also can prove helpful. (See examples of such checklists in Chapter 14.)

Those conducting the inspection should communicate with workers in the area and encourage their assistance and suggestions during the inspection, because the person at a machine is more familiar with it than anyone else. Workers in certain areas are generally more aware of the problems or hazards. Moreover, if a worker's suggestion is accepted, there is a positive reaction. If it is not accepted, the reasons should be explained.

By law, certain inspections will require special expertise and must be conducted by safety personnel or industrial hygienists. They include (*a*) inspections of such equipment as cranes, pressure vessels, fire extinguishers, chains, and ropes and (*b*) industrial hygiene surveys to identify health hazards.

Periodic inspection by supervisors and managers not only reduces the potential for accidents in the workplace, but also demonstrates the company's commitment to safety.

Accident investigation

The investigation of accidents is a critical part of the accident prevention program. It identifies accident causes so that similar accidents can be prevented in the future by mechanical improvements, better supervision, or improved employee training. It determines the causes of human errors. An investigation makes supervisors and employees more aware of particular hazards in the workplace, and it directs attention to safety and accident prevention, in general. Minor injuries and near-accidents (incidents) must be reported as well as lost-time accidents, because nothing is learned from unreported accidents. Near-accidents are often predictors of future serious accidents; potential accident causes can be corrected, and the need for medical treatments of involved workers may not surface until later.

All levels of management are responsible for accident investigation, but the first-line supervisor is best qualified for the following reasons:

1. Supervisors are closest to, and most knowledgeable of, jobs, working conditions, and workers. They should know the details of jobs and job procedures, hazardous environmental conditions in their areas, and any unusual circumstances that must be taken into account. They should also know their workers and their personal characteristics, job experience, and job language.

2. A supervisor's sense of responsibility will be reinforced by accident investigations. After each investigation, supervisors should become more aware of their responsibility for safe procedures and accident prevention.

3. Supervisors learn more about accident causes by investigating accidents. By such investigations, they should learn the unsafe practices

and conditions to look for, how best to train new employees in safety considerations, what hazards to caution employees about, and generally how to prevent accidents.

4. Supervisors are the primary instrument of accident prevention. They can help to prevent the recurrence of accidents as well as prevent minor injury accidents because they are closest to the scene. Information that they obtain during accident investigations will show them whether they should change their instructions on safe job procedures, eliminate unsafe conditions, or correct the sources of such conditions.

Of course, supervisors know how to recognize the causes of accidents. Supervisors will sometimes require assistance from the safety staff, equipment suppliers, and insurance representatives. Management should see to it that supervisors are given the training and assistance that they need.

In investigating an accident, the supervisor's primary problem is in obtaining the cooperation of the employee(s) involved. Employees can be reluctant to cooperate because of any number of fears—of reprimand or discipline, of ridicule from co-workers, or loss of face. Employees should be told that the investigation is being conducted to learn the facts so that future accidents can be prevented. They should be reassured that an investigation is not an attempt to pinpoint blame.

The investigator should also emphasize how much he or she appreciates the worker's cooperation. After the investigator has gained the confidence of the employees, he or she should ask precisely what happened before, during, and after the accident. The interview should close with a general discussion of the accident and a request for the worker's opinion on how it could have been prevented. Those involved in accidents should be encouraged to contact the investigator at a later time if they remember additional facts.

Witnesses are an important source of information, and in the case of a fatal injury, may be the only source of information. Similarly, a witness may be the only source if the involved worker is too badly injured to give a clear account. For minor accidents and near-accidents, witnesses also may be able to give a more accurate account of the situation. The following points should be remembered with respect to interviewing witnesses:

1. *Interview witnesses promptly and individually.* Witnesses should be interviewed as soon as possible after the accident while memories are fresh and before they have been affected by discussions with others. For that reason, witnesses should also be interviewed separately.

2. *Reassure the witness of the investigation's purpose.* The investigator

should explain that ascertaining the basic facts is important in preventing future accidents, and emphasize the important service being provided by the witness in giving honest responses. If company policy permits, the investigator should reassure the witness that his or her testimony will not result in disciplinary action against the injured worker.

3. *Obtain the witnesses' versions with minimal interruption.* Witnesses should be interviewed at the scene of the accident whenever possible. The investigator should ask witnesses to recount the incident in their own words, and not interrupt them unless there is a question of clarity.
4. *Direct specific questions to clarify or amplify the witness account.* The investigator should avoid asking leading questions and should maintain a friendly attitude so that the witnesses will not feel that extended questioning is an attempt to discredit their statements.
5. *Summarize the statements from witnesses and clarify any questionable areas so that the witnesses account and the investigator's understanding of it are in agreement.*

In accident investigations, the first facts to be established are the identity of those directly involved and the time the accident occurred (date and exact time). These facts permit identification of possible witnesses and specify the shift, production period, weather conditions, supervisors in charge, and other factors that may be important. Although dates and times may not seem meaningful to the investigator, they may suggest causes in later readings of the accident report. Once these facts are established, the following questions should be asked:

1. *Where did the accident occur?* The location is important because it can be used to identify high-risk areas. This might raise questions about the safety of other workers in such areas. Usually, a concise descriptive phrase is all that is needed in answering this question.
2. *What is the injured worker's job?* The injured worker's position, occupation, or job title should be determined so that risk hazard by job or occupation can eventually be evaluated.
3. *Was the injured worker physically capable* and did he or she have sufficient experience to perform the work?
4. *What task was being performed* when the accident occurred? This question addresses specific duties within a job or occupation—lubricating a machine, sawing, and operating a grinder, for example. The task involved can be used to identify specific hazards of jobs or equipment and job procedures, as well as worker capability and need for additional training.

5. *What happened?* The answer to this question should be a precise account of the incident itself and a description of the events leading to it. The following information is needed:
   a. See item 1, above. More information on the location of the accident.
   b. What were the conditions at the time of the accident? Recap other general information that might have a bearing on the accident.
   c. Ascertain the injured worker's physical position in terms of his surroundings.
   d. Determine what the worker was doing immediately preceding the accident.
   e. Describe circumstances that might have precipatated the accident.
   f. Describe the accident or incident, including specifics—for example, "struck by," "caught between," "overexerted," "rubbed" or "abraided" against, "contacted temperature extremes," "contacted caustics," or others.
6. *What other elements may have contributed* to the accident or incident?
7. *How can future accidents of this type be prevented?* An accident investigation is not completed until corrective action has been determined. Action to prevent recurrences should be identified on the basis of the direct and indirect causes of the accident under study.

The causes of plant accidents are many and varied; consequently, there will be many types of corrective actions—some already in effect, some pending but not yet implemented, and some still not identified. The accident investigation report must specify which of these applies and, in the case of the last, must offer recommendations for remedial action. These need not be detailed technical descriptions, but preferably should be general descriptions with an explanation of the purpose. If the recommendation is technically detailed, it may be rejected because of some minor technical point.

Figure 8-8 is an example of a supervisor's accident investigation report form.

Job safety analysis

Job safety analysis identifies the hazards or potential hazards associated with each step of a particular job, determines the severity of the hazard, attempts to establish a probability of accident occurrence, and attempts to reduce or eliminate hazards. To be effective, the procedure should include input from workers on how each task is performed and the hazards that they perceive. A description and examples of job safety analysis are given in the Council's *Accident Prevention Manual for Industrial Operations.*[8]

HEARING CONSERVATION PROGRAMS

A hearing conservation program is required (*a*) where the sound level in a working area has not been reduced to 90 dBA or below by engineering means; (*b*) reliance must be placed on administrative controls to limit duration of exposure, or (*c*) relying on hearing protection to reduce the sound level actually reaching the ear. The program should be applied to all employees whose work brings them either steadily or infrequently into areas in which sound levels exceed 90 dBA. The steps in a typical compliance plan might include the following, not necessarily in this order and some usually going on simultaneously:

- A detailed survey of sound levels and sound spectra to determine the sources of excessive sound.
- Initiation of engineering studies to determine methods for reducing sound levels at their sources.
- Planning and initiation of feasible administrative controls, such as modifying production schedules to divide noisy jobs among enough people to bring each below the permissible limit or spreading part-time noisy operations.
- Initial audiograms for employees excessively exposed.
- Installation of a personal protective equipment (PPE) program.
- Follow-up audiograms at appropriate intervals to assess effectiveness of the PPE program and administrative controls.
- Installation of engineering controls, or process changes to reduce noises at their source.
- Repeated noise surveys to measure effectiveness of the engineering changes.

When the compliance plan involves long-term engineering projects, for example, one or two years, it may be revised from time to time as conditions change. The orderly completion, on schedule, of the various phases of the compliance plan, together with other components of the hearing conservation program, are considered adequate compliance with federal requirements. Details of how to start and run a hearing conservation program are given in the Council's *Industrial Noise and Hearing Conservation.*[9]

ACCIDENT INVESTIGATION REPORT

CASE NUMBER

COMPANY ______ ADDRESS ______

DEPARTMENT ______ LOCATION (if different from mailing address) ______

1. NAME of INJURED	2. SOCIAL SECURITY NUMBER	3. SEX ☐ M ☐ F	4. AGE	5. DATE of ACCIDENT

6. HOME ADDRESS	7. EMPLOYEE'S USUAL OCCUPATION	8. OCCUPATION at TIME of ACCIDENT
	9. LENGTH of EMPLOYMENT ☐ Less than 1 mo. ☐ 6 mos. to 5 yrs. ☐ 1-5 mos. ☐ More than 5 yrs.	10. TIME in OCCUP. at TIME of ACCIDENT ☐ Less than 1 mo. ☐ 6 mos. to 5 yrs. ☐ 1-5 mos. ☐ More than 5 yrs.
11. EMPLOYMENT CATEGORY ☐ Regular, full-time ☐ Temporary ☐ Nonemployee ☐ Regular, part-time ☐ Seasonal	12. CASE NUMBERS and NAMES of OTHERS INJURED in SAME ACCIDENT	
13. NATURE of INJURY and PART of BODY		

14. NAME and ADDRESS of PHYSICIAN	16. TIME of INJURY A. ______ A.M. P.M. B. Time within shift C. Type of shift	17. SEVERITY of INJURY ☐ Fatality ☐ Lost workdays—days away from work ☐ Lost workdays—days of restricted activity ☐ Medical treatment ☐ First aid ☐ Other, specify ______
15. NAME and ADDRESS of HOSPITAL		

18. SPECIFIC LOCATION of ACCIDENT ON EMPLOYER'S PREMISES? ☐ Yes ☐ No	19. PHASE OF EMPLOYEE's WORKDAY at TIME of INJURY ☐ During rest period ☐ Entering or leaving plant ☐ During meal period ☐ Performing work duties ☐ Working overtime ☐ Other ______

20. DESCRIBE HOW the ACCIDENT OCCURRED

21. ACCIDENT SEQUENCE. Describe in reverse order of occurrence events preceding the injury and accident. Starting with the injury and moving backward in time, reconstruct the sequence of events that led to the injury.

A. Injury Event ______

B. Accident Event ______

C. Preceding Event #1 ______

D. Preceding Event #2, #3, etc. ______

Figure 8-8. Example of supervisor's accident investigation report.

TOXIC AND FLAMMABLE HAZARDS CONTROL

The tracking down and eliminating of some obvious hazards associated with large concentrations of dangerous substances can be done in a manner similar to that of eliminating purely mechanical hazards. For example, is a section of your plant using a solvent that is explosive when its

22. TASK and ACTIVITY at TIME of ACCIDENT

A. General type of task ______

B. Specific activity ______

C. Employee was working:

☐ Alone ☐ With crew or fellow worker ☐ Other, specify ______

23. POSTURE of EMPLOYEE

______

24. SUPERVISION at TIME of ACCIDENT

☐ Directly supervised ☐ Not supervised

☐ Indirectly supervised ☐ Supervision not feasible

25. CAUSAL FACTORS. Events and conditions that contributed to the accident. Include those identified by use of the Guide for Identifying Causal Factors and Corrective Actions.

______

26. CORRECTIVE ACTIONS. Those that have been, or will be, taken to prevent recurrence. Include those indentified by use of the Guide for Identifying Causal Factors and Corrective Actions.

______

PREPARED BY ______	APPROVED ______
TITLE ______	TITLE ______ DATE ______
DEPARTMENT ______ DATE ______	APPROVED ______
Developed by the National Safety Council	TITLE ______ DATE ______

Figure 8-8. (Continued.)

concentration reaches 500 parts per million in air? Installation of relatively simple monitoring equipment to sound an alarm and rudimentary vents to keep concentrations low can solve the problem.

The most common occupation exposure "incidents" have occurred when a company has switched one chemical for another—perhaps a new solvent was used for a dye process or metal-cleaning operation. Sometimes the switch is made by a supplier who alters existing formula without bothering to tell his customers. The customer—the employer—learns that something is wrong only when the employees begin having allergic reactions to the new toxic agent.

Medical surveillance

Ideally, new employees should receive a careful medical examination to uncover possible existing problems before placement. Such an examination should include tests for allergic reactions to chemicals common in the plant. When equal employment laws and union rules permit, pregnancy tests are advisable if known fetal hazards, such as arsenic or lead fumes, exist. (See Chapter 9 "Women in the Workplace.")

Further, when union rules permit, regular medical surveillance of employees can be useful in uncovering problems before they become serious. Large companies try to keep epidemiological data on illnesses and causes of death among present and past employees. Smaller firms are generally required to keep health records, but usually are not required under OSHA rules to analyze them. Unless a company has hundreds of employees or its hazard has caused a severe health problem, there will not be enough cases in that company to ensure statistical accuracy. But health officials will try to follow up on problems uncovered in large companies by checking their records and combining them with the records of smaller employers. At plants of E.I. duPont de Nemours, for example, employees under 40 years old receive medical examinations every two years; those over 40 are examined annually. An examination usually consists of a health history questionnaire, an examination by a doctor, chest X-ray, vision and hearing checks, a test for lung function, urinalysis, and blood tests. Electrocardiograms are recommended for employees at ages 35, 40, 42, 45, 48, 51, and annually thereafter.

Checklist

A checklist of potential toxic and flammability hazards should address the following questions:

1. Are any chemical agents that are deemed dangerous by OSHA (see 29 CFR 1910.1000, Subpart Z) in use or storage in your plant?
2. Have suspected areas been checked to make sure exposure levels are within tolerances?
3. Are employees wearing protective clothing and/or masks in high-exposure areas?
4. Have employees been screened medically for potential problems (allergies, anemia leading to chemical cyanosis pregnancy)?

5. Are employees allowed to eat or smoke in contaminated areas?
6. Are dangerous substances clearly labeled, and do the labels contain emergency treatment information?
7. Are only specially trained and protected employees allowed to han dle particularly hazardous materials?
8. Are process substitutes allowed into your plant—new solvents, for instance—without proper screening for hazards?

CONCLUSIONS

This chapter has been concerned with the elements of a successful safety program including senior management commitment, safety policy, organization of the safety program, rules and regulations, worker participation, responsibility and accountability, training and motivation, safety inspections, accident investigation, and job safety analysis. Each of these areas is important to the success of a safety program, although not every company will be able to incorporate all of them.

Since all companies differ, it is not possible to specify a single specific program that will meet all safety needs. But there are certain criteria that can be used in developing a safety program or improving an existing program. The criteria are:

- Involvement of senior management
- Involvement of middle management
- Coordination in the workplace by supervisory staff
- Involvement of employees.

If the safety program meets these four criteria, it can be effective. These criteria indicate what the program should achieve; they do not identify the required components of the program (those have been discussed previously). These components will differ between plants, facilities, and departments depending on operations, working environment, geographic and climate conditions, and other factors. Therefore, safety programs must be flexible so that they meet the requirements of organizations and their people.

REFERENCES

1. Bird, F. *Loss Control Management.* Loganville, Ga.: Institute Press, 1976.
2. Diekemper, R., and D. Sparz. "A Quantitative and Qualitative Measurement of Industrial Safety Activities." *ASSE Journal,* December 1970.

3. Heinrich H., D. Peterson, and N. Roos. *Industrial Accident Prevention,* 5th ed. New York: McGraw-Hill, 1980.

4. "Industrial Safety Study." National Safety Council, Chicago: 1967.

5. "Safety Program Practices in Award-Winning Companies," NIOSH, HEW Report, 1970.

6. Peterson, D. *Safety Management—A Human Approach.* Huntington, N.Y.: Aloray Publishers 1975.

7. Petersen, D. *Human Error Reduction.* New York: Garland Publishing, 1981.

8. McElroy, Frank. *Accident Prevention Manual,* 8th ed. *Administration and Programs Volume,* Chicago: National Safety Council, 1981.

9. Olishifski, Julian B., and Earl R. Harford. *Industrial Noise and Hearing Conservation.* Chicago: National Safety Council, 1975.

Chapter 9

Women in the Workplace

by Joseph LaDou and Mary Phares

Women have been employed outside the home in increasing numbers since World War II and today they represent more than 40 percent of the work force. This, however, is not a new phenomenon. Before the Industrial Revolution, women were an integral part of the work force, laboring in the fields and producing goods in the home to meet the needs of an agrarian society. They even dominated some guilds and trade associations, mainly those associated with the production of food, drink,and apparel, and only five out of 500 guilds either were not dominated by women or were formed equally of men and women. In fact, women were so prominently recognized as a major factor in the economy that men working outside the home often were paid only enough for their own support because employers concluded that their wives would support the rest of the family with their cottage industries or guild endeavors. During that time, men and women assumed equal responsibility, not only for the production of goods, but also for household tasks and child rearing.

With the Industrial Revolution and the substitution of mechanical power for human power, the workplace moved from the homes and fields into large, equipment-filled factories. A steady stream of mechanical inventions and new technical processes made hand manufacture obsolete and altered forever living conditions at all levels of society. The new industrialization that brought families into the factories and cities gradually eroded the position of women in the work force. A new middle class developed in which women were no longer producers, but were relegated to household duties or leisurely pursuits. As husbands became more affluent, they hired servants to perform the household chores, and the leisure wife became a status symbol for her husband, attesting to his success in the industrialized society. For financially poor women, the only options for making a living were factory work or domestic service. The trades that women had once dominated were taken over by large manufacturing concerns and steadily became male-oriented as the guilds, and later the trade unions, worked to keep cheaper female labor out. Thus were women eased out of the economic mainstream.

Although many women were no longer part of the mainstream, they continued to work. In the United States in 1870, for example, 1.9 million women were working outside the home, primarily in agriculture, manufacturing, and domestic service. By 1930, the number of employed women had risen to 10.8 million. At the beginning of World War II, women represented 30 percent of the work force, and by the end of the war, their contribution rose to 38 percent. This increase undoubtedly was caused by war production requirements because, within a few years after the war, women's share of the work force dropped to 29 percent.

Since 1950, however, there has been a steady trend upward. In 1980, women represented 46 percent of the work force, and projections for 1990 indicate that 48 million women will be working outside the home, or 51 percent of the work force. Women also have slowly been making inroads into typically male-oriented occupations and can be expected to continue this pattern. Thus, women are working in ever-increasing numbers in light and heavy industry, transportation, community fire and protection services, and, in fact, in almost every job that formerly was held only by men.

WOMEN AT WORK—THEIR SPECIAL PROBLEMS

Because a relatively large percentage of workers today are women, occupational health and safety professionals must be aware of the special problems that women face in the workplace. Physically, they are usually smaller in stature and have less strength than men. Physiologically, they differ from men in one significant aspect—that of carrying the embryo and fetus into the workplace.

A major problem facing occupational health professionals is the difficulty of identifying the job-relatedness of health problems, particularly in handling workers' compensation disability claims. For additional commentary, see the discussion in Chapter 17, "Understanding Workers' Compensation."

Too little is known about the interrelationship of work and health, especially with regard to genetic effects in both men and women. Research in this area is highly complex, often demanding medical histories extending several decades back. (Some effort has been made to require that the occupations of parents be included as part of their children's birth certificates in order to obtain epidemiological information, but this practice has been objected to as an invasion of privacy.)

Acute situations, such as massive exposure to a toxic substance resulting in liver damage, can be readily identified as work-related. But diseases that develop slowly over many years, with symptoms that appear only infrequently and may go unnoticed for long periods, are very difficult to ascribe to the work environment. Moreover, there may be more than one contributing element. For example, both industrial solvent exposure and alcohol abuse can cause liver disease. Which, then, is to be judged the cause in an individual who is exposed to such solvents and is also a heavy drinker? In addition, the individual's personal physician may be of little help in determining the work-relatedness of a health problem since many physicians in private practice are unaware of specific health hazards of the job and consequently cannot relate patient symptoms to the work they might perform.

Within the limitations of the present knowledge of occupational hazards to women, this chapter attempts to provide safety and health professionals with a background of information that will help them (*a*) to understand better the conditions they observe, and (*b*) to seek further enlightenment from the literature and symposia on the particular problems that women face in the work environment. Other chapters that deal with related matters are Chapter 7, "Occupational Stress," and Chapter 10, "High-Risk Workers."

Physical strength and work capacity

Women have less physical strength than men, with some investigators concluding that they have only two-thirds as much strength (see Table 9-A). About one quarter of women are stronger than the weakest quarter of men. Much depends on the height, weight, and physical fitness of the individual. Differences between individuals are important considerations in work assignments. (See Figure 9-1.)

The isometric strength of 25 different muscle groups has been shown to be optimal at 20 years of age in women and 30 years of age in men. With increasing age, there is a gradual decrease in muscle strength which progresses more rapidly in women after 40. Oxygen consumption with exercise, a key indicator of work capacity, also decreases in both women and men with aging. Findings from subjects of both sexes and various age groups tested on bicycle ergometers accompanied by heart-rate recordings have indicated that aerobic capacity and heart rate with exercise are optimal at about age 20 and decline gradually thereafter. (Mean oxygen uptake was 3.6 liters/min for the men and 2.7 liters/min for women. Size and physical fitness differences accounted for some, but not all, of

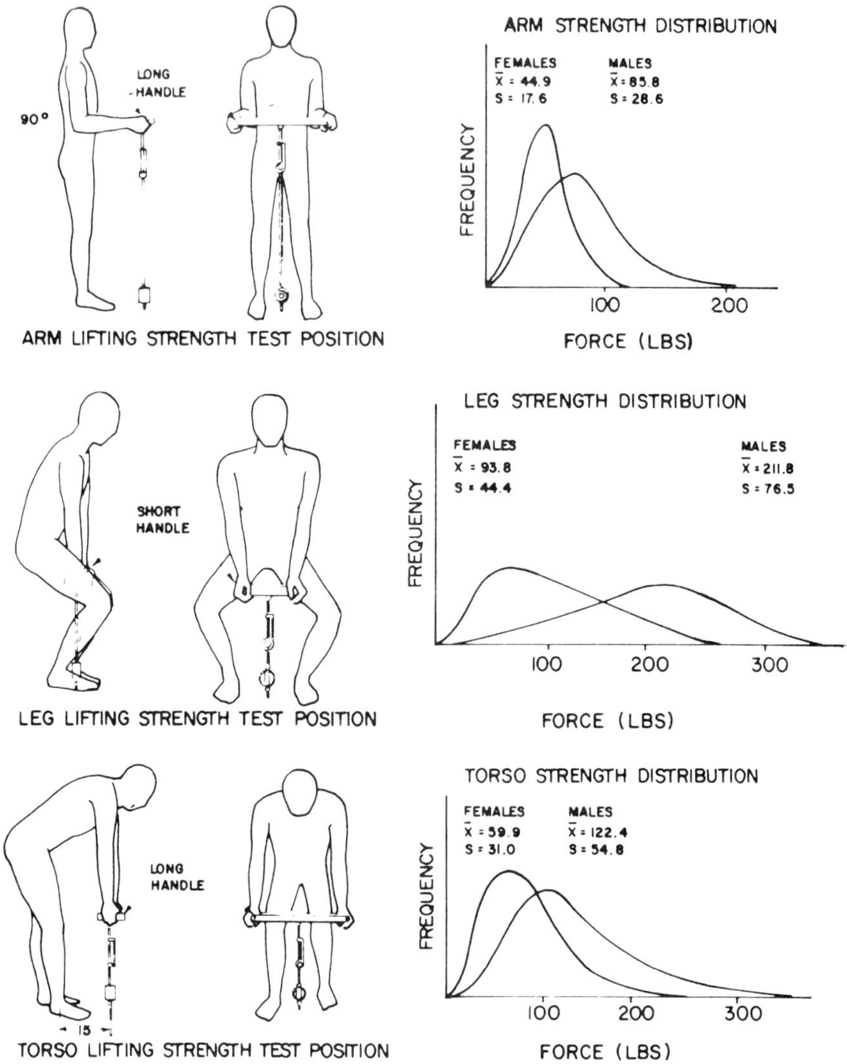

Figure 9-1. Standard strength test postures and distributions.

Journal of Occupational Medicine, vol. 20, no. 6, June 1978.

the divergence, since there is a 10-15 percent deficit in females after correction for the weight difference.)

Women have a higher fat content than men. Since fat tissue has little blood, blood volume for an adult male averages 75 ml/kg; for an adult female, 65. For the same age and body weight, females also have total lung volumes about 10 percent smaller. Women also have hemoglobin

Table 9-A.

PREDICTED STANDARD STRENGTH FOR VARIOUS CASES OF STATURE, BODY WEIGHT, GENDER, AND AGE

		Torso Strength			
		Male Body Weight		Female Body Weight	
Age	Height	100 lb	200 lb	100 lb	200 lb
20 yrs	5′0″	75.3	123.8	50.3	78.8
	6′0″	88.4	155.2 (max)	63.5	105.3
50 yrs	5′0″	56.4	91.0	31.4 (min)	41.1
	6′0″	69.6	117.5	44.6	67.5
		Arm Strength			
		Male Body Weight		**Female Body Weight**	
Age	**Height**	**100 lb**	**200 lb**	**100 lb**	**200 lb**
20 yrs	5′0″	72.5	88.3	41.2	55.9
	6′0″	76.8	96.7 (max)	44.4	64.3
50 yrs	5′0″	64.6	72.4	32.3 (min)	40.0
	6′0″	58.8	80.8	36.5	48.4
		Leg Strength			
		Male Body Weight		**Female Body Weight**	
Age	**Height**	**100 lb**	**200 lb**	**100 lb**	**200 lb**
20 yrs	5′0″	117.5	226.9	82.4	131.8
	6′0″	187.4	246.7 (max)	92.2	151.5
50 yrs	5′0″	151.7	175.4	56.6 (min)	80.2
	6′0″	156.5	184.8	61.3	89.7

Source: *Journal of Occupational Medicine,* Vol. 20, No. 6, June 1978.

content about 10 percent lower than men; hematocrit (the relative amount of plasma and blood cells) is 42 for women, and 47 for males. Arterial oxygen content is 16.7 ml/100 ml versus 19.2 for males. Thus, for submaximal work (oxygen uptake of 1.5 liters/min), women need 9.0 liters of cardiac output to transport 1 liter of oxygen while men require only 8.0.

S. Konz has reviewed the male/female differences and concluded that, "Although the above figures help explain the differences in athletic performance, cardiovascular differences between men and women should have little importance for industrial work since most industrial tasks should not be designed to require maximum physiological output. If very high output is required, then age and training (fitness) are probably more important than sex."[1]

Many strong, physically fit women should be able to perform adequately work tasks that require physical strength. The physical capacity of applicants for such jobs can be determined from tests of aerobic capacity (oxygen uptake) and work capacity in which the individual exercises on a bicycle ergometer to maximum capacity while heart rate, blood pressure, and other important physiological variables are monitored.

Menstruation and pregnancy

Occupational health professionals also may question whether or not women can fulfill jobs demanding strength and endurance during menstruation and pregnancy. Two studies indicate that they can. These studies tested the work capacity of women with dysmenorrhea during various times of the menstrual cycle[2] and pregnant women both before and after birth.[3] The menstruating women exercised on bicycle ergometers while taking mental ability tests and were found to function as effectively during menstruation as they did otherwise. No changes occurred that were significant enough to affect work performance.

The pregnant women also exercised on bicycle ergometers while being monitored for cardiorespiratory responses and were tested both during pregnancy and three months post partum. Their aerobic capacity (oxygen consumption) during specific workloads was similar both before and after birth. During pregnancy, however, their ventilation, heart rate, cardiac output, and stroke volume were higher and their arteriovenous differences lower at each workload. These differences occurred during the first trimester and remained constant throughout the rest of pregnancy. The decrease in arteriovenous oxygen differences could indicate that pregnant women use part of their cardiovascular and respiratory reserves for the pregnancy itself and consequently may have a reduced capacity for additional stress. The rate of increase or decrease in these variables, however, was the same in both the pregnant and nonpregnant states.

The above findings appear to indicate that menstruating and pregnant women are capable of performing their usual work. However, pregnant women in their final trimester probably should be transferred to less physically stressful work because of tendencies for (*a*) greater blood flow to the fetus; (*b*) an increase in blood flow to the legs during exercise and standing; (*c*) and changes in the center of gravity, size, and mobility of the body that occur at this time.

Occupational injuries—low back pain

The California Workers' Compensation Institute has indicated that disabling injuries in women increased steadily over the 1970s, rising by 134 percent, while women's employment rose by only 52 percent. Over the decade, women's injuries rose from 18 percent of total injuries in 1971 to 23 percent in 1979, partially reflecting the shift of women workers into heavier work, for example, the 81 percent increase in women in construction work.

Back injuries are one of the most common complaints in both light and heavy industry and to a lesser degree in clerical and service operations. (See Table 9-B).

TABLE 9-B. DISABLING BACK STRAINS PER 1,000 WORKERS BY INDUSTRY, CALIFORNIA, 1984

Industry	Number of Back Strains per 1,000 Workers
All industries	6.6
Construction	14.9
Transportation and public utilities	13.6
Mining	10.1
Agriculture, forestry, and fishing	8.0
Manufacturing	7.2
Wholesale trade	6.9
Service	6.8
Retail trade	6.3
State and local government	2.2
Finance	2.1

Source: Division of Labor Statistics and Research: *Work Injuries and Illnesses in California,* Quarterly Statistical Summary, August 1985.

Low back pain requiring medical attention occurs in about 50 percent of workers at some time during their working lives, and about 50 percent of these incidents result in lost work time. Often the cause of low back pain cannot be diagnosed. Low back pain of undiagnosed origin ranges from 21 percent to 84 percent of compensable cases; in 12 percent of these cases, the pain could not be traced to a specific accident.[4] (However, cases of malingering are rare.) Each year, low back injuries cost U.S. business and industry about $15 billion, or more than all other industrial injuries combined.

Low back pain occurs most frequently and is most intense in industrial workers in their late 30s and early 40s. There is some question as to whether men or women suffer more lower back complaints, but a number of investigators have concluded that they are equally at risk. Findings

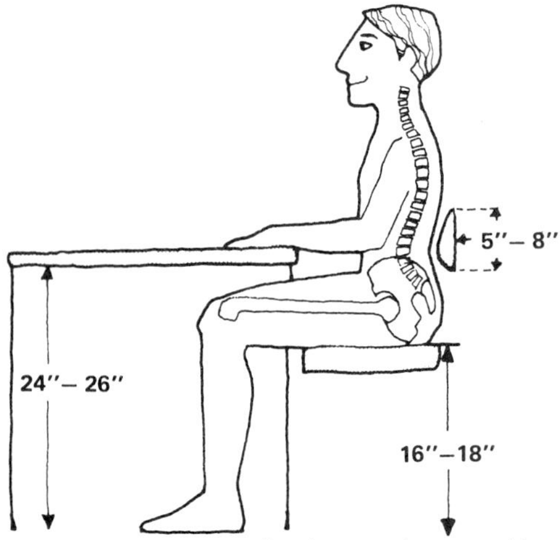

Figure 9-2. Normal spinal curvature should be maintained in the seated posture. Note that the table and chair heights shown are recommended for a male worker and are not correct for a female worker.

from three industry studies indicate, however, that only 16-20 percent of compensable low-back injuries occur in women. The higher incidence of these complaints in males may stem from their work in male-oriented jobs that tend to be physically more stressful. If that is the case, women can be expected to become more at risk as they increasingly enter male-oriented fields.[5]

A worker who sustains the first low-back injury will be more susceptible to future injuries, which can become increasingly severe. Consequently, every effort should be made to avoid that first injury. Occupational health professionals should instruct employees in the importance of physical fitness and proper posture, provide exercise that will strengthen the back and stomach muscles, and show the proper methods of handling heavy loads or performing office tasks that *are stressful* to the lower back—typing, video console work, filing, reaching, or standing for long periods. They should also evaluate the safety and adequacy of such work equipment as chairs, desks, filing cabinets, and lifting devices.

Physical fitness has been found to aid significantly in avoiding compensable back injuries. The lifting capability of the individual is another factor to be considered in avoiding such injuries. Workers tested on isometric simulations of job strength requirements have shown a three times greater risk of back injury when the lifting requirements approached or exceeded their physical strength. Because of their greater strength, men are at an advantage when screening strength tests are a part of preplace-

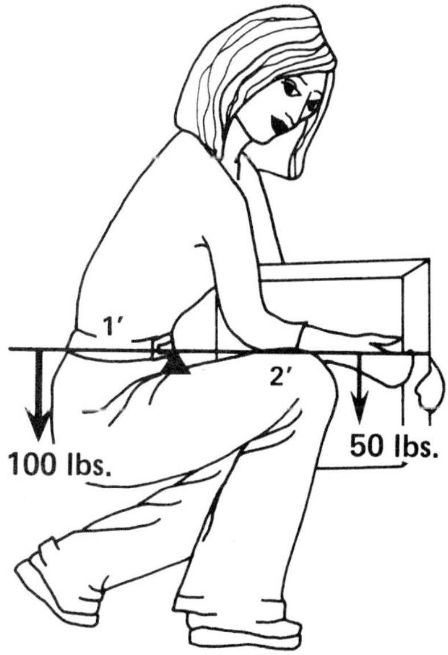

$$f \times d = W \times D$$

$$f \times 1 = 50 \times 2$$

$$f\ 1 = 100$$

$$f = 100 \text{ lb}$$

Figure 9-3. Demonstration on the pressure of intervertebral disks with the lifting of a heavy weight. This simple message is conveyed better to women workers by using the example of a woman lifting.

ment examinations.

Good posture helps in avoiding back pain. With correct posture, the body is in a position that causes the least amount of stress on the joints, bones, ligaments, and muscles, and requires the least amount of effort during standing, walking, sitting, reaching, bending, and lifting. Improper posture contributes not only to back problems, but also to fatigue. The normal spine of the adult human is curved slightly in an S form (Figure 9-2). Any deviation from this curve causes stress on the ligaments, muscles, and disks in the spine, particularly in the lower back. As a result, bending, lifting, pushing, and pulling requirements of a job can exacerbate postural pain into a back injury.

The rules to avoid injuries to the back while lifting are simple—

- Stay close to the work,

- Keep feet apart with one foot ahead of the other,
- Keep the back straight,
- Do not twist the body while moving a work load.

A few simple photographic slides shown during an injury-prevention program could present a clear message to workers on how to avoid back injuries (Figure 9-3). One single presentation, however, is not sufficient, and it must be followed by a practice session and continuing reminders and encouragement.

There are definite limitations to a back-injury prevention program that relies solely on counseling of workers. Although an effort should be made to protect the worker from injury (by instruction in proper positioning of individuals with respect to the weight being lifted), it is of less value without other hazard control methods, such as job redesign and preplacement evaluation of the worker.

Many clerical jobs, particularly those dominated by women, require long hours of sitting at a desk operating business machines, or video consoles, or performing similar tasks. Women working in such jobs should be instructed in proper sitting posture—that is, to keep the spinal curves in the same position as they are during proper standing posture.

Poorly designed work settings can also contribute to back problems, tension, and fatigue in clerical jobs or others that require long hours of sitting. The chair must allow the individual to sit with a minimum of pressure on the thighs to prevent blood pooling in the lower extremities when circulation is impeded. It should be constructed so that it maintains the spine in the proper posture of the lower back. Table or desk height is also important and should be such that the individual does not need to raise up or lean over to work.

Heat exposure

Women do not tolerate heat as well as men. The National Institute for Occupational Safety and Health (NIOSH) has recognized this problem by establishing an allowable heat stress of 79 F (WBGT)* for unacclimatized men and 76 F (WBGT) for unacclimatized women. In environmental heat studies, women exhibit higher skin temperature and pulse rates and lower sweat rates than men under the same heat condition.[6] Men and women exercised on a treadmill in a hot environment exhibit higher sweat rates in physically fit men. In a hot, moist

---

*WBGT or wet bulb globe temperature is a generally accepted index of heat stress.

environment, unacclimatized women exhibit less heat tolerance and higher pulse rates and core temperatures. In a hot dry environment, on the other hand, no difference is found between the heat tolerance of men and women until relative humidity reaches 80 percent. At the higher humidity, core and skin temperatures remain equal, but the sweat rate of women does not increase to the extent that it does in men.

Although both men and women acclimatize to heat equally well, the more physically fit individuals acclimatize best. An individual's aerobic capacity also has an important bearing on heat tolerance, since a smaller person with limited aerobic capacity will use more reserve capacity than a larger, stronger person. Thus, the difference in tolerance to heat may not be so much a question of sex as of size and physical fitness, although women are at a disadvantage determined by oxygen uptake capacity and other measures.

The circulation is the body's first defense against heat. At higher temperatures, blood flow is shifted away from the interior of the body to the skin, thereby creating a loss of body heat by convection or radiation across the skin. The amount of cardiac output delivered to the skin during heat adaptation may rise to 20 percent as compared with 5 percent under normal conditions. With this change, blood pressure drops, blood flow to organs other than the skin decreases, and heart rate increases to increase cardiac output.

The body's second, more efficient, defense against heat is the delivery of perspiration to the skin which can carry with it 580 kcal (kilocalories or large Calories) of heat when evaporated from the skin. In healthy, acclimatized men given adequate salt and fluid replacement, the sweat rate can exceed one liter per hour for several hours. Women have fewer sweat glands than men, and apparently trigger their activity at higher temperature, helping to explain their disadvantage in heat tolerance.

These compensations for heat exposure are often not sufficient to protect some individuals, and heat exhaustion or heat stroke may occur.

- *Heat exhaustion* results from dehydration and inadequate blood flow to the brain and is characterized by nausea, dizziness, and, ultimately, fainting. Usually, it is not a serious medical problem because fainting places the individual in a position in which the blood flow is automatically returned to the brain. If it occurs in those people who are operating machines, equipment, or vehicles, however, there can be serious safety problems.

- *Heat stroke,* on the other hand, is a medical emergency. It is usually accompanied by a high body temperature (42 C or higher) and severe disturbance of the central nervous system. The employee suffering from heat stroke may become irritable or aggressive, exhibiting hysterical or disoriented behavior, or may become completely apathetic. If the individual remains untreated, death ensues. Consequently, heat stroke victims must be identified and treated rapidly. Rectal temperature is the only accurate indicator of heat stroke. If the temperature taken rectally is 105.8 F (41 C) or higher, the employee should be treated for heat stroke. Treatment consists of cooling the body continually for at least one hour, preferably until rectal temperature reaches 100.4 F (38 C). Cooling is best accomplished by spraying a combination of water and air over the naked body. Ice and ice water are not recommended as coolants because vasoconstriction of the peripheral blood vessels can occur, blocking heat transfer and eventually resulting in hypothermic shock when cooling is stopped.

Since women tolerate heat less well than men, those who are to work in hot, moist environments should be acclimatized. Acclimatization consists of exposing the individual to progressively longer periods (one to four hours daily) in the hot environment and can be accomplished over a period of four to seven days. Physically fit individuals will acclimatize more rapidly than those less physically fit. Moreover, working or exercising in heat acclimatizes an employee faster than resting in heat. Finally, body salt and fluid content must be maintained during acclimatization or the process will be ineffectual. Once acclimatization has been accomplished, its effects will last only about two weeks. Thus, if an employee has been on leave for two weeks or longer, he or she will have to be reacclimatized.

THE REPRODUCTIVE FUNCTION AND THE WORKPLACE

Women's unique child-bearing function requires that she carry the embryo and the fetus into the workplace. Although many women work in jobs that are not hazardous to the embryo and fetus, many work in environments that expose them to chemical and physical substances that can be mutagenic, teratogenic, or carcinogenic. Unfortunately, too little is known about the effects of many toxic substances on the reproductive organs of either women or men, and primary attention has been placed on the reproductive function of women in the workplace rather than approaching the problem as a joint question of both male and female reproductive toxicology.

Women have claimed discrimination when refused jobs in areas that are hazardous to reproduction while men are permitted in these jobs. In coping with such claims, it should be remembered that, either through circumstance or preference, not all women will reproduce in their lifetimes. Furthermore, although the child-bearing period usually extends from 15 to 45 years of age, it often does not extend beyond 35 years of age, thus leaving many years for productive work outside the home with, in some cases, less risk to progeny than may be imposed by men workers.

The dearth of information on toxicity to the embryo and toxic responses of the fetus during pregnancy make it difficult for occupational safety and health professionals to counsel pregnant workers or answer questions in this important area. Much additional research will be needed before the task is simplified. Taking into account the limited knowledge at present, it is hoped that the following discussion will offer some insight into the problem. The most serious aspects of the problem are infertility, damage to the embryo, and damage to the fetus.

Infertility and mutagenesis

Exposure to toxic substances may affect either the sperm or egg so that fertilization will not occur or, if it does, the zygote will be abnormal and usually imperceptibly aborted. Infertility may result from either maternal or paternal exposure to certain substances in the workplace. Either type of exposure, acting independently or in combination, can result in lack of fertilization, improper implantation, spontaneous abortion, or production of an abnormal embryo. Investigators of NIOSH studying occupational mutagenesis, teratogenesis, and transplacental carcinogenesis concluded that:

> . . . it is difficult to anticipate the relative sensitivity to germinal mutagenic events in men compared to nonpregnant women. Depending on the action of a particular chemical, the expected effects will differ depending on the sex of the exposed individual. Spermatogenesis involves a continuously replicating cell population with billions of cells; oogenesis is a prenatal event in mammals, and the female germ cell population is finite at birth with no further replication. Therefore, chemicals whose mechanisms of toxicity depend on cell division would have their greatest effect on the male germ cells; chemicals affecting non-replicating cells might have their greatest effect on the limited oocyte population. Chemical exposure which decreases the total number of cells would also have a more prominent effect on the female since damaged cells are not replaced as in the male.[8]

Thus, women are born with all the germ cells (eggs) they will have during their lives. This means that occupational exposures may affect the eggs all during the woman's work life. In contrast, the male germ cells (sperm) are produced in a brief, two-month cycle during which time there is less likelihood of occupational exposure. Although the cells responsible for sperm maturation could be damaged over a long period of time, the reversibility of some exposures that damage sperm cells (DBCP and lead) suggests that men are at less risk than women with exposure to mutagens.

Most abnormal conceptions terminate in spontaneous, unsuspected abortions at a very early stage or are reabsorbed into the female's body. Those that survive may be aborted later, may be still born, or may be born as defective children. It is estimated that early intrauterine abnormalities are responsible for 15–40 percent of spontaneous abortions. Of the birth defects detected in children up to age 10, 6 percent result from known environmental agents and 20–30 percent from major genetic defects. Many of these genetic defects may be caused by toxic exposures in the workplace.

The embryo

During the first trimester of pregnancy, and particularly during the first eight weeks, the embryo is in its most vulnerable state. Even during the first week, before the placenta is in place, toxic agents absorbed by the mother can be rapidly transmitted to the embryo. This susceptibility continues after placental implantation to the end of the first trimester. The substantial effects that external agents can have on the embryo during this period are documented by the thalidomide tragedy in which thousands of severely deformed infants were born to mothers who had taken the drug in early pregnancy and by the administration of diethylstilbestrol (DES) to prevent miscarriages that resulted in a high risk of cervical cancer in female offspring and possible reproductive defects in male offspring. Whereas the mother may not be highly sensitive to toxic exposure, the embryo, because of its more rapid cell turnover, lack of detoxification mechanisms, and undeveloped biologic barriers, is extremely susceptible.

The fetus

In the remaining two trimesters, the fetus is usually considered to be less susceptible to external influences. Nonetheless, adverse influences may result in growth retardation and behavioral or functional disabilities in the infant or child. In the second trimester, the fetus grows at a rapid rate and the placental blood flow reaches a maximum. Under these conditions there can be a considerable transfer of toxic substances from the mother to the fetus. As the fetus progresses toward term, it usually becomes less susceptible to teratogens but more susceptible to transplacental carcinogens. Just prior to term, it is the most susceptible to carcinogenic substances that must be metabolized to have an effect.

The infant

Toxic substances in the workplace can also affect the infant after birth through exposure to contaminants carried home by either the mother or father. Such exposures may negatively affect developmental processes that are not completed until after birth, including physical growth and final development of the central nervous system and some of the endocrine glands. In the very young infant, moreover, some substances cannot be metabolized because of the immaturity of the system.

Occupational health professionals should also be aware that breast-fed infants are more at risk from contaminants than are bottle-fed babies. Substances that can be absorbed through the skin, ingested, or inhaled may affect the quality and amount of the mother's milk, and some potentially dangerous substances can be transmitted in the milk, thus directly exposing the infant. For example, lactating women in Michigan who had ingested food contaminated with PCB showed measurable amounts of the chemical in their breast milk.

The importance of protecting not only the workers, but also their progeny, is emphasized by the primary objective of the Occupational Safety and Health Act of 1970, which is to provide a safe and healthful workplace for every worker, male or female, in the nation and to preserve our human resources.

The mother

Research to date has focused on fetal responses to contaminants in the workplace rather than such responses in the mother. In this area, as in others, there is limited information about women's particular occupational health problems. From available knowledge, however, it appears that pregnant women may be more vulnerable than nonpregnant women to work-related insults to their bodies. During pregnancy, detoxification mechanisms change, together with lung function and blood composition and volume. Changes in detoxification mechanisms and lung function can increase susceptibilty to airborne contaminants and decrease ability to cope with them because a portion of cardiorespiratory reserves is assigned to supporting the fetus. The increased output of the heart to accomplish this may increase susceptibility to toxic inhalants in the mother because a larger respiratory volume comes in contact with the gases. The increase in blood volume during pregnancy also decreases the concentration of iron in the blood to below normal levels, resulting in what is termed the *physiological anemia of pregnancy.* This condition is so common among pregnant women as to be considered normal for pregnancy. The decrease in hemoglobin that occurs with pregnancy, however, may be serious for women who are exposed to lead, benzene, X-rays, and other substances or work conditions that inhibit the production of red blood cells. Considerably more research will be required to understand fully the work-related impact of changes in blood composition during pregnancy.

Although findings seem to indicate that women during pregnancy can continue to perform any work to which they are accustomed, they may have a reduced capacity for coping with more physically demanding tasks. Progressive weight increases, as well as changes in ventilation and arteriovenous oxygen, contribute to the conclusion that pregnant women on the job must learn to pace themselves so as to accomplish the work within their capabilities.

TOXIC SUBSTANCES IN THE WORKPLACE

The effects on women, in particular, of exposure to hazardous substances in the workplace have been largely ignored in epidemiological studies even though many female-oriented industries continually expose their workers to a wide variety of chemicals. The electronic, textile, and jewelry industries, for example, have a preponderance of female employees and they are exposed to hazardous solvents and organic and inorganic compounds daily. In other industries, where men and women have been equally exposed to harmful substances, investigators have concluded that the occupational disease experience of women was comparable to that of men. For example, exposure to asbestos in shipyards during World War II affected men and women equally. But women have the additional problem of carrying their unborn progeny into the workplace where hazardous exposures may occur. The following sections review the substances that are known to be hazardous to reproduction. (Figure 9-4 summarizes these substances, and Table 9-C correlates them with various occupations.)

Lead

Lead is not absorbed into the body through the skin, but by ingestion or inhalation of dusts or fumes. (See the discussion in Chapter 12, "Employee Education.") When ingested, only about 10 percent will be absorbed through the gastrointestinal tract, while the remainder will be excreted in feces. When inhaled, more than half will be exhaled unchanged and the remainder will settle in various areas of the body. Large particles settle on the surface of the bronchi and are returned to the nasopharynx where they are swallowed. Minute particles can reach the lung alveoli where they are absorbed.

The reproductive hazards of lead exposure have been known for a long time. Exposure to lead is also believed to be associated with sterility in males, and possibly in females, as well as abnormal spermatogenesis, stillbirth and infant mortality, lower birth weights, and problems of growth and mental retardation.

During pregnancy, lead compounds pass readily through the placenta. Since lead has a greater effect on the rapidly growing tissue of the embryo and fetus than it does on adult tissue, even minimal amounts may be serious at the embryonic and fetal stages. Moreover, the newborn infant, although removed from the workplace, may still be affected by exposure to lead. The lactating mother who continues to be exposed will transmit the lead to the infant in her milk. If either parent brings lead dust home on the clothing, the home is contaminated and the infant exposed. Postnatal exposures are important because lead can affect physical and mental development and the central and peripheral nervous system—the younger the organism, the greater the effect.

Aside from the reproductive problems associated with lead exposure, it can be damaging to all adults. The systems and organs affected include the central and peripheral nervous systems, the gastrointestinal tract, the hematopoietic (blood forming) system, and the kidneys. Symptoms include fatigue, sleep disturbance, abdominal pain, and headache. Before any of these symptoms can be determined to be lead-related, however, they must be evaluated for other pathological or psychological origins.

How much lead exposure is safe for a female employee? OSHA has set a maximum of 50 mg lead/m^3 of air and 40 mg/100 ml of blood. With adult men, it has been estimated that exposure to 50 mg/m^3 air can result in blood lead levels of 35 to 45 mg/100 ml blood. Women, on the other hand, have been observed to have lower blood levels than men at equal exposures.

To achieve an acceptable margin of reproductive safety, J. Messite and M.D. Bond[7] have concluded that blood levels of female workers should be no higher than 30 mg/100 ml blood, which requires an exposure to airborne lead no higher than 50 mg/m^3 air, with no additional exposure (via smoking or hobbies) either in the workplace or at home.

Benzene

Benzene is an organic solvent that is highly toxic by inhalation, ingestion, and skin absorption. It is used in a wide range of processes including chemical and pharmaceutical production, printing and lithography, and dry cleaning, as well as in the manufacture of adhesives, detergents, petroleum products, and paint. About 2 million workers in the United States are estimated to be exposed to benzene.

CHART OF TOXINS AND THEIR EFFECTS

AGENTS	DEFINITIVE SYSTEMIC TOXICITY: Neurologic	Hepatic	Renal	Hematologic	Cardiovascular	Pulmonary	SPECIAL EFFECTS	SUSPECTED REPRODUCTIVE EFFECTS: Carcinogenesis	Reduced Fertility	Spontaneous Abortion	Mutagenesis	Teratogenesis	SUGGESTED BIOLOGICAL TESTING	AIR STANDARDS*: OSHA (existing)	NIOSH (proposed)
Heavy metals															
Cadmium		•	•	•	•	•		•	○			•	Urine cadmium Quant. low molecular weight proteins	fume 0.1 mg/M³ dust 0.2 mg/M³	40μg/M³
Lead	•		•	•					○	•		•	Blood lead* Blood zinc protoporphyrin Urine ALA Urine coproporphyrin	inorganic 0.2 mg/M³ (Proposed 0.1 mg/M³)	inorganic 0.1 mg/M³ and Blood Lead 60μg/100gms
Mercury	•		•							•		•	Blood mercury Urine mercury	inorganic 0.1 mg/M³ organic 0.01 mg/M³	inorganic 0.05 mg/M³
Organic solvents															
Benzene (benzol)	•			•				•				•	Urine phenols CBC	10 ppm (emergency proposal: 1 ppm)	1 ppm
Halogenated hydrocarbons															
2 chlorobutadiene (chloroprene)	•	•	•	•		•		•	○	•	•		Liver function test	25 ppm	1 ppm
Dibromochloropropane		•	•					•	○				Sperm count Serum testosterone	10 ppb (emergency proposed)	10 ppb
Epichlorohydrin		•	•			•			•				Liver function test	20 mg/M³	2 mg/M³
Ethylene dibromide		•	•			•		•	○					20 ppm	1 mg/M³
Polychlorinated biphenyls (PCB's)		•		•			Chloracne	•	•	•			Blood analysis Adipose tissue Serum transaminase	Compounds with 42% chlorine 1 mg/M³ 54% chlorine 0.5 mg/M³	1μg/M³

Figure 9-4. Chart of toxins and their effects.

Tetrachloroethylene (Perchloroethylene)	•	•				•		•			•	•		100 ppm	50 ppm
Vinyl chloride	•	•			•		Acrosteolysis	•			•		Liver function test	1 ppm	1 ppm
Hypoxic agents															
Carbon monoxide	•			•	•							•	Carboxy hemoglobin	50 ppm	35 ppm
Anesthetic gases															
Halogenated gases e.g. halothane methoxy fluorane	•	•						•	•	•	•	•		not specified	Halogenated anesthetics 2 ppm based on weight of specific gas sampled not >1 hour
Pesticides															
Carbaryl	•								•					5 mg/M³	5 mg/M³
Chlorinated hydrocarbons (e.g. chlordane)				•				•			•		Adipose tissue analysis Blood analysis	varies with specific compounds e.g. Chlordane 0.5 mg/M³	
Chlordecone (kepone)	•	•					Pleuritic and joint pains	•	o				Blood analysis	emergency standard 1 μg/M³	1 μg/M³
Estrogenic compounds															
Diethylstilbestrol					•			•	•			•	Blood analysis	not specified	not specified
Ionizing radiation (Whole body)															
X-rays and gamma rays				•			Gastro-intestinal disorders	•	•	•	•	•	Personal film badge dosimetry Thermoluminsescent dosimetry	NCRP recommendation 1.25 rad/quarter, 5 rad/year Pregnancy: 0.5 rem/full pregnancy	
Miscellaneous substances															
Carbon disulfide	•				•		Retinopathy		•	•		•	Urine iodine-azide test	20 ppm	3 mg/M³
Ethylene Oxide				•		•					•		CBC	50 ppm	50 ppm

* In pregnancy, blood lead less than 40 μg/100 gms is suggested.
o Evidence only male infertility; no data on females
• Animal and/or human data
· Time weighted average 40-hr. week, 8-hr. day (OSHA), 10-hr. day (NIOSH).

From: *Guidelines on Pregnancy and Work.* Chicago: The American College of Obstetricians and Gynecologists, 1977.

TABLE 9-C. EXAMPLES OF POSSIBLE EXPOSURES RELATED TO VARIOUS OCCUPATIONS

OCCUPATION	POTENTIAL EXPOSURES
1. Agriculture	
("Migrant workers' wives work in fields while pregnant and sustain work as long as able)	Multitude of substances and conditions, including physical and biologic. Excessive work hours; intermittent and prolonged heavy work. Chemicals, especially pesticides, herbicides, insecticides, solvents, dusts, fumes, gases, heat and high humidity, infections (bacterial, viral, rickettsial) and vibration.
2. Chemical and pharmaceutical production workers	
a. Concomitant exposures and physical stressors and possible synergistic effects are factors to be considered	See text.
b. Ergonomic/biomechanic deficiencies	See text.
3. Cleaning personnel	
a. Launderers	Contaminated clothing, detergents, enzymes, soaps, heat, humidity.
b. Dry cleaners	Contaminated clothing, heat, perchloroethylene, Stoddard solvent, naphtha, benzene, trichloroethlene and others.
4. Clerical personnel	Carbon tetrachloride and various other cleaners, physical stresses, poor illumination, ergonomic deficiencies.
5. Domestic workers (homes, hotels, motels, office buildings)	Alkalies, bleaches, detergents, heat, cold (physiologic factors), shift work, at times hard work, and various solvents in cleaning agents.
6. Electronics assemblers	Antimony, epoxy resins, lead, methyl ethyl ketone, methylene chloride, tin, trichloroethylene and visual stresses and strain.
7. Hairdressers and cosmetologists	Acetone, aerosol propellants (freons), benzyl alcohol, ethyl alcohol, hair dyes, hair spray resins (polyvinylpyrrolidone), halogenated hydrocarbons and other solvents of a wide variety.
8. Hospital/health personnel	
a. Registered nurses, aides, orderlies, physicians, students, patients	Alcohol, anesthetic gases, ethylene oxide, infectious diseases (bacterial and viral), puncture wounds and X-ray radiation.
b. Dental hygienists	Anesthetic gases, infectious diseases (bacterial and viral), mercury, punture wounds, ultrasonic noise, vibration, and X-ray radiation.
c. Laboratory workers (clinical and research)	Infectious diseases (bacterial and viral), puncture wounds, wide variety of toxic chemicals, including carcinogens, mutagens and teratogens; and X-ray radiation. Many workers are subject to shift rotation.
9. Opticians and lens grinders	Coal tar, volatiles, hydrocarbons, iron oxide, other polishing and grinding dusts and solvents.

OCCUPATION	POTENTIAL EXPOSURES
10. Photographic processors	Bromides, caustics, iodides, iron salts, mercuric chloride, pyrogallic acid and silver nitrate.
11. Plastic fabricators	Acids, acrylonitrile, alkalies, hexamethylenetetramine, peroxide, phenol formaldehydes, styrene, urea formaldehydes, vinyl chloride and vinylidene chloride. High heat and humidity.
12. Printing operatives	Benzene, carbon tetrachloride, ink mists, lead, methanol, methylene chloride, microwave, 2-nitropropane, pigments, trace metals, toluene, and various solvents, including trichloroethylene; ultraviolet, microwave and other radiation sources. Noise and vibration.
13. Sign painters and letterers	Epichlorohydrin, lead chromate pigments, lead oxide, toluene, trace metals and xylene. A wide diversity of other chemicals as solvents and strippers.
14. Textile and related operatives	
a. Textile operative	Asbestos, dyes, flame retardants, formaldehyde, heat, noise, raw cotton dust, synthetic fiber dusts.
b. Sewers and stitchers	Asbestos, cotton and synthetic fiber dusts, flame retardants, formaldehyde, noise and organic solvents.
c. Upholsterers	Same as above.
(Some specific chemicals encountered in the above occupations are benzene, carbon disulfide, chloroprene, perchloroethylene, styrene, toluene and trichloroethylene.)	
15. Transportation	
a. Bus, taxi, truck and other vehicle drivers b. Airlines	Carbon monoxide, polynuclear aromatics, lead and other combustion products of gasoline and jet fuels; circadian dysfunctions; heat; physical stresses; shift changes; and vibration (ultrasonic, noise and hypersonic)

Adapted from *Guidelines on Pregnancy and Work*. Chicago: The American College of Obstetricians and Gynecologists, 1977.

The effects of benzene on human reproduction are not well known. Since so few animal studies have been conducted and information on human responses is almost nonexistent, the findings must be considered inconclusive with much research still needed.

NIOSH has stated that benzene must be regulated as a carcinogen because it is concluded to be leukemogenic and to cause progressive malignancy in the blood-forming organs. The agency has also recommended that exposure to benzene not exceed 1 ppm for a two-hour period.

Other chemical hazards

The effects of lead and benzene have been treated separately in this chapter because of the very large number of workers who are exposed to these hazards. There are, however, other materials in the workplace to which a much smaller number of workers are exposed that may or may not affect the female reproductive system, the pregnant worker, or the fetus. Information on the effects of these substances is also limited because of insufficient research or inconclusive findings. (Hazardous substances used primarily in agriculture are not considered here.)

Carbon disulfide, another suspect chemical, is toxic with inhalation or repeated skin contact. It is used in degreasing operations, rubber vulcanization, dry cleaning, oil extraction, chemical analysis, and grain fumigation. It is also used extensively in the viscose rayon industry where workers are exposed to a combination of carbon disulfide and hydrogen sulfide. Exposure to this combination may cause toxic effects that are additive or synergistic. Most studies of carbon disulfide exposure have been concerned with workers in the viscose rayon industry and have been published in the foreign literature since production of rayon is minimal in this country. In these studies, exposure resulted in sexual dysfunction in males and menstrual problems, spontaneous abortions, and premature births in females. In contrast, animal studies have shown little reproductive impairment in rabbits and a definitive embryotoxicity and weak teratogenic effect in rats. NIOSH has recommended that exposure to carbon disulfide and hydrogen sulfide be controlled. Much additional work, including comprehensive and carefully conducted epidemiological studies, will be needed to determine the effect on female workers of these chemicals, either alone or in combination.

Polychlorinated biphenyls (PCBs) are used to insulate electric capacitors and transformers, in investment casting waxes and molds, and in heat exchange units. They are no longer permitted in telephone line equipment, and older units containing such insulation gradually must be phased out. PCBs are readily absorbed through the skin and by inhalation or ingestion. They are cumulative because they cannot be properly metabolized and can be transferred through the placenta or in mothers' milk.

In the workplace, PCBs remain in the air and on working surfaces long after they have been used and can be carried home on the contaminated clothing of workers. Although human toxicity virtually has been limited to chloracne, care should be taken with these substances, and future research is needed with regard to their effect on reproduction.

Reproductive effects observed in animals have included lowered fertility, increased number of fetal resorptions and abortions, decreased survival of offspring, and teratogenesis. In female Rhesus monkeys fed PCB for two months, fertility decreased for eight months after dosage ceased. Animals fed PCB for six months exhibited irregular menstrual cycles and prolonged excessive menstrual bleeding, lowered fertility, increased fetal resorption, and spontaneous abortion. Live-born offspring were lower in weight than those of control animals, and nursing infants developed PCB poisoning from their mothers' milk in less than two months, with 50 percent of them dying. The most dangerous aspect of PCB is its resistance to degradation and continued persistence in the environment.

Control of chemicals posing a potential embryo-fetotoxic risk

The American Academy of Occupational Medicine has received the following recommendations from a committee composed of occupational physicians in the chemical industry concerning action that should be taken on substances determined to present a risk of embryo-fetotoxicity:

- Employees who may be affected should be informed of the possible consequences of exposure to such substances and appropriate safe handling procedures should be established and communicated.

- Engineering controls should be used to the extent practical to reduce and to maintain at acceptable levels exposure to embryo-fetotoxins. Such controls should be augmented by administrative controls as appropriate.

- Whenever further engineering and administrative controls are not practical to keep exposures at or below acceptable levels, the use of personal protective equipment should be required where appropriate. Employees who are required to use such equipment should be adequately trained in its proper use.

- Where there is potential for exposure to an embryo-fetotoxin for which an acceptable exposure level cannot be set, due to inadequate data, women of reproductive potential should be excluded from the work area.

- Where engineering and administrative controls, augmented as appropriate by personal protective equipment, are determined to be inadequate to ensure acceptable levels of exposure to an embryo-fetotoxic compound, women of reproductive potential should be excluded from the work areas.[8]

For additional information on toxic exposure to mutagens, teratogens, and carcinogens, see Reference 8-11.

WORKING WOMEN AND CORONARY HEART DISEASE

The increase in the number of women in the work force has raised questions in recent years concerning the cardiovascular health of these women. Will they show increasing incidence of coronary heart disease (CHD)? Will they lose their longevity advantage over men? To seek answers, Haynes and Feinleib[12] undertook an extensive psychosocial investigation of subjects, participating in a major study of heart disease, who were undergoing their eighth or ninth biennial medical examinations. This study followed the cardiac histories over an eight-year period of 580 men, 387 women working outside the home, and 350 housewives between the ages of 45 and 64 who were free of CHD at the time of their first medical examinations.

Mortality statistics do not suggest that women have increased their risk of CHD over the past 10 years. In fact, mortality rates from this cause have declined over the period in both women and men, with the women showing a greater decrease. Trends in illness and disability from CHD are less clear, however, since morbidity rates do not provide information indicating whether or not women in the work force have a higher incidence of CHD over time than men. The S.G. Hayes and M. Feinleib study[12] just referred to sought to clarify this question. In this study, the women classified as working women had worked outside the home at least half of their adult lives. Those who had worked less time outside the home were classified as housewives. These two groups did not differ significantly in age or education, but did differ in marital status and number of children. About 20 percent of the working women had never been married, and about 25 percent of those who had married at some time but also had no children. More than one-third of the working women surveyed had worked or were working in clerical or related occupations, primarily as secretaries, stenographers, bookkeepers, bank tellers or cashiers, or sales personnel.

The study found that women working in clerical positions had twice the incidence of CHD as other white-collar or blue-collar workers and that women with children were at greater risk than single women without children. The incidence of CHD also rose with the number of children so that working women with three or more children were more likely to develop CHD (11 percent) than working women with no children (6.5 percent) or housewives with three or more children (4.4 percent) (Figure

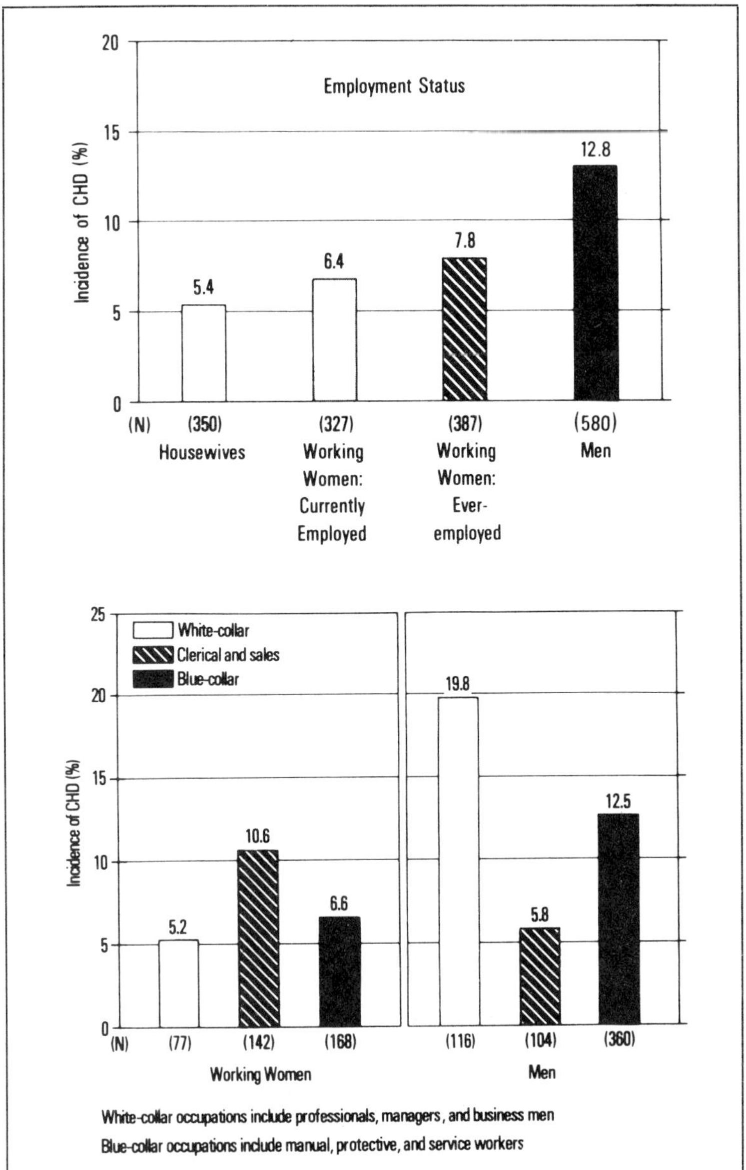

Figure 9-5. *Upper.* Eight-year incidence of coronary heart disease by employment status among men and women aged 45-64 years. *Lower.* Eight-year incidence of coronary heart disease by occupational status among working women and men aged 45-64 years.

From: S.G. Haynes and M. Feinleib. "Women, Work and Coronary Heart Disease: Prospective Findings from the Framingham Heart Study." *A.J.P.H.* 70(2):133, 1980.

9-5). Furthermore, women doing clerical work, married to blue-collar workers, and with three or more children, showed a higher risk of CHD than any of the other groups.

Previous Framingham studies have indicated that two of the most important psychosocial predictors of CHD among both working women and white collar men are Type A behavior and the occupation stressor of suppressed hostility. (More detailed information on Type A behavior and occupational stress is given in Chapter 7, "Occupational Stress.") For present purposes, Type A individuals are competitive, aggressive, governed by the clock, and determined to be first.

The Haynes and Feinleib study indicated that suppressed hostility, nonsupportive superiors, and long periods of time in the same job contributed to the high incidence of CHD among female clerical employees. These investigators concluded that:

> These risk factors may be the product of one or more of the following factors: the particular working environment for clerical occupations, self-selection of certain personalities into the labor force, or economic stress. Whatever the origins of these risk factors, the findings suggest that impersonal relationships, coping styles, and the occupations of some employed women, coupled with family responsibilities, may be involved in the development of coronary heart disease.[12]

Because more than one-third of the women employed in the United States today are in clerical occupations, additional information is needed on those working women. Since economic pressures and child-rearing responsibilities may play a major role in the proclivity for CHD, occupational health professionals should investigate the backgrounds of women clerical workers before determining whether their cardiovascular complaints are job-related or psychosocial problems.

SEXUAL HARASSMENT

Sexual harassment is found throughout U.S. business and industry. The Equal Employment Opportunity Commission (EEOC) has defined sexual harassment as sexual conduct on the part of a superior or coworker that:

1. Requires submission as a condition for obtaining or retaining a job.
2. Affects personal decisions of the harassed woman.
3. Interferes with a person's work performance or creates an intimidating, hostile, or offensive work environment.

The EEOC (despite harassment problems within its own agency) has stated that employers have an affirmative duty to prevent and eliminate such sexual abuse.[13]

Sexual harassment may consist of off-color remarks or jokes, bodily contact when none is necessary, threats of job loss or poor personnel reports for not cooperating, and outright attempts at rape. Women hesitate to speak out against this form of harassment for the following reasons:

1. They are afraid that they will not be believed, will be suspected of inviting the advance, or will be ostracized by their coworkers.
2. They fear publicity.
3. They fear for their jobs.
4. They are unsure whether the act was sexual harassment or was simply an act of human interrelationship.
5. They fear that the act was invited subconsciously.

It is true that such harassment is extremely difficult to prove, but women must be encouraged to report instances of sexual harassment to their superiors because only by this means will it be stopped.

Occupational health professionals should be alert to sexual harassment among women employees because it can result in decreased work performance, symptoms of mental illness, and psychosomatic problems. Sexual harassment affects not only the individual but the company as well. It is important to resolve and prevent these problems.

CONCLUSIONS

Because increasing numbers of women are being hired for more physically demanding jobs, problems of strength and endurance take on greater importance. Depending on their height, weight, and physical fitness, many women can successfully accomplish most male-oriented work. Some problems, however, may be encountered by women working in hot, moist environments since women have a lower tolerance to heat than men. With heat acclimatization, however, this obstacle can often be overcome. Menstruation and pregnancy do not appear to deter performance in jobs that women are accustomed to, but women are generally at a disadvantage in lifting jobs.

The greatest area of risk for women in the workplace is related to child-bearing. Many substances and processes commonly used today are becoming known as mutagenic, teratogenic, and carcinogenic. Women working in industries or services using these substances or processes may be subject to infertility, embryo resorption, spontaneous abortion, fetal malformation, or growth and development problems in live-born children. Occupational safety and health professionals working with women employees must be aware of the risks that these women face when they are determining where and in what jobs they should work.

Although working women, in general, have not exhibited an increase in coronary heart disease despite their increasing numbers in the work force, women working in clerical and associated occupations have shown a higher incidence of CHD than other working women, housewives, or men. The causes are not entirely understood because psychosocial factors may play an important role. Until additional information is available, occupational health professionals should take into account the potential for coronary heart disease in women in these occupations.

Finally, occupational safety and health professionals should be alert to problems of sexual harassment that may occur within their purview. They should encourage women to voice their objections to these conditions and support them by creating a better working environment for all workers.

ADDITIONAL READINGS

Cummings, D. C., and R. W. Rebar. "Exercise and Reproductive Function in Women," *Am. J. Ind. Med.,* 4:113, 1983.

Landrigan, P. J., *et al.* "Reproductive Hazards in the Workplace," *Scand. J. Work Environ. Health,* 9:83, 1983.

Marbury, M. C., *et al.* "Work and Pregnancy," *J. Occup. Med.,* 26(6):415, 1984.

"Women at Work: Special Issue." *Ergonomics,* 27(5), 1984.

REFERENCES

1. Konz, S. *Work Design.* Columbus, Ohio: Grid Publishing, 1979.
2. Gamberale, F., *et al.* "Female Work Capacity During the Menstrual Cycle," *Scand. J. Work Environ. Health,* 1:120, 1975.
3. Guzman, C. A., and R. Caplan. "Cardiorespiratory Response to Exercise During Pregnancy," *Am. J. Obstet., Gynecol.,* 108:600, 1970.
4. Snook, S. H. "Low Back Pain in Industry," A. A. White and S. L. Gordon, eds. American Academy of Orthopedic Surgeons, *Symposium on Idiopathic Low Back Pain.* St. Louis: C. V. Mosby, 1982.
5. Mital, A. "Comprehensive Maximum Acceptable Weight of Lift Database

for Regular 8-Hour Work Shifts," *Ergonomics,* 27(11):1127, 1984.

6. Shapiro, Y., K. B. Pandolf, and R. F. Goldman, "Sex Differences in Acclimation to a Hot-Dry Environment." *Ergonomics* 23(7):635-642, 1980.
7. Messite J., and M. D. Bond, "Occupational Health Considerations for Women at Work," in C. Zenz, ed. *Developments in Occupational Medicine.* Chicago: Year Book Medical Publishers, 1980.
8. Karrh, B. W., *et al.* "Guidance for the Evaluation, Risk Assessment and Control of Embryo-Fetotoxins," *J. Occup. Med.* 23:399, 1981.
9. *Guidelines on Pregnancy and Work.* Chicago: The American College of Obstetricians and Gynecologists, 1977.
10. Levy, B. S., and D. H. Wegman, eds. *Occupational Health.* Boston: Little, Brown, 1983.
11. Mattison, D. R., ed. *Reproductive Toxicology,* New York: Alan R. Liss, 1983.
12. Haynes, S. G., and M. Feinleib, "Women, Work and Coronary Heart Disease: Prospective Findings from the Framingham Heart Study," *A.J.P.H.* 70(2):133, 1980.
13. Collins, E. C. G., and T. B. Blodgett, "Sexual Harassment...Some See It ...Some Won't," *Harvard Bus. Rev. 59(2):76-95, 1981.*

Chapter 10

High-Risk Workers

by Edward J. Calabrese

It is generally recognized that individuals differ widely in their susceptibility to environmentally induced diseases. Consequently, considerable efforts have been made over the past several decades by occupational physicians (*a*) to study the differential susceptibility of workers to toxic substances; (*b*) to determine a biochemical explanation for causes of these enhanced risks; and (*c*) to develop simple, accurate, and cost-effective means to identify persons with such vulnerabilities.[1]

The use of the knowledge of high-risk groups in occupational health policy is controversial. As soon as the term "high-risk group" is used, often polarization occurs over how this knowledge will be used and how it might affect individual workers. A series of articles in the *New York Times* in February of 1980 by Richard Severo[2] about the potential use of this knowledge by government and certain industries clearly illustrates the sensitivity of the issue.

Although this chapter does consider some of the policy questions associated with high-risk groups in the occupational setting, it is primarily intended to provide the biomedical basis for differential susceptibility in the population and the role that knowledge plays in the regulatory process.

The chapter emphasizes the role of genetic factors in occupationally induced disease. However, detailed examples of developmental processes affecting susceptibility are given along with a general discussion of nutritional variables and a discussion of the important implications of smoking and asbestos in lung cancer development. (Those interested in a more detailed discussion, especially of other risk factors such as predisposing diseases, refer to *Pollutants and High Risk Groups*, Edward J. Calabrese, 1978, John Wiley and Sons.)

Factors affecting differential susceptibility

A variety of biological factors affects the susceptibility of individuals to environmentally induced diseases. They are developmental processes (aging), genetic or hereditary background, nutritional status, the presence of preexisting diseases, and finally, lifestyle considerations. Table 10-A lists specific conditions under each of the five above-mentioned factors that may enhance an individual's susceptibility as well as the types of toxins to which each subgroup may be at increased risk, and the estimated number of people in the United States exhibiting such specific behaviors. The table, pp. 226 to 231, indicates that there are a number of biological factors that may enhance risk to environmental toxins.

Developmental/aging factors are obviously conditions that are biologically unavoidable, representing periods of time when everyone is at enhanced risk. Other biological factors affecting susceptibility are not as universal. For example, black American males have a high incidence of erythrocyte glucose-6-phosphate dehydrogenase (G-6-PD) deficiency. It is estimated that as many as 13-16 percent of all U.S. black males have this deficiency. It is also important to recognize that many risk factors are not independent of each other. For example, a young child of Mediterranean origin may have an erythrocyte G-6-PD deficiency, as well as a nutritional status considered less than adequate for vitamin E. It is possible that the presence of such multiple factors may act in a totally independent way in affecting toxicity, or that these factors may interact in such a manner as to increase or magnify the toxicity of a given agent or combination of such toxins. An example of the latter may be the concomitant presence of a G-6-PD deficiency and inadequate vitamin E status that enhance the toxicity of such an environmental/occupational oxidant as ozone.

DEVELOPMENTAL/AGING PROCESS

Developmental processes as factors affecting an individual's susceptibility to environmental toxins are not usually associated with occupational medicine since they are a province of the young. Consequently, government agencies with such a broad protective mandate as that of the Environmental Protection Agency (EPA) are most concerned in identifying developmental processes that affect susceptibility.

Although it is true that the occurrence of developmental patterns in the young and their role in affecting susceptibility would not typically fall under the regulatory frame work of the Occupational Safety and Health Administration (OSHA), there is considerable concern over the exposure of pregnant women to industrial toxins. This concern involves both the pregnant person and the developing embryo and fetus. The health risk of the unborn as a result of their mother's being exposed to toxins while on the job has been a highly controversial area. This is true because some employers have required that the women who work in certain high-hazard areas (where fetal toxins such as lead are present) be reproductively sterilized. The ethical issues surrounding this issue are indeed profound, but it is important to realize that the biological contribution is found in the principles of differential susceptibility of the very young. (Also see Chapter 9, "Women in the Workplace.")

Occupational medicine is also concerned with the potential enhanced risks of the elderly to various toxic agents. Unfortunately, this group is generally overlooked as a high-risk group. However, geriatric medicine has established that a wide range of medicinal agents are metabolized differently in older persons. Typically, the older person is less efficient in metabolizing drugs and requires doses that are adjusted to account for this phenomenon. Since most pollutants are metabolized along the same pathway as drugs, it is only reasonable to assume that older persons will likely have a more difficult time detoxifying and excreting industrial toxins than younger workers. In fact, this potential problem has never been addressed. Most of the standards designed to protect industrial workers do not single out older persons for special treatment. It is likely that this potential problem may become a more prominent issue if efforts that succeed in making the Social Security System solvent also extend the retirement age.

Some of the specific developmental/aging factors that have been identified are as follows:

1. **Immature enzyme development.** The classic example of the occurrence of an immature enzymatic pathway is that of glucuronidation. Glucuronidation is a biochemical process that results in the detoxification and excretion of a wide range of compounds, especially secondary and tertiary compounds of a phenolic nature. This process involves the addition of a derivative of glucose, that is, glucuronic acid, to the agent(s) in question. From a developmental perspective, the process of glucuronidation does not become functional in the adult sense until 6-8 weeks after birth. This developmental immaturity is thought to contribute to the occurrence of neonatal hyperbilirubinema (the presence of bilirubin in the blood in excess of the normal quantity) since bilirubin, a breakdown product of hemoglobin, is thought to be excreted in part via glucuronidation. Moreover, there are numerous substances such as polychlorinated biphenyls (PCBs) that are known to be excreted with glucuronic acid. Presumably, the very young may be at increased risk to such compounds.

2. **Rapid cell division.** It is well established that the very young are at increased risk to radiation-induced cancer because of more rapid cell divi-

sion. Knowledge of this fact has led to educational programs that inform pregnant women to avoid exposure to X-rays so as to protect their developing children. The enhanced risk of the child may be best explained by the higher proportion of cells in a cell division cycle than in adults.

3. Developmental immaturity of Immunoglobulin A (IgA). Adult levels of IgA are not achieved until the individual reaches approximately 9-10 years of age. IgA plays an important role in protecting secretory areas of the body, especially the respiratory tract. Consequently, it has been speculated that the well-enhanced susceptibility of the very young to upper and lower respiratory tract infections may be related to this temporary IgA deficiency. This hypothesis has been supported by studies in adults with a genetically determined IgA deficiency who have been shown to have a significantly greater risk to develop both upper and lower respiratory tract infections.[3,4]

4. Nitrite-induced methemoglobinemia. The very young go through a transitory period in which they display a relatively high gastric pH. The significance of this somewhat elevated pH is that nitrate-reducing bacteria are capable of converting the generally innocuous nitrate to the much more toxic nitrite. Nitrite is a potent oxidizing agent that may convert hemoglobin to ferrihemoglobin, which is more commonly referred to as methemoglobin. In contrast to hemoglobin, methemoglobin is not capable of irreversably transporting oxygen. Consequently, individuals in whom the levels of methemoglobin have been increased from a normal of about 1 percent to about 10 percent would likely begin to experience the onset of clinical symptoms of cyanosis.

Besides their enhanced ability to convert nitrate to nitrite (because of the favorable environment for nitrate-reducing bacteria in the stomach), infants have several other biological factors that enhance their ability both to form methemoglobin and to reduce their capability to reduce or convert it back to hemoglobin. They are able to form more nitrite-induced methemoglobin because the very young have significantly greater concentrations of fetal hemoglobin, a form of hemoglobin that is markedly more susceptible to oxidation than normal hemoglobin. Once methemoglobin is formed, infants experience difficulty converting it back to hemoglobin, since they have developmental immaturity of the enzyme methemoglobin reductase.

GENETIC FACTORS

Table 10-A shows that approximately 30 genetic conditions are either known or hypothesized to enhance one's individual susceptibility to environmental and occupational pollutants. The role of genetic factors in explaining the differential susceptibility to environmental and occupational pollutants has generated much public attention. As mentioned earlier, a series of *New York Times* articles in 1980, directed considerable attention to (*a*) the issue of genetic susceptibility to industrial toxins; (*b*) the identification of persons with such characteristics; (*c*) the reliability of the data; (*d*) the use by industry of this knowledge; and (*e*) the extent of informal and formal OSHA policy toward the concept of genetic screening. This series of articles and subsequent hearings caused Congress to ask the Office of Technology Assessment to evaluate the role of genetic testing in the prevention of occupational illness.

It has long been suspected that genetic factors may affect susceptibility to occupational illness. In 1938, geneticist J. B. S. Haldane suggested a possible genetic role in the occurrence of bronchitis among potters and even endorsed the possibility of eliminating the genetically predisposed person from potential unhealthy work environments. More specifically, Haldane[5] stated:

> I should be interested to know whether, to take a single example, the death rate among potters from bronchitis is still eight times that of the general population. If so, it would not be unreasonable if a certain proportion of the funds devoted to pottery research at Stoke-on-Trent were spent on research on potters rather than on pots. But while I am sure that our standards of industrial hygiene are shamefully low, it is important to realize that there is a side to this question which has so far been completely ignored. The majority of potters do not die of bronchitis. It is quite possible that if we really understood the causation of this disease, we should find that only a fraction of potters are of a constitution which renders them liable to it. If so, we could eliminate potters' bronchitis by rejecting entrants into the pottery industry who are congenitally disposed to it. We are already making the attempt to exclude accident prone workers from certain trades. The principle could perhaps be carried a good deal further. There are two sides to most of these questions involving unfavorable environments. Not only could the environment be improved, but susceptible individuals could be excluded.

It is uncertain to what extent the sentiments of Haldane encouraged research on the issue of genetic factors as determinants of differential susceptibility to industrial toxins. It was not until the early 1960s that various researchers began to publish the biochemical foundations of genetically related differential susceptibility to industrial toxins.[6-8] This notion of inherited susceptibility to certain industrial agents was generally adopted by the American Conference of Governmental Industrial Hygienists (ACGIH). ACGIH leaders were the principal formulators of the concept of genetically related high-risk groups. Nevertheless, the federal government has paid little attention to the phenomenon, either with respect to encouraging research or in policy formulation.

However, strong leadership in the high-risk area was provided by the EPA, especially in the development of ambient air quality standards in 1969-70, and in the development of rationale to support the promulgation of primary drinking water standards. EPA's concern with high-risk groups was typically directed only to selected developmental factors. In no case was a subgroup of individuals identified as being genetically at increased risk. *(Text continues on page 231.)*

TABLE 10-A
THE IDENTIFICATION AND QUANTIFICATION OF HIGH-RISK GROUPS

High-Risk Groups	*Estimated Number of Individuals in United States Affected*	*Pollutant(s) to Which High-Risk Group Is (May Be) at Increased Risk*
	Developmental Processes	
Immature enzyme detoxification systems	Embryos, fetuses and neonates to the age of approximately 2-3 months	pesticides PCBs
Immature immune system	Infants and children do not reach adult levels of IgA until the age of 10-12	respiratory irritants
Deficient immune system as a function of age	Progressive degeneration after adolescence	carcinogens respiratory irritants
Adolescents, especially carriers of prophyria genes	See population frequency under Porphyrias in this table	chloroquine hexachlorobenzene lead variety of medical drugs
Differential absorption of pollutants as a function of age	Infants and young children	barium lead radium strontium
Retention of pollutants as a function of age	Individuals above the age of 50	fluoride
Pregnancy	Approximately several million females per year in the United States	anticholinesterace (pesticides) carbon monoxide lead

High-Risk Groups	*Estimated Number of Individuals in United States Affected*	*Pollutant(s) to Which High-Risk Group Is (May Be) at Increased Risk*
Circadian rhythms including phase shifts	All people have certain periods of the day when they are more susceptible to challenge	hydrocarbon carcinogens and probably most other pollutants
Infant stomach acidity	Infants	nitrates
	Genetic Conditons	
Albinism: Tyrosinase-positive	1 per 14,000 blacks; 1 per 60,000 Caucasians; very high frequency in American Indians	ultraviolet radiation
Tyrosinasenegative	1 per 34,000 or 36,000 blacks and Caucasians; 1 per 10,000 or 15,000 in Ireland	
Catalase deficiency:		ozone
Hypocatalasemia	5,000,000 heterozygotes	radiation
Acatalasemia	16,000 homozygotes	
Cholinesterase variants	Highly sensitive homozygous and heterozygous individuals of European ancestry have a combined frequency of about 1 per 1250; moderately sensitive genotypic variants of European ancestry have a frequency of 1 per 15,000	anitcholin-esterase insectides
Crigler-Najjar syndrome	Few individuals live to adulthood	PCBs
Cystic fibrosis	8,000 heterozygous for CF, 100,000 homozygous for CF; most common among Caucasians of European ancestry	ozone respiratory irritants
Cystinosis	Most common among Caucasians of European ancestry; unknown frequency	cadmium lead mercury uranium
Cystinuria	1 per 200-250 individuals although asymptomatic, are affected; 1 per 20,000-100,000 are homozygous	cadmium lead mercury uranium

Table 10A (Continued)

High-Risk Groups	*Estimated Number of Individuals in United States Affected*	*Pollutant(s) to Which High-Risk Group Is (May Be) at Increased Risk*
Glucose-6-phosphate dehydrogenase deficiency	1,600,000 American black males (11 percent); Mediterranean Jews 11.0 percent, Greeks 1-2 percent, Sardinians 1-8 percent	carbon monoxide lead nitrate ozone
Glutathione deficiency	Few cases reported	lead ozone
Glutathione peroxidase deficiency	Partial deficiency in some neonates; only a few adult cases reported	lead ozone
Glutathione reductase deficiency	Unknown, but thought to be rare	lead ozone
Gilbert's syndrome	6 percent of the normal, healthy adult population	PCBs
Immunoglobin A deficiency	500,000 homozygotes	respiratory irritants
Immunologic hypersensitivity	2 percent of some worker populations	isocyanates
Inducibility of aryl hydrocarbon hydroxylase	Approximately 45 percent of the general population are at high risk; 9 percent of the 45 percent are at very high risk	polycyclic aromatic hydrocarbons
Leber's optic atrophy	Thought to be rare	cyanide
Methemoglobin reductase deficiency	Found in many ethnic groups; gene frequency is uncertain but thought to be rare; most a temporary deficiency	nitrates nitrites ozone vanadium
Phenylketonuria	1 per 80 in the United States is a carrier; 1 per 25,600 has the disease	ultraviolet light
Porphyrias	1.5 per 100,000 in Sweden, Denmark, Ireland, West Australia; 3 per 1,000 in South African whites; rare in blacks	chloroquine hexachlorobenzene various drugs including barbiturates, sulfonamides, and others

High-Risk Groups	*Estimated Number of Individuals in United States Affected*	*Pollutant(s) to Which High-Risk Group Is (May Be) at Increased Risk*
Serum $alpha_1$ antitrypsin	Approximately 4.0-9.0 percent individuals of northern European descent are heterozygotes; approximately 160,000 homozygotes, based on frequency found in Norway and Sweden	respiratory irritants smoking
Sickle cell trait (heterozygote) anemia (homozyzygote)	7-13 percent of American blacks, heterozygotes Homozygotic condition highly fatal	aromatic amino nitro compounds carbon monoxide cyanide
Sulfite oxidase	Unknown	sulfite sulfur dioxide
Thalassemia (Cooley's anemia)	Homozygous condition highly fatal; heterozygote frequency high (0.1-8.0 percent) in certain people of Italian, Greek, Syrian, and African origin	lead organic chemicals such as benzene and its derivatives ozone
Tyrosinemia	Frequency in general population unknown; frequency of heterozygous carriers in the Chicoutimi region of northern Quebec is for every 20-31 people	cadmium lead mercury uranium
Wilson's disease	400,000 heterozygotes, 200 homozygotes, most frequent among Jews of Eastern European ancestry and non-Jews from the Mediterranean regions, especially Sicily	lead vanadium
DNA Repair Diseases	Homozygotes—rare Heterozygotes—1.2%	ultraviolet radiation chemical carcinogens
	Nutritional Deficiencies	
Vitamin A	25 percent of children between 7 and 12 have lower than recommended dietary allowance (RDA) or slightly higher percent among the lower income groups	DDT hydrocarbon carcinogens PCBs

Table 10A (Continued)

High-Risk Groups	*Estimated Number of Individuals in United States Affected*	*Pollutant(s) to Which High-Risk Group Is (May Be) at Increased Risk*
Vitamin C	10-30 percent of infants children and adults of lower income groups receive less than the RDA	arsenic cadium carbon monoxide chromium DDT dieldrin lead mercury nitrites ozone
Vitamin E	7 percent of the general population are "physiologically deficient"	lead ozone
Calcium	65 percent of children betwen the ages of 2 and 3 receive less than the RDA	lead
Iron	98 percent of children between the ages of 2 and 3 receive less than the RDA	hydrocarbon carcinogens lead manganese
Magnesium	Most United States males have a partial deficiency	fluoride
Phosphorus	Deficiency in people with various kidney diseases	lead
Selenium	Unknown; deficiency thought to be rare	cadmium mercury ozone
Zinc	Unknown, deficiency in association with various diseases, but not thought to be widespread	cadmium
Riboflavin	30 percent of women and 10 percent of men aged 30-60 ingest less than two thirds of the RDA	hydrocarbon carcinogens lead ozone
Dietary protein	10 percent of women and 5 percent of men aged 30-60 ingest less than two thirds of the RDA for protein	DDT and other insecticides
Methionine	Unknown; but thought to be more frequent in individuals following certain vegetarian diets	DDT

High-Risk Groups	*Estimated Number of Individuals in United States Affected*	*Pollutant(s) to Which High-Risk Group Is (May Be) at Increased Risk*
	Diseases	
Kidney disease	Relates to genetic disease (see cystinosis, cystinuria, tyrosinemi), bacterial and virus infections, and hypertensive disease	fluoride lead other heavy metals excessive sodium in diet
Liver disease	Relates to genetic disease (Gilbert's syndrome) and virus infections	carbon tetrachloride DDT and other insecticides PCBs
Asthmatic diseases	4,000,000-10,000,000 of the general population	respiratory irritants: nitrogen dioxide ozone sulfates sulfur dioxide
Chronic respiratory disease	6,000,000-10,000,000 of the general population	respiratory irritants: nitrogen dioxide ozone sulfates sulfur dioxide
Heart disease	15,000,000 of the general population	cadmium carbon monoxide fluoride respiratory irritants sodium
	Behavioral Activities	
Smoking	Widespread	cadmium hydrocarbons lead nickel radioactive compounds
Alcohol consumption	Widespread	lead pesticides PCBs
Drug taking	Widespread	pesticides PCBs numerous potential substances

G-6-PD deficiency

Glucose-6-phosphate dehydrogenase (G-6-PD) deficiency is a sex-linked genetic condition. It is principally characterized as a biochemical defect in erythrocytes by which glucose is oxidized in a series of reactions to provide the erythrocyte with reducing power in the form of nicotinamide dinucleotide phosphate (NADPH) and reduced glutathione (GSH), which are necessary, among other things, to maintain erythrocyte membrane integrity.[9,10] The enzyme G-6-PD catalyzes the first reaction of the pentose monophosphate shunt, and those individuals with G-6-PD deficiency have lower levels of reduced GSH. Since G-6-PD initiates the only reaction by which NADP is reduced to NADPH, it is suggested that a decrease in NADPH can account for the low level of GSH in G-6-PD deficient individuals, because NADPH is the necessary coenzyme for the reduction of oxidized glutathione (GSSH) to GSH.[9,11] Under hemolytic stress conditions, deficients are unable to maintain proper erythrocyte membrane integrity.

Individuals with the G-6-PD deficiency were first recognized as medically important when they exhibited extreme hemolytic sensitivity of erythrocytes on administration of certain antimalarial drugs—primaquine and pamaquine. Numerous other substances of an industrial and medicinal nature subsequently were definitely shown to cause clinically significant hemolytic anemia in G-6-PD deficients.[12] Theoretically, any chemical compound or its metabolite that can accept hydrogen in the defective red blood cell may be suspected as a possible hemolytic agent.[6,7,13]

With respect to industrial practices it has been two decades since the first suggestion was made that G-6-PD deficiency could help to explain worker susceptibility to chemically induced anemia.[6,14,15] In 1963, Stokinger and Mountain[7] proposed a partial list of 37 industrial chemicals known to cause hemolysis to which those individuals with a G-6-PD deficiency would be at enhanced risk. They further suggested that screening tests to identify such deficient people be conducted as part of the preemployment medical examination before job placement. By 1967, Stokinger and Mountain[8] reported that more than 15 industries, research centers, and health-oriented groups were either using the G-6-PD test or had inquired into its use. Not surprisingly, these investigators noted that industries that found the test useful were manufacturers of dyes and dyestuff intermediates, metals, especially lead, and drugs.

Of relevance to the development of occupational exposure standards was a subsequent report by Linch[16] noting that G-6-PD-deficient workers were more susceptible to chemical (i.e., aromatic nitro and amino compounds) induced cyanosis than normal individuals. As a result of the recognition that such G-6-PD-deficient individuals were at increased susceptibility, The E. I. DuPont Company adopted a screening program to detect these persons.

Despite the concern that G-6-PD-deficients may be at greater risk to over 30 hemolytic agents used in industry, the documentation of the

ACGIH Threshold Limit Values (TLVs®) states that for only two substances (carbon monoxide and naphthalene), the TLV may not be sufficient to protect those with heredity blood disorders (those with a G-6-PD deficiency). No evidence was presented by the governmental industrial hygienists (ACGIH) as to why those with hereditary blood disorders may be at potentially increased risk at or below the TLV.

Sickle-cell anemia and sickle-cell trait

The sickle-cell genetic conditions result from the presence of an abnormal hemoglobin molecule, hemoglobin S (HbS), located in the erythrocytes of affected persons. Hemoglobin S differs from the normal adult hemoglobin A (HbA) only by the exchange of the amino acids glutamine by valine, at a single location in the hemoglobin chain, which consists of 287 amino acids. Hemoglobin S is considerably less soluble than HbA under conditions of low oxygen tension. The decreased solubility may result in the formation of a gel within red blood cells, distorting them and causing sickling to occur. An individual with sickle-cell anemia is homozygous for HbS, while one with sickle-cell trait is heterozygous for HbS. In general, the homozygous person has nearly 100 percent HbS, while the heterozygous person has from 20 to 40 percent. The person with sickle cell trait will experience sickling only when blood O_2 tension is greatly reduced (about 10 mm Hg from a normal of 95 mm Hg).[13,17]

Although those with sickle-cell anemia are known to have a considerably reduced lifespan, health hazards have been considered minimal, if not nonexistent, under normal circumstances.[18] For example, life expectancy, mortality, and hematologic status have been reported normal. Situations where problems were found in those with the sickle cell trait are: hypoxia (lack of adequate oxygen) in unpressurized airplanes, during skin diving, mountain climbing, or possibly during anesthesia. The overwhelming majority of those individuals having the sickle cell trait in the United States are not expected to have health problems associated with this genetic condition.

However, Stokinger and Scheel[13] have suggested that persons with sickle cell abnormalities may be at increased risk to (*a*) anemia producers (benzene, lead, cadmium); (*b*) methemoglobin formers (aromatic amino and nitro compounds); and (*c*) blood oxygen tension reducers (carbon monoxide and cyanide). However, these researchers emphasized that they were not aware of any reports describing such enhanced risk. Reinhardt[19] reported that as a service, DuPont "routinely offers all black employees a test for sickle cell trait." This type of screening was initiated in 1972. At DuPont's Chambers Works Plant, those with sickle cell trait and a hemoglobin value of 14 g/100 ml are not permitted to work with hemolytic agents. DuPont officials presented no evidence to support the premise that persons with sickle cell trait and a low hemoglobin value were at increased risk to such hemolytic chemicals as nitro and amino compounds.

Paraoxanase polymorphism

Human serum has been found to contain an enzyme (paraoxanase) that hydrolyzes the cholinesterase inhibitor paraoxon which is the oxidized metabolite of the organophosphate insecticide parathion. Paraoxanase displays considerable interindividual variability while its activity remains constant within a given subject.[20] Geldmacher von Mallinckrodt and his associates[21] have reported a diallelic model with the occurrence of a trimodal distribution of enzyme activity with a simple mode of inheritance. Persons classified in the low activity group have about 25 percent of the activity of the high group. The gene frequency of the low activity allele (a different form of a given gene) is determined to be 0.7. According to von Mallinckrodt, an individual with low paraoxanase activity would be expected to be at increased risk to parathion toxicity, even though there is no substantiation of this hypothesis.

Pseudocholinesterase variants

There are two types of cholinesterase (an enzyme): acetylcholinesterase (ACHase) and pseudocholinesterase (pACHase). ACHase inactivates acetylcholine (ACH) produced at the neuromuscular junction during neurotransmission. The pACHase is found in many tissues as well as in plasma. Although its function is unknown, pACHase may hydrolize certain cholinesters which inhibit ACHase.[22] Whereas most people have identical pACHase, a number of pACHase variants exist. Most individuals with variant forms are typically symptomless, but some may exhibit an extreme sensitivity to the muscle relaxant, suxamethonium, since they cannot hydrolyze such substrates as efficiently as the normal pACHase.

Screening large numbers of humans[23,24] has revealed that the development of atypical or variant types of pACHase is under genetic control. Gene frequencies have been determined for some of the variant genotypes. The most common "atypical" homozygous variant (the dicubaine variant) occurs with a frequency of 1 in 2,800 healthy Canadians of European ancestry[23] and has been found to be extremely sensitive to carbamate substances.[22] This finding is particularly significant in light of the widespread use of carbamate insecticides. It should be pointed out that 3 to 4 percent of the Canadians tested were found to be heterozygote carriers (normal and dicubaine genes together) of intermediate sensitivity.[23]

In terms of public health, the data indicate that individuals have differential sensitivity to the activity of various neuromuscular-acting drugs and carbamate-like chemicals. Differences in sensitivity are directly related to the occurrence of cholinesterase enzyme variants and the diminished ability to inactivate the drug or insecticide analog. Individuals with such pACHase variants should be considered potentially at high risk to anticholinesterase insecticides. It should be emphasized that not all drug and insecticide-like compounds act with greater sensitivity in atypical pACHase variants.[24] For example, two organophosphate insecticides (tetraethylpyrophosphonate and diisopropylfluorophosphonate) do not inhibit differentially among pACHase variants and would not cause a higher risk to those with an atypical variant. In a limited study of organophosphate (OP) poisonings, Tabershaw and Cooper[25] found no increased susceptibility of heterozygous dicubaine carriers.

Serum alpha 1-antitrypsin (SAT) deficiency

Data from animal and human studies strongly support the hypothesis that an intermediate deficiency of SAT is a contributing factor in the development of emphysema. However, it is now being recognized that emphysema is a multi-causal disease and that the heterozygote state, by itself, is not a major predisposing factor. However, in combination with other predisposing factors, for example, high leukocyte protease (an enzyme that digests protein), the heterozygote condition is likely to be at significantly higher risk of developing emphysema. The most reasonable position, at present, would be to try and evaluate the relative contribution of these variables in the development of emphysema.[26,27] If the above associations prove true and are amenable to widespread screening, then it is likely that screening would entail an assessment of two or three important factors.

AHH inducibility and susceptibility to lung cancer

One of the more exciting, yet controversial, areas of ecogenetics concerns the hypothesis that susceptibility to polycyclic aromatic hydrocarbon (PAH)-induced lung cancer can be predicted by the differential ability to induce the synthesis and activity of the enzyme aryl hydrocarbon hydroxylase (AHH). This enzyme, which is found in most mammalian tissues, is known to catalyze the first step in the metabolism of benzo(a)pyrene (BAP) and other PAHs. Being an inducible enzyme, AHH displays increased activity following administration of a number of agents such as PAHs, various drugs, steroids, and insecticides among others.

A number of investigators have shown that there is considerable variation in the extent of AHH induction in cultured leukocytes and that it is tightly under genetic control.[28] According to Kellermann and co-workers,[29] the normal Caucasian population can be divided into three distinct groups with low, intermediate, and high degrees of inducibility with the following respective phenotypic frequencies: 53 percent, 37 percent, and 10 percent.

Since the inducibility of AHH was found to be under genetic control and exhibited wide variation in the population, Kellermann and others[29] sought to evaluate whether or not the AHH inducibility could explain differential susceptibility to lung cancer, presumably caused by PAHs that may have been bioactivated by AHH to carcinogenic products. The researchers selected 50 patients with bronchogenic carcinoma (58 percent squamous cell carcinoma, 24 percent adenocarcinoma, 18 percent oat cell carcinoma), as well as 85 healthy controls and 46 tumor control, and compared their capacity to have 3-methylcholanthrene (MCA) induce the synthesis of AHH in leukocytes. They reported that the healthy and tumor control groups did not differ from each other and displayed phenotypic frequencies nearly identical with that noted above. However, the lung cancer patients displayed a marked shift to the right as only 4 percent were low inducers, 66 percent were moderate, and 30 percent were high. The authors concluded that the risk of lung cancer for the groups with intermediate and high inducibility was 16 and 36 times, respectively, greater than that of the low-inducibility group.

Not all the reports, however, supported the original association between AHH inducibility and susceptibility to bronchogenic carcinoma. For example, Paigen and co-workers[30] reported that they were unable to measure AHH induction in 6 of 12 lung cancer patients who had been operated on but not given subsequent chemoradiation therapy. Because of this lack of response, they compared AHH inducibility in the progeny of lung cancer patients with controls, who were the spouses in each case. This approach was justified on the basis of similarity of age, ethnic background, and environment (diet and home neighborhood) and also because AHH inducibility had been shown not to be affected by the sex of the individual. The results indicated that the progeny of the patients did not differ from the matched control population in induced AHH activity levels or in basal AHH levels.

The genetic basis affecting susceptibility to PAH-induced cancer is overwhelmingly documented in animal studies[31,32] and supported by human epidemiological investigators.[33] However, the identification of a precise and reliable marker or predictor of risk to lung cancer such as AHH inducibility, remains unresolved.

The theory of Kellermann and co-workers that susceptibility to PAH-induced lung cancer is partly a function of the ability to induce AHH remains to be unequivocally established. The negative reports of Paigen[30] and others have clearly revealed that current methods used to establish the above relationship are of limited value. However, the fact that at least nine different studies, involving patients in different parts of the United States and Europe, have supported the hypothesis that lymphocyte AHH inducibility is associated with enhanced risk to PAH-induced lung/laryngeal cancer, along with the impressive support from animal studies,[31,32] strongly support the concept that genetic determinants of PAH metabolism are somehow linked to susceptibility of lung and laryngeal cancer in humans.

NUTRITIONAL FACTORS

Like genetic factors, nutritional status has long been suspect as a factor in industrially related disease. During World War I, it was speculated that TNT-induced adverse health effects were markedly enhanced by inadequate diets.[34] Over the past 50 years, there has been continuing interest in the idea that certain vitamin and mineral deficiencies could exacerbate the effects of pollutants, while the addition of certain vitamins and minerals in the diet would alleviate them to some extent. Considerable research has been conducted in this area because of earlier promising findings and the possible commercialization of the findings.[35,36] For example, the international pharmaceutical company, Hoffman-LaRoche, is a manufacturer of vitamins and is actively studying how certain vitamins affect the body's capability to detoxify and excrete a wide variety of toxic substances. Furthermore, the U.S. National Cancer Institute has established a chemoprevention program in which various nutrients, such as vitamin A and some of its synthetic analogs, have been tested for their anticarcinogenic capability. Along these same lines, a recent report of the National Academy of Sciences[37] concluded that diet has a significant role in affecting the individual's susceptibility to an array of environmentally induced diseases.

Therefore, the evidence concerning the interaction of dietary factors with environmental and industrial toxins is substantial. In fact, specific nutrition programs are now being advocated by the National Academy of Sciences.[37]

The information suggests that consumption of certain foods may alter the occurrence of pollutant-induced toxicity and carcinogenicity is of considerable potential interest to the field of occupational medicine. The term "potential" is used because most of the available data are from animal models whose relevance to the human condition remains to be more precisely defined. The number of industrial studies including workers in experimental studies is limited compared to the animal data. It is clear, however, that research should be undertaken to evaluate to what extent the use of vitamin and mineral supplements affects the body burden of various pollutants and the capacity of the body to detoxify them.

Perhaps the most important reason for encouraging aggressive research along these lines is that nutritional status is at least one factor where human intervention might have a decided and immediate impact. For despite the knowledge that developmental/aging and genetic factors may affect susceptibility to environmental and industrial pollutants, there is little that can be done with the obvious exceptions of either removing workers from the hazardous environment or reducing the level of the toxin(s). It should be emphasized that improving the worker's nutritional status does *not exclude* reducing the exposure to potentially harmful pollutants. As a corollary to this statement, giving workers vitamin supplements should not be used as an excuse for maintaining an unhealthy workplace. However, as stated above, improving one's nutritional status and reducing the level of pollutants in the workplace are not mutually exclusive.

BEHAVIORAL ATTITUDES

Smoking, asbestos, and lung cancer

The most important cancer associated with asbestos workers is cancer of the lung. Approximately 20 percent of all deaths of asbestos workers are caused by lung cancers.[38] Factors affecting the individual's response to the carcinogenic activity of asbestos include age, duration of exposure, and smoking behavior. In 1968, Selikoff and co-workers found that the incidence of lung cancer was not significantly increased among asbestos workers with no history of cigarette smoking. Yet when such history was present, the incidence of lung cancer increased considerably over that expected among other cigarette smokers, without asbestos exposure. The research team concluded that an asbestos worker who smoked cigarettes had a strikingly greater risk of dying of lung cancer, when compared with cohorts who did not smoke cigarettes.

Other studies[39] involving larger samples have confirmed this finding—nonsmoking asbestos workers had few lung cancers, but those who smoked had significantly more lung cancers. It was determined that cigarette-smoking asbestos workers have approximately 8 to 12 times more risk of developing lung cancer when compared with nonsmoking asbestos workers.

Finally, the relationship of smoking to coal miner's pneumoconiosis should be mentioned. Considerable evidence now exists to suggest that among coal workers there is a significantly higher incidence of emphysema and heart disease in those that smoke as compared with nonsmokers.[40]

ROLE OF HIGH-RISK GROUPS IN THE REGULATORY PROCESS

Environmental Protection Agency

The EPA has attempted to consider the special sensitivities of those at increased risk to pollutants in the standard-setting process for air and water pollutants (see Table 10-B). It is interesting to note that EPA has specifically considered both developmental and preexisting disease conditions—that is, the increased susceptibility of the very young to nitrate-related formation of methemoglobinemia, the increased sensitivity of children to certain aspects of lead toxicity, and the incresed risks of those with various cardiovascular diseases to respiratory tract irritants.[41]

TABLE 10-B

HIGH-RISK IN THE DERIVATION OF U.S. ENVIRONMENTAL HEALTH STANDARDS

	Drinking Water Standards
Substance	*High-Risk Conditions Considered*
Arsenic	None
Barium	No specific groups; but a safety factor of two incorporated to account for variation (or increased susceptibility) within the human population.
Cadium	None
Fluoride	Children—prevent mottling of teeth
Lead	Children—prevent neurological disorders
Mercury	Chose humans who exhibited toxicity at the lowest levels of exposure from a group of mercury poisoned adults.
Nitrate	Infants—protect against methemoglobinemia
Selenium	None
Sodium (no standard)	Those with ear and kidney disease
Chlorinated hydrocarbon insecticides (non-carcinogenic)	None
	Air Standards
Carbon monoxide	Heart disease
Hydrocarbons	None
Lead	Children—prevent blood disorders
Nitrogen dioxide	Young children—respiratory infections
Ozone	Asthmatics
Particulates	Aged, those with cardiovascular disease
Sulfur dioxide	Aged, those with cardiovascular disease

Source: Reference 41

Despite the fact that the EPA had made a good-faith effort to incorporate protection for high-risk groups in setting certain standards, a more specific and consistent policy toward utilizing the knowledge of high-risk groups is needed in environmental health policy, in general, and standard setting, in particular.

OSHA Standards and high-risk groups

Soon after OSHA was created, it adopted all the Threshold Limit Values (TLVs®) of the American Conference of Governmental Industrial Hygienists (ACGIH) for noncarcinogens as its new standards.[42] It is important to realize that the TLVs of the ACGIH were intended as guidelines for industry to use as an indication of good workplace hygiene practice. The preface of the TLV booklet of the ACGIH specifically states that these TLVs may not offer full protection to those at enhanced risk.[43] Consequently, it follows that by adopting the TLVs of the ACGIH as *de facto* standards, OSHA decided specifically not to take into consideration those who may be at increased risk.

GENETIC SCREENING AND INDUSTRIAL HIGH-RISK GROUPS

According to Stokinger and Scheel,[13] there are five prerequisites for hypersusceptibility testing: (*a*) the specific genetic effects should be identified; (*b*) an inexpensive and simple test indicating the presence of the genetic defect should be available; (*c*) the defect should have a relatively high occurrence in the worker population; (*d*) it should concern pollutants usually encountered in industry; and (*e*) the genetic factor should be compatible with an apparently normal life until industrial exposure occurs.

These investigators have also reported that five disorders satisfy the prerequisite for the above-stated hypersusceptibility tests. They are G-6-PD deficiency, carbon disulfide sensitivity, SAT deficiency, reagenic antibodies to allergic chemicals, and hemoglobin S in sickle cell anemia. Other genetic conditions such as atypical cholinesterase, pyruvate kinase deficiency, Wilson's disease, and abnormal porphyrin metabolism do not satisfy all the criteria given above. For example, detection of subnormal amounts of cholinesterase activity has not been associated with adverse health in pesticide workers, and the remaining three diseases are considered too rare for genetic screenings. However, a closer look at the five genetic conditions which, according to Stokinger and Scheel, meet the criteria for screening, reveals that a total SAT deficiency is very uncommon ($<1/1000$). Certainly that does not meet even the most liberal definition of a relatively high occurrence in the worker population. Sickle-cell anemia is also not frequent in the black population (0.5 percent)[17] with an even smaller percentage being able to seek employment as adults because of the manifestation of overt illness before adulthood. However, the heterozygous condition (i.e., sickle cell trait) is estimated to affect between 7 to 13 percent of American blacks of African origin. Yet, despite this relatively high percentage, there are no data which indicate that those with the trait would be at increased risk to toxic substances. As observed earlier, Stokinger and Scheel[13] suggested that persons with sickle cell abnormalties may be at increased risk to anemia producers (e.g., benzene, Cd, Pb), methemoglobin producers (i.e., aromatic amino and nitro compounds), and cyanide and carbon monoxide which reduce the blood oxygen tension. Without evidence to indicate such enhanced susceptibility, however, it appears that the inclusion of sickle cell trait or anemia in screening programs would not be appropriate.

The genetic condition that most closely fulfills the criteria of Stokinger and Scheel[13] is G-6-PD deficiency because of its high frequency of 13 percent in the American black males and limited industrial documentation of hemolytic crises in persons exposed to oxidant stressor agents such as the aromatic nitro compounds.[7,13] But despite the concern that G-6-PD deficient persons may be at greater risk to more than 30 hemolytic agents used in industry,[13] the documentation of the ACGIH TLVs[43] states that for only two substances (i.e., carbon monoxide and naphthalene) the TLV may not be sufficient to protect those with hereditary blood disorders (i.e., those with a G-6-PD deficiency). No evidence was presented as to why the ACGIH considered that those with blood disorders may be at potentially increased risk at or below the TLV value, nor was any mention ever made in documentation of TLVs as to whether persons with G-6-PD deficiency were ever evaluated either *in vitro* or *in vivo* with respect to enhanced susceptibility to any of the hemolytic compounds commonly used in industry.

At this point it is appropriate to quote the reasonable assessment of Cooper:

> What is the current state of tests of hypersusceptibility? There is insufficient epidemiologic evidence to support the use of them as criterion for employability, without many qualifications. On the other hand, there is ample scientific evidence to support wider testing. Premature assumptions as to the necessity for such tests, or over-optimistic claims for their benefits, can actually impede testing. On the basis of what we now know, no employer should be regarded as liable or derelict for not choosing to screen his employees. If he screens all employees, he

> would have to consider whether he would be regarded as liable to criticism for using a positive test to deny employment, or conversely, for jeopardizing the health of an individual permitted to work with a positive test. If it is clearly understood that the appropriate application of tests of hypersusceptibility is still moot, still on trial, then progress can be made in studying them. To clarify this point, it would be well if NIOSH developed positions, including consideration of their possible value when appropriate in criteria documents developed for control of occupational exposures.[17]

The inadequacies of the present data base make genetic screening programs for job placement or denial premature. Companies that are currently screening for persons with G-6-PD deficiency, sickle cell trait, and other genetic conditions which are thought to enhance susceptibility to toxic chemicals should only do so in a research sense and not for job placement or denial.

It should be emphasized that present evidence evaluating the effects of pollutants on proposed human high-risk groups with genetic conditions is totally inadequate on which to base policymaking and industrial screening programs. That does not mean that such persons are not at enhanced risk. They may well be. Until such time as further studies are conducted, much uncertainty does exist. This uncertainty not only possibly compromises the health of such workers, but also leaves policymakers in a quandry about how to deal with theoretical risks that seem plausible, such as G-6-PD deficients being at increased risk to naphthalene exposure below the TLV. The federal government and industry should cooperate in an effort to establish well-designed studies to ascertain whether there are definable high-risk groups within industry who need special consideration with respect to more protective regulations.

IMPLICATIONS OF HIGH-RISK GROUPS FOR OCCUPATIONAL MEDICINE

The knowledge of high-risk groups has the potential to affect a broad range of issues associated with occupational medicine. These issues are likely to include the development of (*a*) a better understanding of the causes of occupationally related diseases; (*b*) a better definition of dose-response relationships; (*c*) more accurate quantification of risk within the workforce; (*d*) development of policy toward preemployment medical screening with respect to job acceptance and placement; (*e*) development of policy with respect to the economic impact of differential susceptibility on medical costs; and (*f*) establishing ethical frameworks to facilitate acceptable ways in which the knowledge of high-risk groups may be used within a democratic society.

One of the more difficult recurring biomedical issues is the occurrence of differential susceptibility to pollutant-induced toxicity or carcinogenicity, or both, when individuals have presumably been exposed to the same toxic agents. For example, why does one person develop a smoking-related lung cancer while another smoker of the same intensity and duration does not? This is the essence of the study of high-risk groups—that is, to explain such differential susceptibility. Acquisition of such knowledge may be of considerable use in the development of dose-response relationships. If it is possible to identify specific underlying factors associated with such enhanced susceptibility, it may be possible to quantify the number of such affected persons in the workplace. The development of precise dose-response relationships provides a basis from which government regulatory personnel can begin to make rational decisions affecting the standard-setting process.

Although the knowledge of high-risk groups may ultimately have its greatest and most positive impact on the regulatory process, the most vocal discussions in the occupational sector have focused on the role of such knowledge in the prevention of disease via a process of genetic screening. It has been argued that it is not economically possible to eliminate occupational exposure to all toxins and, at least for the near future, levels of pollutants to which certain subgroups within the worker population may be at increased risk will continue. It follows from this argument that if these high-risk subgroups could be detected during a preplacement physical, the person could be appropriately placed in a less stressful work environment and the occupational disease would be avoided.

As reasonable as this argument may seem on the surface, numerous concerns have arisen concerning this issue. Some of these have revolved around statements such as those that suggest that removing high-risk workers gives industry an excuse for allowing the workplace to be dirty; and those that suggest that since many inherited diseases associated with specific races—for example, G-6-PD deficiency and sickle cell trait—have a high prevalence in blacks as compared to Caucasians, genetic screening for these traits may be a subtle form of racism.

SUMMARY

Biochemical individuality is a part of human existence. Wide human variation exists for hundreds of traits such as height, skin color, and others. Overwhelming evidence has emerged from animal and human studies that individuals also differ widely in their susceptibility to toxic agents. The knowledge of differential susceptibility has long been exploited by the medical profession in making sure that persons are not at increased risk to certain medications. For example, persons who are "fast acetylators" of isoniazid are known to be at increased risk to the development of hepatitis. Such people are now easily and readily identified by prior testing, and thereby much pain and suffering has been avoided.

Individuals are differently susceptible not only to medicinal agents, but also to numerous environmental and industrial toxins. The principal policy question concerns how this knowledge should be used by society. The EPA has incorporated the knowledge of high-risk groups in deriving both air and drinking water standards. But occupational health standards have not really incorporated the knowledge of responses of high-risk groups to chemical agents.

While there is a preoccupation with policymaking, frequently the available biomedical data are insufficient to sustain rational policy decisions. For example, even though it is well established that human erythrocytes with a G-6-PD deficiency are at increased oxidative risk to an array of industrial substances, there is a total lack of adequate dose-response data normally used in the standard-setting process. Consequently, although qualitatively it is known that such persons are at increased risk to industrial oxidants, data on quantitative relationships are lacking. Thus, no one really knows to what extent such groups are protected by current OSHA standards. This lack of quantitative dose-response relationships holds as well for other known genetic high-risk groups. Without defined and quantitative dose-response relationships, any policy directives in this area are premature.

It must be emphasized that a variety of factors affect the occurrence of differential susceptibility. Any program dealing with phenomena must consider aging, genetics, nutrition, disease conditions, and lifestyle in a collective sense, because no risk factor acts in a vacuum, but combines with others to enhance or reduce an adverse effect.

REFERENCES

1. Calabrese, E. J. *Pollutants and High Risk Groups*. New York: John Wiley and Sons, 1980.
2. Severo, R. "Genetic Tests by Industry Raise Questions on Rights of Workers." *New York Times,* Feb. 3-6, 1980.
3. Koistinen, J. "Studies of Selective Deficiency on Serum IgA and Its Significance in Blood Transfusion." Doctoral dissertation, Univ. of Helsinki, 1975.
4. Koistinen, J. "Selective IgA Deficiency in Blood Donors." *Vox Sang.* 29:1,

5. Holdane, J. B. S. *Heredity and Politics.* London: George Allen and Unwin, Ltd., pp. 179-180, 1938.
6. Jensen, W. N. "Hereditary and Chemically-induced Anemia." *Arch. Environ. Health* 5:212-216, 1962.
7. Stokinger, H. E. and Mountain, J. T. "Tests for Hypersusceptibility to Hemolytic Chemicals." *Arch. Environ. Health* 6:495-502, 1963.
8. Stokinger, H. E. and J. T. Mountain. "Progress in Detecting the Worker Hypersusceptible to Industrial Chemicals." *Occup. Med.* 9:537-542, 1967.
9. Beutler, E. "Glucose-6-phosphate Dehydrogenase Deficiency." In: *The Metabolic Basis of Inherited Disease,* 3rd ed. J. B. Stanbury, J. B. Wyngaarden, and D. S. Fredrickson (eds.), New York: McGraw-Hill Book Co., 1972, pp. 1358-1388.
10. Beutler, E., R. J. Dern, and A. S. Alving. "The Hemolytic Effect of Primaquine. VII. Biochemical Studies of Drug-sensitive Erythrocytes to Primaquine." *J. Lab. Clin. Med.* 45:286-295, 1955.
11. Paniker, N. V. "Red Cell Enzymes." *CRC Crit. Rev. Clin. Lab. Sci.* p. 469, April, 1975.
12. Beutler, E. *Hemolytic Anemia in Disorders of Red Cell Metabolism.* New York: Plenum Medical Book Co., 1978, pp. 23-167.
13. Stokinger, H. E. and L. D. Scheel. "Hypersusceptibility and Genetic Problems in Occupational Medicine: A Consensus Report." *J. Occup. Med.* 15:564-573, 1973.
14. Breiger, H. "Genetic Bases of Susceptibility and Resistance to Toxic Agents." *J. Occup. Med.* 5:511-514, 1963.
15. Zavon, M. R. "Modern Concepts of Diagnosis and Treatment in Occupational Medicine." *Am. Ind. Hyg. Assoc. J. 23:30, 1962.*
16. *Linch, A. L.* "Biological Monitoring for Industrial Exposure to Cyanogenic Aromatic Nitro and Amine Compounds." *Am. Ind. Hyg. Assoc. J.* 35:426-432, 1974.
17. Cooper, W. C. "Indicators of Susceptibility to Industrial Chemicals." *J. Occup. Med.* 15:355-359, 1973.
18. Motulsky, A. G. "Screening for Sickle Cell Hemoglobinopathy and Thalassemia." In: *Genetic Polymorphisms and Diseases in Man.* B. Ramot, *et al.* (eds.) New York: Academic Press, 1974, pp. 215-223.
19. Reinhardt, C. F. "Chemical hypersusceptibility." *Occup. Med.* 20:319-322, 1978.
20. Playfer, J. R., *et al.* "Genetic Polymorphism and Interethnic Variability of Plasma Paraxonase Activity." *J. Med. Genet.* 13:337, 1976.
21. Geldmacher-v-Mallinckrodt, *et al.* "Genetisch determinierter Polymorphismus der menschlichen serum-paraoxonase." *Humangenetik,* 17:331, 1973.
22. Lehmann, H., and J. Liddell. "The Cholinesterase Variants." In: *The Metabolic Basis of Inherited Diseases,* 3rd ed., J. B. Stanbury, *et al.* (eds.) New York: McGraw-Hill Book Co., 1972, p. 1730.

23. Kalow, W. and R. O. Davies. "The Activity of Various Esterase Inhibitors Towards Atypical Human Serum Cholinesterase." *Biochem. Pharmacol.* 1:183, 1958.

24. Kalow, W. and D. R. Gunn. "Some Statistical Data on Atypical Cholinesterase of Human Serum." *Ann. Hum. Genet.* 23:239, 1958.

25. Tabershaw, I. R. and W. E. Cooper. "Sequelae of Acute Organic Phosphate Poisoning." *Occup. Med.* 8:5-20, 1966.

26. Kazazian, H. H. "A Geneticist's View of Lung Disease." *Am. Rev. Respir. Dis.* 113:261-266, 1976.

27. Mittman, C. "The PiMZ Phenotype: Is It a Significant Risk Factor for the Development of Chronic Obstructive Lung Disease?"*Am. Rev. Respir. Dis.* 118:649-652, 1978.

28. Kellermann, G., M. Luyten-Kellermann, and C. R. Shaw. "Genetic Variation of Aryl Hydrocarbon Hydroxylase in Human Lymphocytes." *Am. J. Hum. Genet.* 25:327-331, 1973.

29. Kellermann, G., C. R. Shaw, and M. Luyten-Kellermann. "Aryl Hydrocarbon Hydroxylase Inducibility and Bronchogenic Carcinoma." *N. Engl. J. Med.* 280:934-937, 1973.

30. Paigen, B., *et al.* "Distribution of Aryl Hydrocarbon Hydroxylase Inducibility in Cultured Human Lymphocytes." *Cancer Res.* 37:1829-1837 1977.

31. Nebert, D. W. "Genetic Control of Drug Metabolism." In: *Pharmacological Intervention in the Aging Process.* J. Roberto, *et al.* (eds.) *Advances in Experimental Medicine and Biology,* Vol. 97:55-83, New York: Plenum Press, 1977.

32. Nebert, D. W., R. C. Levitt, and O. Pelkonen. "Genetic Variation in Metabolism of Chemical Carcinogens Associated with Susceptibility to Tumorigenesis." In: *Carcinogens: Identification and Mechanisms of Action.* A. C. Griffin and C. R. Shaw (eds.) New York: Raven Press, 1979, pp. 157-185.

33. Tokuhata, G. K. Familial Factors in Human Lung Cancer and Smoking. *Am. J. Publ. Health* 54:24-32, 1964.

34. Viscount-Chetwynd "In Discussion on the Origin, Symptoms, Pathology, Treatment Prophylais of Toxic Jaundice Observed in Munitions Workers." *Proc. R. Soc. Med.* 10:6, 1917.

35. Calabrese, E. J. *Nutrition and Environmental Health, Vol. 1. The Vitamins,* New York: John Wiley and Sons, 1980.

36. Calabrese, E.J. *Nutritional Environmental Health, Vol. 2. The Minerals and Macronutrients.* New York: John Wiley and Sons, 1981.

37. National Academy of Sciences. "Diet, Nutrition and Cancer," Washington, D.C., 1982.

38. Selikoff, I. J., E. C. Hammond, and H. Seidman. "Cancer risk in insulation workers in the U.S." In: *Biological Effects of Asbestos.* IARC, Lyon, 1973, pp. 209-216.

39. Hammond, R. S. and I. J. Selikoff. "Relations of Cigarette Smoking to Risk of Death of Asbestos-associated Disease Among Insulation Workers in the

United States." In: *Biological Effects of Asbestos*. IARC, Lyon, 1973, pp. 312-317.

40. Naeye, R. L., J. K. Mahon, and M. S. Dellinger. "Effects of Smoking on Lung Structures of Appalachian Coal-workers." *Arch. Environ. Health* 22:190-193, 1971.
41. Calabrese, E. J. "Role of High-Risk Groups in Deriving Environmental Health Standards." *Reviews in Environ. Health* 3(2):131-147, 1979.
42. Calabrese, E. J. *Methodological Approaches for Deriving Environmental and Occupational Health Standards*. New York: John Wiley and Sons, 1978.
43. American Conference of Governmental Industrial Hygienists. *Documentation of TLVs*. Cincinnati, Ohio, 1976.

Part IV

Administration of the Occupational Health and Safety Program

Chapter 11

Occupational Health and Safety Information Management Systems

by Adrienne A. Whyte

The following situations could occur in many organizations.

• A retired employee who worked for Company A for 30 years has cancer. Her physician says it is likely that her disease was caused by exposure to chemicals in the workplace. This employee sues Company A and is awarded $500,000 by the jury. During the trial, Company A was unable to produce records listing all of the employee's work areas, the chemicals that were used in those areas, and the exposure profiles for those chemicals during her employment.

• The federal government has just issued a new regulation with medical and workplace monitoring, training, labeling, and exposure limits for carcinogen X. The vice-president for human resources of Company B, a large, multi-plant organization, wants to know if the company uses carcinogen X in any of its plants, and if so, where it is located, how much is used, and whether air concentrations exceed the new exposure limit. The corporate occupational health manager thinks that carcinogen X is being used, but he cannot provide data to answer the vice-president's questions. He surveys the plant for the information. Plant personnel must perform their own surveys to generate the data. The vice-president gets an answer six weeks after asking the questions.

• An occupational injury is reported at Company C's health clinic. Health clinic staff members must complete and route a total of eight different forms to document the injury within the organization. Of course, additional forms must be completed for workers' compensation insurance. A total of four hours of recordkeeping is associated with the initial reporting of each occupational injury.

• Citing the federal access to exposure and medical records regulation (CFR 1910.20), an employee requested all of his medical and exposure records from Company D. He also requested the exposure records for all other welders in the department. The employee had been with the company for 15 years. To give him the requested information within the required 15 days, one nurse spent three days locating and copying medical records and an industrial hygienist spent a week locating and compiling all of the exposure records. The regularly scheduled work of these two health professionals was set aside while they fulfilled the employee's request for information.

Each of the above examples illustrates the critical importance of information management to occupational health and safety programs. Without adequate information systems, occupational health and safety professionals cannot efficiently perform their functions to protect the organization, the employee, and the environment. With proper information management, Company A might have won the suit by providing inventory and survey records showing that the employee was never exposed to a known carcinogen in her workplace; Company B's vice-president could have had the requested information the same day; Company C's health clinic staff would have completed one form for each reported occupational injury, saving about three and one-half hours of work per injury; and Company D could have complied with the employee access regulation with a simple computer run, saving eight days of work.

The importance of information management to occupational health and safety programs did not become critical until the 1970s. The passage of the Occupational Safety and Health Act in 1970, followed by passage of such environmental laws as the Toxic Substances Control Act and the Resource Conservation and Recovery Act, contributed to a regulatory environment that quickly overwhelmed the resources of existing occupational health and safety programs. In the late 1970s and early 1980s, right-to-know laws and regulations added to the problems. All of these new regulations imposed complex procedural, reporting, and record-keeping requirements on industry and, by executive order, on government agencies. Not only did medical, industrial hygiene, and safety surveillance requirements increase, but also the requirements for documentation, reporting, and analysis increased.

As public awareness of workplace hazards began to grow, litigation involving occupational illnesses and deaths also began to have its own impact. Employers were asked to prove in court that employees' diseases were not caused or aggravated by exposures in the workplace. Unfortunately, many employers found their records to be incomplete or unavailable. A similar situation arose with workers' compensation and disability cases. Organizations found that their records were inadequate to respond to the increasing number of questionable injury claims.

These factors, combined with the need for better internal records management, led many organizations to look for ways to improve their systems of records. Most of them chose a common solution—computerized occupational health and safety information systems. Today, it is known that computers can solve many of the information management problems faced by occupational health and safety professionals.

ADVANTAGES OF AUTOMATED INFORMATION SYSTEMS

One of the major advantages of an automated information system is, of course, improved records management. Part of this results from the computer itself, and part from the improved reporting and recordkeeping procedures that normally precede or accompany the institution of a new system. Before a new information system can be designed and implemented, a requirements analysis must be conducted. This analysis, which is discussed later, is a survey of an organization's information needs. Any organization must decide, among other things, what kinds of information need to be computerized; how the system will be developed; and what procedures will be used in data collection, retrieval, and analysis.

These studies are really an exercise in data management; *before a computer system can be set up, the manual system must be improved.* It is pointless to impose a computer system on top of an inefficient manual system of records. The requirements analysis gives potential users ability to define the flow of data from collection to analysis and to eliminate duplication. After the analysis has been completed, there should be little ambiguity left in reporting and recordkeeping procedures. The computer, by automating storage, retrieval, and analysis functions, is the tool used to implement the improved system of records management.

Here are the major advantages of an automated system.

- A well-designed computer system should provide easy access to all elements in the data base. For example, an employer might want to know whether any employees who were exposed to benzene have had abnormal hematology results. A time-consuming search for this information would be necessary with a manual system. With an on-line computer system, this search should be possible in a matter of minutes. With a batch-processing system, the results could be available the next day. The important point is that the data are there, and a computerized information

Figure 11-1. A single terminal takes the place of large patient record files.

Division of Occupational Medicine, Dept. of Medicine,
University of California, San Francisco.

system can provide ready access to them.

• The next advantage of such a system is avoiding duplication. For example, in a manual system, it may be necessary to keep records of occupational injuries in three offices: safety, medical, and personnel. With a shared information system, one comprehensive file on each injury can be kept in the computer. Naturally, the time spent in maintaining duplicate records will be reduced.

• A related advantage is the improved capability for interoffice coordination. With most manual systems, statistics generated by one office often do not match statistics generated by another. Again, take occupational injuries as an example. Unless this issue has received particular attention, the number of lost-time injuries reported by the safety department may not match the number reported by the medical department. A good information system should solve this problem by making discrepancies apparent long before the information is needed.

• A well-designed system will provide data standardization and accuracy. Standardization is particularly important in a large, multi-facility organization because it enlarges the data base that can be used for analysis or decision making. Accuracy of the data can be vastly improved by including input screens and editing routines in the system. An information system can also be used to ensure the completeness of records. For example, the system can be designed to alert the medical department that part of a required physical examination protocol was not conducted. The system can also allow an organization to keep up-to-date records.

• An occupational health and safety information system should provide its users with the capability to correlate workplace exposure and medical records, leading to improved medical surveillance. Correlating these records can be difficult. With most manual systems, it is prohibitively time-consuming. With an effective computer surveillance system, it is possible to routinely match these records and search for work-related diseases or abnormalities. Trends in employee health and safety can become apparent in time to act to prevent injuries and illnesses.

• The abilities to retrieve data easily and rapidly and to correlate workplace records with medical records lead to yet another advantage of information systems—the capability to conduct research on the health of an organization's work force. Once a data base is built, the computer simplifies the once arduous task of data collection and analysis. Naturally, the problem of anticipating the type of data that will be needed must be addressed during system design.

• A computer system, if it has on-line capabilities, can also be used to provide information in emergencies. For example, information on emergency procedures to be followed after exposure to toxic substances can be made available immediately at terminals located in the workplace, at first aid stations, or in health clinics.

• One last advantage is probably the hardest to document. Computerized information systems will save an organization money above and beyond the costs of system development and operation, but the amount of these savings will depend on the system's efficiency.

There are many different types of occupational health and safety information management systems in existence and on the drawing boards. The most comprehensive of these systems can provide all of the above advantages. Less sophisticated systems may provide only some of them. The most important thing to remember is that an organization's information requirements should determine the type of system chosen or designed.

HEALTH AND SAFETY SYSTEM FUNCTIONS

The following sections list the major functions of today's occupational health and safety information systems.

Monitoring employee health and safety

Monitoring employee health is a key function of comprehensive occupational health and safety information systems. The protection of employees is, after all, one of the major goals of occupational health and safety programs. This function is used primarily by occupational medicine specialists to manage and store records related to employee health surveillance. These records, shown in Table 11-A, result from both

TABLE 11-A
OCCUPATIONAL MEDICINE RECORDS

- Demographic Records
- Physical Examination Records
    - Personal and Family Medical History
    - Occupational History
    - Physical Measurements
    - Blood and Urine Laboratory Tests
    - Clinical Evaluations and Diagnoses
    - Audiometry Results
    - Spirometry Results
    - Electrocardiogram Results
    - X-Ray Results
    - Other or Special Test Results
- Immunization Records
- Clinic Visit Records
- Occupational Injury and Illness Records
- Disability Data
- Workers' Compensation Records
- Biological Monitoring Results
- Sickness and Absence Records
- Mortality Records

scheduled and unscheduled events. Scheduled health events include all physical examinations (preemployment and preplacement, periodic, certification, and termination examinations and biological monitoring). Unscheduled health events are those that cannot be anticipated. It is necessary to maintain a record of both scheduled and unscheduled health events to compile a historical profile of an employee's health.

In addition to maintaining health records, this function (or an independent but related function) must also maintain data that identify and describe individual employees and maintain a history of job assignments and locations. The identification of employees is usually accomplished by use of one or more of the following—name, social security number, employee number, and badge number. Descriptive data include such demographic data as sex, date of birth, race, address, identification of nearest relative, and so forth. These are normally found in personnel information systems.

Job assignments and location tell where an employee works and what he or she does. These data are a necessary bridge between the employee health records and the workplace and environmental records. They allow health professionals to correlate medical records with workplace records and are an integral part of comprehensive occupational health and safety systems.

Monitoring workplace conditions

This function, monitoring workplace conditions, manages data that relate to the work environment. Each unique workplace must be identified and described. One way is to use such preexisting identifiers for locations as building and room numbers or plant column or bay numbers. Another method assigns zone or grid numbers to facility area that have common exposure profiles. Whatever method is used, it should be one that can be maintained consistently throughout the organization and, if possible, for the life of the information system.

Industrial hygiene, health physics, environmental, and other sampling and inspection data that describe and quantify conditions in the workplace are managed by this function. Industrial hygiene sampling includes both personal and area sampling, and the system should be capable of accommodating both types.

Monitoring environmental agents

Environmental agent monitoring includes preparation and maintenance of local and organizational inventories of environmental agents, toxicology data, and material safety information on toxic agents within the workplace. These agents include chemical, biological, radiological, and physical hazards. As with the health and workplace functions, environmental agents must be identified and described. Since there is no one standard coding system that encompasses all types of environmental agents, several systems are often necessary. Chemical Abstracts Service (CAS) numbers or the National Institute for Occupational Safety and Health (NIOSH) *Registry of Toxic Effects of Chemical Substances (RTECS) numbers* are used frequently to identify chemicals and chemical mixtures. Other codes, unique to an organization (e.g., company material numbers), are sometimes used to identify environmental agents.

Records of environmental agents must be kept to identify and locate the toxic substances to which employees are actually or potentially exposed and to maintain reference information on their properties, use, storage, and disposal. Reference information is usually maintained on Material Safety Data Sheets (MSDS) or the equivalent. MSDSs are provided automatically and on request by most manufacturers of toxic substances. Storage of MSDS on on-line systems makes reference data on environmental agents available to all who have access to terminals and the authority to retrieve information.

Monitoring of protective measures

The protective measures monitoring function maintains information on measures taken to protect employee health and safety, including engineering controls, such as hoods for ventilation; administrative controls for health hazards; and personal protective equipment. This is an integral part of the system. Health and safety training records should also be included.

In order to identify protective measures, it is usually necessary to design an arbitrary system of identification, since no universally recognized coding systems are available. The descriptive data for protective measures permit a list of requirements for inspection, maintenance, certification, and training that contributes to management of a program. This function allows the monitoring of protective measures needed to facilitate and document regulatory compliance and historical documentation of the measures taken to protect employees.

Monitoring regulatory, administrative, and action items

The fifth system function maintains information on regulatory requirements, including OSHA permissible exposure limits or action levels, or internally adopted standards, such as the American Conference of Governmental Industrial Hygienists (ACGIH) Threshold Limit Values. This function can also be used to track ongoing events or problems that require follow-up, for example, hazard abatement plans and schedules and accident reports.

Scheduling

The computerized system can be used to schedule physical examinations, workplace monitoring or inspections, inspections of personal protective equipment, training, or any other event that is documented in an occupational health and safety system. The schedules can be generated by system logic, or they can be generated by users and maintained by the system.

Standard and ad hoc reporting

Reporting is the most important function of an automated system.

Standard reports are predefined and used routinely, and they are usually preformatted. They can be prepared with one or few instructions (often a choice of a menu of reports). Examples of typical standard reports are physical examination profiles of individual employees and lists of occupational injuries by workplace. All systems should have a standard reporting function.

Ad hoc reports cannot be predefined or preformatted, but are designed by users for special analyses and reporting. There are several methodologies for producing ad hoc reports. Two of them, the question-and-answer and fill-in-the-blanks approaches, prompt the system user to name the variables for study, indicate how they are to be manipulated, and specify the format for the report. Another method requires the use of a simple computer language for queries. The method chosen should be easy to learn and simple to use. An adequate query capability is one of the most important features of an occupational health and safety system, since nothing is more frustrating than to find out after years of data entry that some data cannot be retrieved in the manner required.

Statistical analysis

The statistical analysis function of the occupational health and safety system is a necessity if the data base is to be used for epidemiological research. Commercial software is available for this function, but it is often not occupational health and safety software. Statistical packages such as SPSS (Statistical Package for the Social Sciences) and BMDP (Biomedical Computer Programs) are frequently used. These packages give users the ability to apply descriptive and inferential statistics to their data.

DEVELOPMENT AND IMPLEMENTATION

The development and implementation of an occupational health and safety information system can be a difficult project. Often, health or safety professionals who have little or no experience with large computer information systems must direct, participate in, or budget for the project. The following section gives some guidelines that may be helpful to the newcomer to information system project management.

System project team

Any organization considering an occupational health and safety information system should first establish a system project team. The formation of this team is critical to the success of the project. Individual team members should be capable of directing the project from the initial requirements study through implementation and evaluation. Continuity in the project is very important.

The team should be multidisciplinary, reflecting the disciplines of all users of, and contributors to, the system. Ideally, team members should have expertise in their own technical areas and should be familiar with other specialties. Their motivation is also important; they must be interested in the success of the project.

The project team should include the following types of individuals.

The project manager should be an individual who has first-hand knowledge of the requirements for the system. If the system is to encompass various disciplines, such as occupational medicine and industrial hygiene, the project manager optimally should have technical expertise in, or at least a thorough understanding of, these disciplines.

The project manager must understand computers and have had some experience with them. He or she must be more than a figurehead and must have enough time to participate actively in team meetings, decisions, and projects; to implement team decisions; and to represent the team both within and outside the organization. If such a person cannot be found within the organization, the project manager should at least be a system user who has good management skills and is devoted to the success of the project.

The other individuals on the team should represent the various users. If a comprehensive system involving all aspects of health and safety is planned, representatives should include physicians, nurses, records technicians, industrial hygienists, toxicologists, epidemiologists, and safety professionals. All must be familiar with the information requirements of their own offices.

The team should also include a representative from the organization's information systems or computer department. This person must have had experience with implementing and using information systems and should be familiar with state-of-the-art hardware and software options for systems design. If the organization does not have a computer systems group, a systems consultant should be included on the team.

Requirements analysis

Once a project team has been established, a requirements analysis should be conducted. It can be conducted by team members or by an experienced consultant (from inside or outside the organization). Often, it is quicker, less costly, and more efficient to have the requirements study accomplished by a consultant.

The requirements analysis encompasses all aspects of system development prior to actual system design. It is a careful assessment of the needs that a system will fulfill. It must say *why* a system is needed, based on current and foreseen conditions. It must say *what* system features will satisfy the needs. It must also say *how* the system is to be constructed.

To provide the why, what, and how of system definition, the requirements analysis must address three subjects:

1. *Needs analysis.* The reasons *why* the system should be created. How the system will solve a problem. Why certain technical, operational, and economic feasibilities are the criteria for the system.
2. *Functional specification.* A description of *what* the system will be in terms of the functions it must accomplish. How the system will fulfill its role. Why certain design components should be considered.
3. *Design constraints.* A summary of conditions specifying *how* the system is to be constructed and implemented. What will compose the system. Why particular designs are feasible.[1]

Each of these subjects must be fully documented during the requirements analysis. Collectively, these components of the analysis should provide all of the information needed for system design.

Make-versus-buy analysis

After the requirements and specifications for a system have been defined, the organization should have enough information to conduct a make-versus-buy analysis. Organizations have three basic alternatives: (*a*) purchase commercially available software; (*b*) design and develop software, in-house; (*c*) assign the project to an outside software development firm.

Since 1977, occupational health and safety software has been commercially available. Today, several good packages are available, and any group planning to implement a system should determine whether existing software meets their requirements. Table 11-B shows the commercially available comprehensive occupational health and safety software available in 1984. The advantage of purchasing software is the time and expense that can be saved. The disadvantages are that extensive modifications and work practice adaptations are sometimes necessary if generalized software is used.

The major advantage of developing customized software is the exact fit between the requirements and the system. Little, if any, accommodation should be required by users. Today, new hardware and software

(Text continues on page 264.)

TABLE 11-B. GUIDE TO COMPREHENSIVE OH&S INFORMATION MANAGEMENT SYSTEMS

VENDOR	SYSTEM NAME	SYSTEM COMPONENTS/FUNCTIONS
American Health Sciences Corporation 4811 N. 7th Street Phoenix, Arizona, 85014 (602) 995-1441	SENTRY	Medical, Hearing Conservation, Vision, EKG, Audiometry, Spirometry, Direct Interface to Clinical Test Instruments, Industrial Hygiene, Work History, Personnel Safety, Analytical, Trend Analysis
Biometric Research Institute 1401 Wilson Boulevard Suite 400 Arlington, Virginia 22209 (703) 276-0400	HEALTHTRACK	Health Module (Health History, Routine Examinations, Clinic Visits, Hearing Conservation, Special Monitoring Programs), Industrial Hygiene Module (Materials Inventory, Exposure Data), Workers' Compensation, Personnel Module (Job History), Reporting Module (Individual, Summary and Epidemiological Reports)
Creative Socio-Medics Corporation 16 East 32nd Street New York, New York 10016 (212) 696-3500	COMET	Personnel Module, Medical Module, Industrial Hygiene and Safety Module, Toxicology Module, Environmental Hazard Module, Report Generator, Interactive Medical Records Management System (optional), Clinical Management and Scheduling (optional)
Diamond Shamrock Corporation 717 North Harwood Street Dallas, Texas 75201 (214) 659-7376	COHESS	People (Demographic, Health Evaluations, Health Incidents, Problem File), Places (Workplace Monitoring, Environmental Monitoring), Things (Environmental Agents)
Digital Equipment Corporation Medical Systems Group Two Iron Way PO Box 1003 Marlboro, Massachusetts 01752 (617) 467-6645	DEChealth	Employee Services Module (incl. Worker Tracking), Area Module, Agents Module, Health Services Module, Industrial Hygiene Module, Scheduling, Reporting Module (incl. Medical and Environmental Reporting and Intergrated Data Search), Dictionary Services Module, System Management Module, TLV Reference File (approx. 500 agents), Electronic Mail
Flow General, Inc. 7655 Old Springhouse Road	FLOW GEMINI	Medical Surveillance, Industrial Hygiene and Safety Surveillance, Scheduling, Material Safety Data Sheets, User-Modifiable Data Dictionary, Report

McLean, Virginia 22102 (703) 893-5915		Program Generator, Query Language, Statistical Packages, Interfaces to Clinical Instrument and External Hardware and Software Systems, FLOW GEMINI Environmental Information System
Industrial Health and Hygiene Group 633 Upland Avenue P.O. Box 79 Elk River, Minnesota 55330 (612) 441-6537	BIOTRAK	Employer Information (Industrial Hygiene Demographic, Hazard Exposure, Job Classification, Work History), Employer Health Claims Expense Module (Health Benefits, Disability, Workers' Compensation, Cost Containment), Health Status (Epidemiological and Statistical Studies), Safety and Accident (Hours Worked and Lost), Health Fitness (Health Risk Appraisal, Recordkeeping)
OHIS, INC. 218 West Patrick Street Frederick, Maryland 21701 (301) 694-6504	OHIS SYSTEMS	All custom-tailored: Medical Examination Protocol Generation and Scheduling, Medical Protocol Management (incl. all scheduled and unscheduled exams), Medical Records and Forms Management, Environmental Agent Management (incl. Material Safety Data Sheets, Toxicology, Inventory), Environmental Health/Industrial Hygiene/Safety program Scheduling and Management, Employee Tracking Support, Protective Measures Management, Administrative Support (incl. Workers' Compensation and Resource Management), Reporting and Analytical Support (incl. Statistical Packages) Interfaces to Systems and Instruments for Automated Data Collection
SIS SUNDATA 2 Glenhardie Corporate Center 1285 Drummers Lane Wayne, Pennsylvania 19087 (800) 441-4203	SunHealth	Employee Work History Profile, Medical Records, Industrial Hygiene Sampling, Occupational Illness and Injury, Materials/Agents Inventory, Safety and Substance Training, Scheduling, Material Safety Data Sheets, Audiometrics, Interface to SAS, Nine Environmental Functions
Stewart-Todd Associates, Inc. No. 9 Valley Forge Executive Mall 580 East Swedesford Road Wayne, Pennsylvania 19087 (215) 687-2030	ETHOS	Personnel/Administration Module, Industrial Hygiene Module, Interface to Industrial Hygiene Instrumentation, Medical Module, Audiometric Module, Safety/Accident Module, Material Safety Data Sheets Module **Compiled by Adrienne A. Whyte, PhD, OHIS, Inc., Frederick, Md.**

Note: Some of these vendors may not offer these systems today, and others have enhanced their system functionality since this table was prepared.

From *Occupational Health & Safety* magazine, vol. 53, No. 4 (April 1984). Used with permission.

technologies make custom software development a viable alternative.

When comparing the available alternatives, the project team should remember two cardinal rules. First, the alternative should be judged on the basis of criteria established by the requirements analysis. A commercially available system cannot be evaluated for use in an organization unless the requirements for the system have been thoroughly defined. Second, life cycle costs should be used to compare the costs of the considered alternatives. Life cycle costs are all costs associated with system design, development, implementation, and operation over the life of the system. A system's life is the amount of time the software/hardware configuration can be expected to perform without major modification.

System design and production

Once a set of requirements has been prepared, and if a purchase of commercially available software has been ruled out, the next step is to establish the structure of the system. This activity, often called architectural or general design, consists of defining the subparts of the system and the interfaces among them. In the detailed design step that follows, the precise algorithms (procedures for accomplishing the goal) and data structures are defined. Detailed design may involve several iterations (repetitions). Production of the system involves programming, testing of individual pieces, integration of the various pieces, system testing, documentation, and the initial performance evaluation. All of the activities associated with system design and production must be directed or accomplished by systems professionals. The project team's role in this phase of development should be one of support and acceptance testing.

System implementation

Implementing the new information system should be a joint effort between the system developers and the project team. While there are many activities associated with implementation, the most important one is user training. This training should establish realistic expectations among users about the system's capabilities and requirements. Any changes required in recordkeeping procedures and potential transitional difficulties should be thoroughly explained.

User training should not be left to programmers or systems analysts. Ideally, it should be carried out by user representatives from the project team. If a system has been purchased, some training will, by necessity, have to be conducted by the system supplier. However, a core of system users should be trained by the supplier, and then, if possible, these users should train others.

Training should be conducted within the everyday work routine. Naturally, some general training sessions must be held, but the most effective type of training is "on-the-job" training, with health and safety professionals learning to use terminals for data input and output in the context of their actual work.

Training should be timed so that there is no gap between training and actual use of the system. When something entirely new is learned, most people need to practice it to remember it. The longer the gap between training and system use, the less retention there will be.

Follow-up to training is also important. Periodically, the project team should return to the users to determine whether their expectations for the system are realistic and whether they are using the system correctly. The system manager should study error rates. If error rates are too high, the reasons should be identified and the users retrained. Follow-up training is just as important as initial training is to the success of a system.

The user's manual is an important aspect of training. During the development process for any information system, system documentation is written. Much of this documentation is for the use of systems analysts and programmers. However, a user's manual is also prepared, and great care must be taken in its preparation. Both systems analysts and user representatives must participate in writing the user's manual. It must be simple to understand and direct in instructions for system use. If it is properly written, it will be a permanent, valued reference for all users. If it is not well prepared, it either will not be used or will promote errors.

System operation

Every system should have a manager or data base administrator. This person has the ultimate responsibility for the day-to-day operation of the system, including: (*a*) management of system security; (*b*) supervision of the data base content; (*c*) problem solving; (*d*) coordination of changes to the system; (*e*) archiving of data; (*f*) data quality; and (*g*) planning for future needs and applications. The system manager should know the application well and work closely with users to make sure that the system meets user expectations.

Evaluation

After an information system has been implemented, it should be evaluated annually. Estimating an information system's worth to an occupational health and safety program can be difficult for the same reasons that conducting a cost-benefit analysis is difficult. However, evaluation is an important follow-up, and the following factors should be considered in evaluating a system:

- Completeness

- Timeliness
- Reliability
- User acceptance
- Costs
- Improved availability of information
- New capabilities.

If problems are identified in any but the last area, corrective steps should be taken. Problems may result from human, hardware, or software elements of the system.

OTHER FACTORS TO CONSIDER

Anyone faced with the task of developing an occupational health and safety information system must consider many factors during the design and implementation phases. This section presents some of the more important ones.

User acceptance

User acceptance must be considered at every stage in the development of an information system. For a system to work, it must be easy to use, and its benefits must be obvious to all.

There are several ways to promote user acceptance. First, users should be involved in the system design from the beginning. This is most easily accomplished by making them members of the project team. These user representatives should form working groups within their own departments to discuss how the system should work, what kinds of inputs would be acceptable, and what staff members should expect from the system. If the system is designed at an organization's headquarters, it is sometimes difficult to involve users from the field. In that case, field personnel should be kept aware of all findings and decisions throughout project phases. They should be surveyed for special information requirements during the requirements analysis, and they should review all documents produced during system design.

Data input

There are many different ways to get data into an information system and each has advantages and disadvantages. This section reviews the most frequently used methods.

Data can be written on machine-readable forms as they are collected. These forms can then be put through a special terminal that "reads" the data and transmits them to the computer. Mark-sense forms are the most frequently used machine-readable forms. The forms can be arranged so that they are simple to complete, requiring only a pencil mark in the appropriate box. When mark-sense forms are used, they replace conventional forms on which data collectors (like physicians or nurses) write notes, enter numerals, or check boxes. After mark-sense forms are used for machine input, they can be kept as permanent written records.

The advantage of such forms is that they do not require a special coder or records technician to transcribe the data from a handwritten form to a computer terminal. However, there are some disadvantages. The terminals that read and transmit them can be expensive, and unless there is a heavy volume of mark-sense forms to be entered, they are probably not cost effective. Another disadvantage is that they are somewhat unreliable for data entry. Extraneous marks, traces of coffee spills, or bent forms will either be misread or rejected by the terminal.

A more common data entry method uses cathode ray or CRT terminals. CRT terminals resemble a television screen and usually are equipped with a typewriter keyboard. Data are entered into the computer by typing on the keyboard. As alphanumeric characters are typed in, they appear on the screen.

CRT terminals have many uses in health and safety information systems. They can be used by special records technicians who take forms received from medical personnel and type the data into the terminal. Depending on how the information system is set up, technicians may enter the information and review it for accuracy before they transmit it to the computer. Frequently, a formatted screen is used to guide data into the terminal. Basically, this screen looks like a form, and the records technician fills in the blanks. Editing programs, which can be made part of the system's software, can alert the technician when an error has been made. In this type of entry, errors are easily corrected. Although having a trained records technician use a CRT terminal is much more reliable than entry by "keypunchers," it still has the disadvantage of an extra step between data collection and input. Another potential disadvantage involves coding. When records technicians must make decisions about physicians' notes and code physical findings and diagnoses, errors and lack of standardization can result. For example, two technicians may code the same finding in slightly different ways. In this case, the number of technicians entering data should be kept to a minimum, and they should be trained thoroughly in procedures for coding data.

The most powerful way in which CRT terminals can be used is source data entry. In this approach, the source of the information (the employee or the nurse) enters the data directly into the terminal. For example, after a brief instruction period, an employee might be asked to sit down at a terminal and give a medical history. Each question appears on the screen, and the employee then types in the answer.

Light-pen terminals can also be used for data entry. They are CRT terminals with a small, pen-sized instrument attached. When the light pen is beamed at a choice on the screen, that choice is recorded for input to the computer. A keyboard is also attached so that narrative and descriptive information that cannot be broken down into choices can be entered into the computer. The light pen allows data collectors to enter data very quickly and verify them before they are permanently recorded.

Similar terminals have been developed that require users to touch with their fingers the data on the screen to be transmitted to the computer. These terminals have the same advantages as light-pen terminals—they are fast, efficient, and accurate.

Voice data entry is a relatively new input method. With voice data entry, the user speaks into a microphone to enter the data. The system, having been taught previously to recognize the user's voice, "understands" the commands and responds appropriately. The major disadvantage of voice data entry is the limited vocabulary, since the system will only respond to words that have been programmed. The advantages are that voice entry is efficient and frees the user from the keyboard. As the technology of voice data entry is refined, it will become more widely used.

Two frequently used types of data entry are tape and direct machine-to-machine entry. Tapes, made by one computer and then loaded into another, are especially efficient when two information systems share data. For example, many occupational medicine departments send blood and urine to commercial laboratories for processing. These laboratories often have their own computer systems. When they can provide a tape of laboratory results, rather than printouts, both time and money can be saved. Direct entry from one machine to another is also efficient. For example, it is possible to send audiogram results directly from a microprocessor-based audiometer to the system computer. With a little signal conversion, this can be done quite efficiently.

Because the data processing field is constantly changing, new methods are constantly being introduced. Even now, there are other ways to enter data into computer systems. For that reason, the team's systems analyst should survey the available technology and report on the advantages and disadvantages of the options. It may be that a combination of methods will be the most appropriate.

Data quality control

Quality should be an important consideration in data collection and input procedures. Source data entry—where the individual who is the source of the data enters them into the system as just described—is one method for decreasing errors and information loss. However, additional precautions should be taken. A good information system should have built-in programs for checking the data. For example, a system can automatically check an employee ID for accuracy by verifying the name. Validation programs can check the valid limits of laboratory test results or physical measurements. They can also catch errors like the assignment of Pap smear results to a male employee. However, these validation programs, unfortunately, will not catch all errors. For example, a blood pressure of 100/60, because it is within normal limits and would be accepted as a valid blood pressure, could be entered into a system even though the correct blood pressure was 120/80. But programs can be designed so that the individual who enters the data must verify the entry (by looking at the screen and signifying that it is correct) before the entry is recorded permanently.

Another quality control function is to scan the screen for missing items. For example, the computer can signal when a required data element from a physical examination is missing. The quality control function is a necessity in occupational health and safety information systems, and it should not be overlooked during the design process.

System configurations

The types and placement of the computers to be used for an occupational health and safety information systems must be determined during the requirements study. Today, there are many alternatives for computers, ranging from large mainframe computers to personal computers, and information processing can be accomplished on one or many of these.

Advancements in microcomputer technologies have made personal computers a part of our personal and professional lives. Personal computers offer ideal tools for information management because they are easy to use and relatively inexpensive. Many companies are making personal computers available to managers for information management, word processing, spreadsheet manipulation, and electronic mail. They are part of an overall strategy for office automation. Occupational health and safety professionals can use personal computers for "local" information management and tie them into central mainframes or large minicomputers for long-term storage, cross-functional analyses, and corporate data base management. This kind of system configuration that distributes system tools mimics the decentralized organizational structure of many corporations. It is ideally suited to seemingly opposing needs for local records management and control and corporate-wide data base management.

Other system configurations are being used in the development of health and safety systems. They include: (*a*) use of a mainframe or large minicomputer for all system functions with terminals distributed to users and communication over telephone lines; (*b*) use of large computers for regional data base management with terminals or microcomputers used for data input and communication among the larger machines for corporate-wide analyses; and (*c*) use of networked microcomputers for all system functions.

There are many factors that will influence an organization's choice and location of computer hardware for an information system, including currently available hardware, corporate standards for computer hardware, telecommunications, and software, and the information requirements of system users. All system options and constraints should be studied during the requirements study so that an economical, efficient, and integrated system can be developed.

Batch versus interactive access

The type of access and processing, batch versus interactive or on-line, is an important consideration. Generally, the philosophy on the use of the system will determine which alternative is best.

Systems developed for batch processing are usually intended to provide support for health surveillance, epidemiological research, and health risk analysis without replacing the traditional manual system of records. On the other hand, systems developed to be interactive or on-line give users immediate access to data. Thus, with appropriate safeguards, they can actually replace many of the paper records traditionally maintained by occupational health and safety professionals. The capability to lend operational support to occupational health and safety records management has made on-line systems popular.

System developers need not choose only one of these methods. Interactive access can be provided wherever required and batch access and processing can be offered where processing requirements are not time-critical.

Confidentiality

As the use of computers has increased, individual privacy has become a hotly debated issue. The question of the privacy of medical records is an integral part of this issue. Thus, confidentiality of medical records in computerized occupational health and safety systems must be given careful attention. Occupational health programs must be concerned with two areas of confidentiality: (*a*) the doctor-employee relationship and the issue of confidentiality in general; and (*b*) confidentiality of automated medical records. The first is very important to the practice of occupational medicine. Confidentiality in the work environment is also discussed in Chapter 18. All organizations should have written guidelines for maintaining the confidentiality of occupational medical records.[3]

When a new, computerized system is being designed, protecting the confidentiality of medical records should receive careful attention. Although many states have enacted laws regarding medical confidentiality, few have been concerned with the privacy of computer medical records. However, since legal and ethical requirements protect medical information in general, it has been assumed that computer medical records are similarly protected. Data processing personnel who have access to medical records are expected to accept the same responsibility for privacy protection as has been practiced by physicians and nurses.

The remainder of this section discusses the security of automated medical records, with some ideas from two books on medical records.[4,5] Compare the accessibility to a computer with that of a medical chart. All employee medical records are usually stored in a central records room that is supervised by a nurse or records technician. Under this arrangement, confidentiality can be violated in three ways: (*a*) by deliberate disclosure of information by authorized users, (*b*) by accidental disclosure of information by authorized users, and (*c*) by deliberate removal of charts from restricted areas by unauthorized individuals.

In a computer data base, records are written on such machine-readable media as disks or magnetic tapes. A set of programs on the computer provides access to these records. Remote terminals linked to the computer by telecommunications are often used for entry and retrieval of data. In a computer system, data security may be threatened in four ways: (*a*) by deliberate or accidental disclosure of information by authorized users, as in the manual system; (*b*) by accidental disclosure because of hardware, software, or human failure; (*c*) by deliberate passive monitoring; or (*d*) by deliberate active interrogation.

There are no systems precautions to prevent the disclosure of information by authorized users. As with a manual system, only legal and ethical restraints apply. The second security failure—accidental disclosure through software, hardware, or human error—can be avoided. Memory protection schemes, improvement in hardware reliability, back-up systems, and careful software planning can avert accidental disclosures of information. Care must be taken in software development to incorporate procedures to reduce the risk of accidental disclosure.

Operators must be trained so that careless disposal of computer printouts or routing of confidential information to the wrong terminal or computer does not occur. Passive monitoring refers to wiretapping of transmissions between the computer and a peripheral device, such as a terminal or printer. Active interrogation is the most sophisticated means of violating computer security. It can occur when an individual who is authorized to use some parts of a record attempts to obtain unauthorized parts, or when an unauthorized individual, with the help of a password, key, or familiarity with retrieval programs, gains access to records. It may also occur when special terminals or software modifications are used to bypass control and protection programs.

Many protective measures have been developed to protect computerized records from passive and active interrogation. The use of leased or private telephone lines, rather than dial-up lines, affords some protection against passive monitoring.

Central control of remote terminal functions places control of terminal retrieval capabilities in the hands of the computer operator rather than the terminal operator. For example, an industrial hygienist could be prevented from using his or her terminal to call up a medical record in a central computer.

Access control must also be designed into the system. Legitimate users must be identified, authorized, and given a method for identifying themselves to the system. Sometimes, machine-readable plastic identification cards are used. Another method requires password entry into the keyboard. If dial-up lines are used, password identification of the terminal may be added as an extra precaution.

Central monitoring procedures can also be used. That is, the computer can keep track of which person requests what information from where and when, so that any attempts to perform unauthorized requests are recorded.

Processing restrictions can be imposed on the data base management system to monitor access. Scrambling devices or routines can be used to code and uncode data automatically between the computer and the terminal. A popular method in occupational health systems is to scramble social security numbers, making difficult the identification of individuals in computer records or in group reports. Physical security of all disk and tape files also can be controlled by locking all files in storage vaults. Software can be written so that it is impossible for someone without access to programs to read an individual's record from a file. In other words, one medical record might be spread over 20 separately stored files on a disk.

With the appropriate combination of these and other safeguards, computerized medical records are actually more secure than most manual records since, in many cases, manual records are not even securely locked. In some organizations, anyone can walk into a records room, open a file cabinet, and pull out a handful of records. But anyone wanting a record from a properly secured computer system will have to do a lot of advanced planning and, in most cases, must have the technical knowledge to retrieve it.

In research programs, confidentiality can be protected easily if individual identities are not required with the data. Scrambled social security numbers or other computer-generated numbers can be used on computer printed reports to differentiate between, but not identify, individuals.

In addition to guarding the confidentiality of medical records, these procedures will also guarantee confidentiality of the other records in the system, whether they are industrial hygiene records, research data, or accident reports.

SUMMARY

Computerized occupational health and safety information systems have been implemented in many organizations to improve information management. Although the existing systems differ in size, capabilities, and approach, they all have similar functions. Occupational health or safety professionals planning an information system should acquaint themselves with the steps required to implement and develop a system and with factors that are unique to occupational health and safety systems.

ADDITIONAL READING

An excellent series of papers on occupational health information was published in the following book: "Computerized Occupational Health Record Systems," *Annals of ACGIH,* vol. 6. Cincinnati, Ohio: American Conference of Governmental Industrial Hygienists, 1983.

REFERENCES

1. Ross, D. T., and K. E. Schoman. "Structural Analysis for Requirements Definition," in *Software Design Techniques* (P. Freeman and A. I. Wasserman, eds.), 3rd ed. Long Beach, Calif.: IEEE Computer Society, 1980.
2. Whyte, A. A. "A Guide to Comprehensive Occupational Health and Safety Information Management Systems," *Occupational Health & Safety,* 53:4, 1984.
3. Belair, R. R. "Employee Health Record Privacy and the Occupational Physician," Chapter 7 in *Occupational Health Law* (J. LaDou, ed.). New York, N.Y.: Marcel Dekker, 1981.
4. Davis, L. S., and J. F. Terdiman. "The Medical Data Base," in *Hospital Computer Systems* (M.F. Colln, ed.). New York, N.Y.: John Wiley & Sons,

1974.

5. Springer, E. W. *Automated Medical Records and the Law.* Pittsburgh, Pa.: Aspen Systems Corp., 1971.

Chapter 12

Employee Education

by Joyce A. Spencer and Joyce A. Simonowitz

Employee education has always been a major component of occupational health practice. This chapter focuses on that dimension of safety and health education concerned with work-related risks and appropriate controls, legal and ethical reasons for training workers, responsibility for providing education, teaching methods and content, resources, and anticipated outcomes of such training. There is no question about the value of health education activities, and the occupational setting is also an ideal place for hypertension screening, weight control programs, smoking abatement, and breast self-examination clinics. But there is some concern that these health promotion programs in industry may actually take the place of bona fide occupational safety and health programs.[1]

With two million workers reporting that they were severely or partially disabled from an occupationally related disease, and approximately 700,000 employees who have long-term total disabilities, there are forceful reminders that workplaces still pose substantial risks to the health of American workers.[2] The point is that industrial health programs must not focus on improving health and personal lifestyles at the expense of programs directed at the effects of work itself. In light of the voluminous information now available on toxic materials common to workplaces, and technologies available to control many of these hazards, the primary focus of education should be to arm workers with enough information for self-defense.

LEGAL AND ETHICAL NOTIFICATION

Why train workers? The major motivation for today's employers is that the law requires employee training. The 1970 Occupational Safety and Health Act (OSHAct) laid the basic groundwork by requiring that (*a*) a standard be developed that requires labels or other forms of warning for hazardous materials (*b*) workers be made aware of relevant symptoms of overexposure and emergency treatment, and (*c*) they be made to understand appropriate precautions.

The act also directed that workers receive notification and the right to review records of exposure to regulated toxic substances, results of medical or biological tests, opportunities to observe workplace monitoring for regulated substances, and notice of OSHA violations and subsequent citations. Employers are also to inform employees of their rights under the act by posting notices in the workplace. Thus, OSHA states that workers have a basic right to information about workplace conditions that potentially may affect their lives, health, and ability to remain productive members of society. Finally, the act makes the employer responsible for fulfilling workers' rights to such information.

Since those basic provisions of the original act, additional educational provisions have been added to some standards, such as those for lead, vinyl chloride, and coke oven emissions. There has been a progression from the requirement for general information (and perhaps the assumption that employers would apply the requirements liberally) in the early standards, to very specific and detailed hazard information in more recent standards. The progression of one such standard is examined later in this chapter.

Most toxic substances, however, were not covered by specific regulatory standards requiring employee training. With the Federal standards promulgated in the early 80s called the "Right-to-Know" laws, composed of the "Access to Employee Exposure and Medical Records" (29 CFR 1910.20) and the "Hazard Communication" (29 CFR 1910.1200), it is now the duty of employers to provide a wide variety of information.

Manufacturers and importers must assess the hazards of chemicals and all employers must provide information to their employees concerning hazardous chemicals. This should be done by instituting hazard communication programs which include the use of labels, Material Safety Data Sheets (MSDS), training, and access to written records. The "Access to Employee Exposure and Medical Records" provides for employee, designated representative, and OSHA access to employer-maintained exposure and medical records.

These two regulations direct that workers, who receive notification of exposure and the exposure level, be allowed to copy results of medical or biological tests, have the opportunity to observe workplace monitoring for toxic substances, and receive training to understand the labeling and use of the MSDS. This final provision of the hazard communication regulation is to be enforced in May 1986 and is effective in all states.

Furthermore, administrative law judges of the National Labor Relations Board, in several cases, have upheld union rights to generic names of chemicals and related safety procedures.[3] The employer's duty to warn employees of potential harmful exposure is evolving in other courts, too. One of the major arguments in the many asbestos victim cases concerns failure-to-warn issues. Although most of those cases are still in the courts, G. A. Peters[4] contends that asbestos manufacturers and users are still not meeting common law requirements for warnings, while J. D. Blum[5] suggests that employers' failure to warn about a workplace health hazard should be seen as an intentional tort. Clearly, the law is still being tested, but the trend and the law lie in the direction of the individual's right to information.

Beyond the legal mandates and motivations, however, there are ethical considerations for employee education. With 39,000 chemicals in use and only 24 OSHA standards for hazardous substances, it is evident that regulations do not exist for most of the serious health hazards. The 1978 "Registry of Toxic Effects of Chemical Substances," compiled by the National Institute of Occupational Safety and Health, listed approximately 34,000 chemicals with toxicity. That figure included 7,500 new listings that did not appear in the 1977 issue. When the 1979 issue was compiled, an additional 5,000 new entries were included, maintaining a pattern of several thousand new additions to the Registry annually. The point is that workers may not know about the toxicity of the chemicals with which they work. What will be done about informing at-risk workers in circumstances in which a questionable substance has no standard, and, therefore, no educational or notification requirement? See p. 401.

The concept of "informed consent" as it is applies to hospital patients, though not entirely applicable to the health risks of the workplace, does provide some practical rationale for telling workers what the hazards are, and how they are being protected. Under the "Patients Bill of Rights" of the American Hospital Association, the sick person has the right to determine whether or not to accept treatment or procedures prescribed by the treating physician. The patient has rights to information about procedures, alternatives, prognosis, complications, sequelae, discomforts, costs, inconveniences, risk, and experience. The individual also has the right to the information in language he or she can understand, and must be given enough clear information to make a personal decision about health care and what will be done to his or her body.[6]

In the work setting, the consent principle is inappropriate because overexposure simply is not acceptable, regardless of information and/or consent.[7] But if the sick person has rights to information and determination regarding his or her body, should the well, healthy employee or working person have any less right to the same consideration when there are health implications of exposure on the job? If the sick person is entitled to know about procedures and prognosis, discomforts and inconveniences, risk and experience of proposed treatment, surely working persons should have similar rights to information about the nature and toxicity of the substances with which they work, controls and their effectiveness, personal discomforts and inconvenience of both hazards and controls, morbidity and mortality data, and the relative risk.

The prevention of occupational disease depends largely on nonpersonal factors.[8] The ways in which work is completed can influence the degree of risk, and it is only the informed individual who can be expected to work responsibly, to observe safe work practices, and to use and care for protective equipment properly.

Although some employers have rationalized that health hazard information would frighten workers, or that "all the data were not in" on toxicity, or that trade secrets would be revealed, there is no ethical rationale for withholding from workers information about potential health hazards associated with workplace chemicals or processes.

Responsibility for training

The OSHAct actually assigned a great deal of responsibility for training to the Occupational Safety and Health Administration (OSHA) and to the Secretary of Labor. The Secretary of Labor, in consultation with the Secretary of Health, Education, and Welfare (now, Health and Human Services), was to "provide for the education and training of employers and employees in the recognition, avoidance and prevention of unsafe or unhealthful working conditions." House committee notes on the law reflected concern that the secretary should specifically conduct employee training that would promote greater awareness of health hazards in the workplace.[9]

OSHA originally responded to this mandate by providing elementary educational materials, for example, pamphlets on worker rights and responsibilities and specific hazards and by entering into limited training programs around the country. It became apparent, however, that OSHA itself could not meet the obligation of trying to assume responsibility for employer and employee training. To cope more realistically with this problem, during the latter part of the 1970s three new program items were developed that were relevant to training. First, financial grants were made to labor, employer and voluntary organizations, and academic institutions potentially able to provide training; secondly, extensive and specific training requirements were included in standards promulgated by OSHA; and lastly, the "Right-to-Know" regulations were promulgated.

By the development of these programs, OSHA has given a clear signal to labor unions and other organizations that they could and should also develop a framework for training members. And, at the same time, OSHA has reinforced the individual employer's ultimate responsibility for an informed body of employees. Response to this delegation of responsibility has been varied. Many labor organizations that received government funding have incorporated occupational safety and health specialists. These unions have engaged in surveys and epidemiological studies among members. They have developed training materials suited to the needs of the trade or skill, and have conducted educational programs for leaders, as well as members, across the country. Health and safety committees have become stronger; newsletters and hazard-alert bulletins have been distributed to thousands of union members. Several major international unions have established occupational health and safety as a permanent part of their programs, to be funded in a continuing basis by the membership.[7,8]

Meanwhile, employers, who have received relatively little of the training grant monies, but all of the legal responsibility, have been less enthusiastic about the training requirements. A few employers have been able to work cooperatively with union representatives to share information and concerns and have even experimented in joint training. Most, however, retain strong adversarial positions and attempt to provide employee training solely through a company program.

One of the problems that arises out of the labor-management standoff is the question of trust. Unfortunately, the occupational health arena is a natural place for animosity, and the two sides continue to be suspicious and critical of each other's integrity. Whereas management claims that labor's criticisms are politically motivated, the intensity of feeling on the part of labor is based on accounts of management manipulation of data, reluctance to provide toxicological information, or issuing actual misleading material. Management has been guilty, according to the unions, of the inappropriate use of controls contraindicated by medical standards and deliberate deception. Regardless of feelings about the validity of the charges or their universal application, the concerns of labor must not be ignored, and management should not attempt to be the sole source of or to oppose union efforts to provide occupational health information for workers.[5] Furthermore, health professionals hired by employers are not immune from the skepticism of organized or unorganized labor.[10,11]

In planning an information program for workers then, it is recommended that representatives from a broad spectrum of agencies be asked to participate. Utilizing the specialized areas of knowledge from different disciplines, expertise should be drawn from labor, academia, government, and public service agencies, as well as management.

Practical applications

To meet the legal and ethical requirements of educating the work force, the information given must be sufficiently precise to answer questions about the substance, the kind and degree of exposure, the controls in use and the degree of their effectiveness, any personal discomfort or inconvenience involved, morbidity and mortality data—both animal and human—and the relative risk. The information provided must serve as a warning of possible danger and enable workers to avoid the hazard or minimize the risk.[4] They must be informed not only about emergency procedures for the sudden acute exposure, but also of chronic illness associated with long-term, low-level exposure and length of latency periods.

Information should be shared with workers about whether or not there is solid, confirmed evidence, incomplete research with suspicious findings, or disagreement in the scientific community regarding the validity of data.[11-13] Although controversial, this premise coincides with the basic concept that workers have the right to information about possible or potential hazards. If evidence indicates that no hazard exists, or less danger than was originally suspected, the revised information, together with the substantiating evidence, can be shared with workers who will respect the honesty with which it is given. When evidence is withheld, working people become hostile and develop a mistrust of any information they receive.

THE LEAD STANDARD AS A MODEL

As mentioned earlier, recently developed standards have incorporated increasingly comprehensive training requirements. The lead standard (Title 29—Labor, *Code of Federal Regulations,* Section 1910 1025) can be used as a model, because it attempts to incorporate many of the dimensions of ethical information-sharing. The lead standard requires "each employer who has a workplace in which there is a potential exposure to airborne lead at any level" to inform employees in lay language.

By developing a training program on the basis of potential exposure at any level, workers who have even a remote chance of exposure will receive information that may be important in an emergency, a temporary change in job assignment, or simply when mingling with friends working in an area of higher exposure. When information is limited to those who work in an area of predetermined exposure levels, temporary changes in routine may lead to a breakdown in the information/control system, and all workers may question the rationale and integrity of the company policy.

The training requirement of the lead standard goes on to require the employer to assure that each employee is informed of the following specifics.

a. "The content of the standard." Making workers aware of a legal standard for a substance with which they work, what the standard covers, how it protects them, and what is required of the employer allows the worker to participate in the monitoring of the work environment, its compliance, and its protection. Implied here, too, is the need to know whether there is a permissible exposure level (PEL) without a comprehensive standard, or even an unofficial recommendation for an exposure limit, because this information will help the worker to evaluate the risk, and to decide how carefully one must work with the substance.

b. "The specific nature of operations that could result in exposure above the action level." It is important for workers to know exactly how and where a hazardous exposure could occur. Those working in a hazardous area need to be able to recognize exposure and know how to react to the threat. Others who work where such exposure is unlikely should have an understanding of other potential exposure(s).

c. "The purpose, proper selection, fitting, use and limitations of respirators." Perhaps one of the more misunderstood components of many hazard control programs is the respirator. Any time workers must wear respiratory protection from any hazard, a comprehensive training program is imperative. Often managers, supervisors, and foremen are not knowledgeable about respirator uses and limitations. Outside consultants (not sales persons) may be needed to assist with training. Included in the training should be a frank discussion of why respiratory protection is needed—the specific exposure, the level, reasons why engineering or other controls are not adequate, and at what times the device should be worn. The type of respirator is important—why it will offer protection and why it is different from others that may be used elsewhere for another kind of exposure. There are other relevant questions. What will the respirator not do? How often should it be changed? Will it work for a rescue in a confined space or a low-oxygen environment? If it is issued for the paint booth, is it also good protection for sanding the family car? Face-piece size, proper fit, strap placement, cleaning and maintenance for valve operations are important. Personal discomfort and inconvenience are well worth discussing. The more a person understands about the use and limitations of a respirator, the better the possibility of optimal protection with it. The misuse of a respirator, or the false security it may give, can create a more immediate and serious hazard than the original threat.

d. "Purpose and description of the medical surveillance program . . . including the adverse health effects associated with excessive exposure to lead (with particular attention to the adverse reproductive effects on both males and females)." The worker has a right not only to know what the hazardous substance or agent is and where it appears in the workplace, but also to have explicit information on the possible effects on the human body by an exposure. Information on chronic and long-term effects, as well as short-term acute disability, should be provided together with symptoms of both. Animal and human research results, as well as morbidity and mortality data, should be discussed. In the case of lead, women have long been excluded from lead exposure areas because of the fetal toxicity of lead, but men often have not been informed that lead can also have adverse effects on the male reproductive system.

When there is a potential for a physical effect from a known exposure, medical or biological monitoring, or both, are important controls. Such monitoring can be done ethically only when employees understand what tests are being conducted and why, and what the results might be and their meaning.

e. "Engineering controls and work practices associated with the employees' job assignments." Employees who know what engineering controls are in effect, and who understand their value in health protection, are in a position to assist in monitoring the controls, making certain that exhaust ventilation is on and working, keeping enclosures in place, and positioning baffles appropriately. Occasionally, such controls add to the complexity of a job. Workers who understand the function and importance to themselves will be far more tolerant of the added complexity than those who are left to think that it is just a management ploy to make the work harder.

The ways in which people do their jobs can contribute also to the aggravation or control of a hazard. Where work practices are an important component of a health hazard control program, employees must understand the nature of the health hazard and the contribution of the specific work practices to the control plan. For lead workers, vacuuming is far superior to sweeping in controlling lead dust, but an individual who does not understand the concept may find it more convenient to use the broom. Similar applications can be found for most other types of exposures. Not only do workers have the right to know what the employers are doing for their protection, but they also have the right to know about the interrelationships between engineering, work practices, and personal protective devices in the total control program.

f. "The contents of any compliance plan in effect." The lead standard implies that employees have the right to know whether or not their employer is in compliance with existing state or federal OSHA regulations, and if not, what plan the employer has for achieving compliance. The broader application, however, might be that employees are able to contribute to the planning for hazard reduction, if management communicates its concerns about recognized or suspected hazards. It is reasonable to expect that once workers know about potential hazards associated with their workplaces, they will want to know what the company will do. This part of the lead standard indicates that employees have a right to this information, and that companies should plan to share it with them.

METHODS OF TRAINING

Adults generally develop an interest in learning when confronted with something that is of importance to them personally.[14,15] They learn most effectively at times when they are actively involved in the teaching/learning process, seeking out information that they perceive to be important and contributing out of their own practical experience.[16]

Successful education combines lectures, written materials, visual aides, and demonstrations. Although a program may include all of these, learner participation in discussions, exploration, research, sharing, and role playing is invaluable. Information given in a classroom lecture, a supervisor safety talk, or a tape recording is likely to go in one ear and out the other, if offered as the sole source of information. If in addition to their ears, the learners also use their eyes, such as with posters, diagrams, film, and written material, the information will be more deeply impressed. The ultimate goal, however, is to involve learners in the process so that they interact with and internalize the information. With the internalization of occupational health and safety information comes the concern that is so vital to responsible work and participation in the hazard control program.

In more practical terms workers who might be exposed to lead should be given some nontechnical literature on the physical effects of lead and asked to work together in small groups to make a list of all the possible health effects and symptoms of overexposure. As they read, discuss and develop their lists, their experiences as lead workers will create added interest. Moreover, the fact that they are allowed to discover for themselves, from impartial literature, will help to dispel suspicions of management manipulation.

Rather than telling the group what control methods are in use for a particular exposure in the plant, employees can be given information on control principles—substitution, personal rotation, engineering, personal protective devices, and the like. Workers should, by all means, be allowed to discover the controls in place, and then discuss them together. A film might be used to present the advantages and limitations of a given control method, and the group could then discuss how to maximize the controls in use in the plant. Thus, if an exhaust ventilation is used, what do the affected workers need to know about it for optimal effectiveness? Is it practical to expect individual workers to remember to turn on the exhaust equipment before beginning the job?

For toxic substances with an OSHA standard or PEL, workers should become familiar with the standard and its provisions. They may need help with some of the technical aspects. Workers who have the opportunity to wrestle with these issues are more likely to gain the level of understanding needed to make decisions about their own work, to work responsibly where there are potential health hazards, and to participate actively in the hazard control program.

Ideally, occupational health training begins with the new worker, whether young and inexperienced, or simply new to a particular plant environment. Understanding a given employer's commitment to occupational health from the outset will enhance the overall program. But the motivated worker who encounters older, experienced employees with less understanding and involvement in the program will soon lose that early interest and concern. All affected employees must, therefore, be involved at the outset of a new hazard control program and periodically in refresher sessions. Updated programs, in conjunction with relevant periodic medical evaluations (or biological tests), are appropriate for reviewing hazard recognition and control methods. Any time there are to be new processes or new or different materials, training should be initiated before the change. New information on materials and processes already in use should be shared with workers at the earliest possible time. Training sessions for affected employees should not be optional and should be held on company time.

A detailed discussion of training methods is given in the Council's book *Accident Prevention Manual for Industrial Operations Administration and Programs* volume.

MATERIALS AND RESOURCES

A wealth of material is available for the occupational health educator. There is probably no single resource or package of material that will adequately meet the needs of a given plant or group of workers. Instead, a variety of resources and types of materials should be considered, and that which is most appropriate be used accordingly. The most desirable approach would be an individualized program planned by management in conjunction with labor representatives using appropriate materials from several resources, as well as original material ideas.

The materials available generally fall into one of two basic categories.

- *Motivational materials* give some information about a hazard or group of hazards, but the major thrust is the employees' responsibility to do something personally for their own health and safety, for example, wear earplugs, lift heavy items correctly, or use respirators. Such materials may by useful if behavior modification for compliance with a health and safety program or an OSHA regulation is the sole goal of training. But if the ethical responsibility to help workers make informed choices in their work and personal lives is to be fulfilled, motivational material alone will fall far short of meeting the need.

- *Hazard-recognition programs* that have as a primary goal teaching workers how to identify possible hazards, present basic principles of hazard evaluation, risk assessment, and control methods. Rather than assuming that the company is paternal and will care for the employees, these programs assume that the worker is interested in his or her own welfare and is capable of learning and making decisions relating to work. Recent "Right-to-Know" laws require this type of information to be given.

Required materials for the hazard training program can be obtained from the following sources. (Also see Chapter 15, "Information Sources for Safety and Health Professionals.")

Manufacturers and importers. All suppliers are required to supply Material Safety Data Sheets when selling products or raw materials.

National Institute of Occupational Safety and Health. NIOSH is the research arm provided by the OSHAct and has the most current information on regulated substances and on chemicals for which no standard exists. Basic training materials are available; also, technical information from which teaching materials can be developed.

Nonprofit health agencies. Although many of these agencies emphasize health promotional materials (stop-smoking programs, cancer detection clinics) for use in industry, some recognize the workplace contribution to disease and are developing programs and materials to address these issues. The American Lung Association is one such agency that is developing educational materials on occupationally related lung diseases.

National Safety Council. For many years this not-for-profit organization has been a spokesman for industry on workplace injuries and, more recently, occupational illness. For member companies, a vast library of films and other training materials is available. Many Council materials are predominantly motivational.

Private educational firms. As standards for employee education have been developed, private enterprise has seen the opportunity to develop and sell training programs "to meet OSHA requirements." These are often packaged in modules that require little input from either the employer or the employee.

Public-interest groups. Such groups as the Urban League have produced some useful literature on occupational health hazards. Some communities have organized Coalitions for Occupational Safety and Health (COSH), groups that can provide professional expertise and materials.

Trade organizations. Many trade groups have responded to federal requirements by developing information packages and training programs to help member companies comply with regulations. The materials are primarily motivational.

Trade unions. Many international unions, as well as the AFL-CIO, have developed extensive training materials for their members. They are focused almost exclusively on hazard recognition, control principles, legal rights and responsibilities. The materials are available to other groups engaged in labor education. Employers with union plants who are able to work jointly with their labor representatives in their occupational health training should be encouraged to use these materials.

Academic programs. Specialty training for health professionals has been developed in NIOSH-sponsored programs across the country. Technical material and expertise developed in these programs can be of help in the development of training programs. Less technical information and training programs have been developed by labor education centers and industrial relations programs of major universities. Hazard recognition materials can be obtained from the Labor Occupational Health Program at the University of California (Berkeley); the School for Workers of the University of Wisconsin (at Madison); and the Ohio State University Labor Education and Research Service. Other educational institutions produced helpful material, and local educational institutions should be investigated for possible assistance.

State and federal OSHA offices. OSHA has printed materials which may supplement information from other sources. OSHA consultation offices provide a variety of educational materials and services.

Anticipating worker response

The notification and training presented here has the primary purpose of providing tools with which workers can shape their lives. The ways in which individuals and groups use those tools may go beyond the major employer interest in compliance, work practice, use of protective devices, and general cooperation with the control program. It is realistic to anticipate that the more information workers gain, the more they will expect of the employers, applying pressure on management to fulfill the legal obligation for a healthy and safe work environment. Employers are expected to seek out hazardous information, share it with the workers, and correct the hazard. If employees know about engineering controls, they will not be easily persuaded to wear uncomfortable protective devices. Knowledgeable workers will be monitoring their own workplaces for potential hazards and appropriate controls, to the mutual benefit of employers and employees.

Workers who know they have legal right to file OSHA complaints may very well exercise that right if they feel that management is not fulfilling its obligations. Concerned and responsible employers who have good working relationships with their employees will recognize this, take employee concerns seriously, seek neutral consultation on disputes, and involve workers in these consultations. Where there is no satisfactory resolution, knowledgeable workers can reasonably be expected to exercise their legal rights to file complaints and to follow each step in the investigation, appeals, and hearing process allowed by law.

Increasingly, health and safety are becoming major collective bargaining issues. As labor gains knowledge about occupational health problems, they will use the bargaining table to achieve advances in workplace health and safety policy. In existence now are contract agreements for engineering controls, independent health consultants to be approved by both management and labor, expedited grievance and arbitration procedures for health and safety problems, and joint health and safety committees with significant investigative and policy making functions.[17]

Motivated by knowledge, workers (and their employers) should become involved in the standard-setting process, in encouraging an expanded research effort to close gaps in toxicity data, and in changing the system to provide fair compensation for victims of uncontrolled occupational health hazards.[12]

In summary, occupational health education, as discussed in this chapter is frank communication with workers about workplace environmental factors that can create personal health risks, and workplace changes that can reduce those risks. The information is given in a way that facilitates personal decisions and actions to reduce personal risk. The right of workers to have information about occupational health hazards is supported by law and ethical standards. The law now makes the employer responsible for the health and safety of the workplace. Knowledgeable workers can also contribute, in tangible ways, to the success of a good health and safety program. The information offered needs to be consistent with worker self-interest and in accord with recognized adult education methods.

Information cannot be a substitute for preventive measures, however. It remains the employers legal and ethical responsibility to provide a safe and healthful workplace.[18] Education will be accepted only if workers feel management is doing its part in changing the risk by changing the environment. Placing unwarranted responsibility on the employee while allowing hazards to continue will destroy the credibility of management. As workers become more knowledgeable, they will become more sophisticated in their expectations. Collective action to compel environmental changes will result, and legal and political pressure for better protection will increase. Employers, therefore, should take seriously their responsibility to create a workplace as safe and healthful as is technologically possible, then to provide comprehensive education and training for their employees, enabling the workers to become responsible partners in workplace health and safety.

REFERENCES

1. Whorton, M. Donald, and Morris E. Davis. "A Perspective on Health Intervention Programs in Industry," in Rebecca S. Parkinson and Associates, eds. *Managing Health Promotion in the Workplace.* Palo Alto, Calif.: Mayfield Publishing, 1982.
2. "An Interim Report to Congress on Occupational Diseases," Washington, D.C.: U.S. Department of Labor, Government Printing Office Document No. 0-311-416/4007, June 1980.
3. *Occupational Safety and Health Reporter,* Bureau of National Affairs, 9(24):790-91, January 24, 1980.

4. Peters, George A., and Barbara J. Peters. *Sourcebook on Asbestos Diseases.* New York: Garland Publishing, 1980.
5. Blum, John D. "Revealing the Invisible Tort: The Employer's Duty to Warn," in R. Egdahl and D. Walsh, eds. *Health Services and Health Hazards: The Employee's Need to Know,* New York: Springer-Verlag, 1978.
6. Curran, William J. "Ethics and the Law," in Jean Mayer, ed. *Health,* New York: Litton Educational Publishing, 1974.
7. Brooks, Wayne: "Safety and Health on the Job: A Challenging Responsibility for Employers," in Judson MacLaury, ed. *Protecting People at Work,* Washington, D.C.: U.S. Government Printing Office, 1980.
8. Ashford, Nicholas A. "Economic Issues I" in *Crises in the Workplace: Occupational Disease and Injury.* Cambridge, Mass.: MIT Press, 1976.
9. *Occupational Safety and Health Act of 1970: Law and Explanation,* Chicago: Commerce Clearing House, 1971.
10. Whorton, M.D., and M. E. Davis. "Ethical Conduct and the Occupational Physician," *Bull. N.Y. Acad. Med.* 54:733-741, 1978.
11. Goldsmith, Frank, and Lorin E. Kerr. *Occupational Safety and Health: The Prevention and Control of Work-Related Hazards.* New York: Human Sciences Press, 1982.
12. Friedland, John. "The Challenge of Informing Workers of Job-Related Health Hazards," in R. Egdahl and D. Walsh, eds. *Health Services and Health Hazards: The Employee's Need to Know,* New York, Springer-Verlag, 1978.
13. Lehmann, Phyllis. *Cancer and the Worker.* New York: New York Academy of Sciences, 1977.
14. Brown, Mary Louise. *Occupational Health Nursing: Principles and Practices.* New York: Springer Publishing, 1981.
15. Bellanca, Joseph J. *"Motivating Managers and Workers," Occup. Health and Safety,* 51:15-16, 55, 1982.
16. Ingalls, John D. *A Trainer's Guide to Androgeny,* rev. ed. Washington, D.C.: U.S. Government Printing Office, 1972.
17. Wood, Michael. *The Safety Committee and Collective Bargaining: New Concepts.* Kansas City, Mo.: The International Brotherhood of Boilermakers, Iron Ship Builders, Blacksmiths, Forgers and Helpers, AFL-CIO, 1979.
18. Van Den Eeden, Stephen K. "The Right-to-Know" and the Occupational Health Nurse," *Occup. Health Nurs.,* 33:281-285, 1985.

Chapter 13

Work Scheduling

by Joseph LaDou

Work scheduling, shift work, and alternative work schedules, and their physiological effects concern occupational health and safety professionals because of the possibility that employee health, performance, and safety may be affected. Scheduling will, of course, differ among companies and industries, depending on their particular operations and economic demands. In certain processes, such as steelmaking, pulp- and papermaking, and petroleum refining, plants must operate 24 hours a day with back-to-back shifts. In other operations, economics require that costly capital equipment be operated continuously. Finally, severe competition for worldwide markets can be the reason for 24-hour production.

When continuous production is required (without the payment of excessive overtime) a number of scheduling alternatives are available. The most common is the continuous shift, in which three crews of employees work a specified number of the day, evening, or night shifts, and then are off for two days, regardless of when those days fall. A second option, called the discontinuous shift, uses four crews—three work Mondays through Fridays, with the weekend off—and the fourth crew covers the weekend. Another way of scheduling shifts uses two crews to work a total of 80 hours a week, during both the day and evening shifts, or during both the day and night shifts. A twilight shift also is used in which part-time workers, primarily women, are employed to work four hours at the end of the normal working day. If overtime payments are not a concern, scheduling is more flexible because discontinuous three-shift work (three crews, Monday through Friday) can be extended into part of the weekend, or two shifts can be extended to 12 hours to provide continuous coverage over a five-day week. Finally, split shifts are used by public transportation, carriers, hotels and caterers, and hospitals. Employees work so many hours on one shift and then return to work for so many hours on another. One area of particular concern with regard to shift work is the rotating schedule, in which workers serve so many days on one shift, so many days on another, and are continually moved from shift to shift. Since most shift systems rotate, occupational health and safety professionals should pay particular attention to this area.

Alternative work scheduling has developed in recent years, because of greater emphasis on the psychological needs of workers and also the availability of transportation in the community. Flexible scheduling allows workers to work and leave on their own schedules, generally giving them a more satisfactory life-style. Also, companies in heavily populated areas allow employees to arrive and leave work at various times to avoid traffic and bus congestion. Details are given toward the end of this chapter.

THE BIOLOGICAL CLOCK

One of the more important questions occupational health and safety professionals face is whether or not shift work, rotational scheduling, and other alternative scheduling will disturb the internal rhythms of the body, that is, the human biological clock.

It has been determined that all organisms function according to internal rhythms and that these rhythms are synchronized by such factors as light/dark, summer/winter, clock time, or social stimuli. These factors are termed *Zeitgebers* (time givers). For plants and animals, light (or the absence of it) is the primary synchronizer. For example, certain animals procreate and hibernate at specific times related to their rhythms, and plants germinate, bloom, and fruit according to rhythms.

Light is also a synchronizer for humans, but a relatively weak one. Important synchronizers for humans are awareness of clock time and social interaction. Human biological rhythms are disturbed, for example, when flying across many time zones, hence, the occurrence of jet lag. Another important time-giver for humans is social interaction, and when they are denied interaction with families and others, their biological clocks become desynchronized (out of phase). In humans and most animal and plant species, the rhythms usually follow a near 24-hour pattern. For that reason, they are called circadian rhythms—from the Latin *circa dies* meaning "about a day." The present knowledge of these rhythms has developed through studies with plants, animals, and humans in which subjects have (*a*) been removed from the light cycle entirely; (*b*) exposed to a lengthening of either cycle; (*c*) had their sleep/wake cycles inverted, or (*d*) been removed from association with their own species.

The rhythms are synchronized by the time-givers, and, when the latter are altered, the rhythms take on patterns of their own and become desynchronized or free-running. They follow a different time pattern. It has been determined from considerable study that the following functions fluctuate or change rhythmically—body temperature; the endocrine system; urine flow; and the cardiovascular, hematopoietic (blood forming), respiratory, and autonomic nervous systems. Moreover, the absorption rate, metabolism, effect, and excretion of drugs are strongly rhythmic, so that a medication taken at one time of the day may not be as effective as when taken at another. When the phase of time-giver is shifted—for example, when the light/dark cycle is shifted—the body rhythms either shift with the time-giver or become desynchronized from the external environment (light, time, society) and tend to become desynchronized from each other. For example, rhythm of body temperature, low in the early hours and high later on, may run on an entirely different schedule and become desynchronized from other systems as well.

Biological clocks often run faster with advancing age so that the free-running period gradually shortens. Free running of rhythms occurs when conditions are constant, but are typically entrained to the 24-hour period in the natural environment. Consequently, circadian rhythms respond to environmental time cues. These rhythms develop in a species during the fetal stage, since it has been learned that animals later removed to foster mothers continue the rhythms of their natural mothers.

When human rhythms become desynchronized, they must be reentrained to avoid physiological problems. Humans and animals are reentrained by removing other time cues, adjusting the circadian rhythm to parallel the period of the environmental cycle, and maintaining a stable and reproducible relationship between the timing of the observed rhythm and the advent of the time cue. For reentrainment to be successful, when the time cue is removed, the rhythm must begin to free run according to the environmental cycle, not the previous rhythm.

It should be noted that environmental temperature, whether hot or cold, can suppress some circadian rhythms, such as locomotor activity. Thus, the two might seem to be in accord when bodily activity actually was stimulated by the desire to keep warm or cool as the case might be, which would mask the real rest/activity rhythm. To avoid this problem, environmental factors must be reduced insofar as possible. Similarly, if the time cue extends over a period that is too short or too long, reentrainment will be limited because the time-cue is not strong enough and the rhythm will free run.

Other factors that affect reentrainment include the availability and timing of food intake, water availability, and social cues. Food and water availability and intake have been studied primarily in animals. Studies of

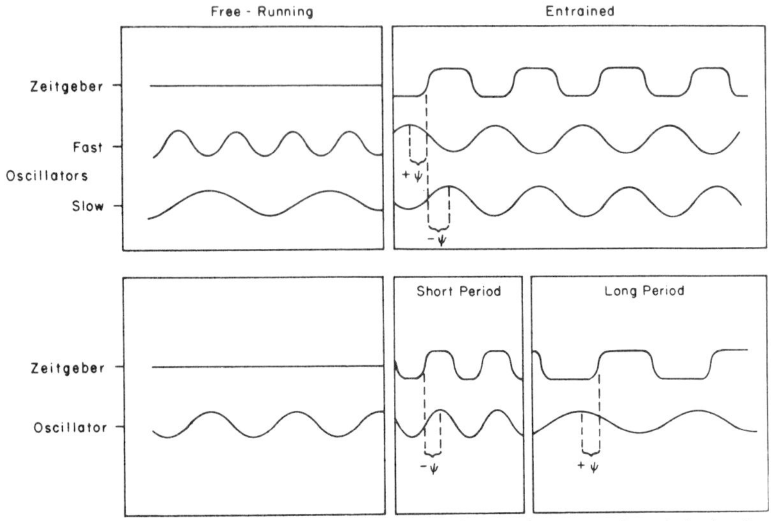

Figure 13-1. Schematic drawing of the influence of zeitgeber period and rhythm free-running period on the phase relationship (ψ) between rhythm and zeitgeber during entrainment. *Top left,* free-running rhythms of fast and slow oscillators. After entrainment to a zeitgeber cycle *(top right),* both rhythms show a period identical to that of the zeitgeber, but the fast oscillator is phase-advanced relative to the zeitgeber cycle, whereas the slow oscillator is relatively phase-delayed. *Lower panels,* oscillator with an intermediate free-running period entrained to a short-period and a long-period zeitgeber. With a short-period zeitgeber, the oscillator is relatively phase-delayed with reference to the zeitgeber, whereas with the long-period zeitgeber, the oscillator is relatively phase-advanced. Whether an actual phase lag or phase lead is seen depends on the phase reference points used. If a later phase reference point such as the maximum of the zeitgeber cycle were picked, no phase-lag rhythms would be seen. The same relative directional effects on the phase relationship would have occurred, but the oscillator reference phase would always have been phase-advanced with respect to that of the zeitgeber.

Source: Moore-Ede, Martin C., Frank M. Sulzman, and Charles A. Fuller. *The Clocks that Time Us.* Cambridge, Mass.: Harvard University Press, 1982, p. 75. Used with permission.

cycles of eating and fasting, with food available only at certain times of the day, have shown that food availability and intake can either desynchronize or reentrain circadian rhythms. In contrast, the availability of drinking water apparently does not affect circadian rhythms.

Social cues, although more difficult to isolate and quantify than physical cycles, play an important role in entraining circadian rhythms of some animals and all humans, but the nature of this entrainment remains unclear, because it can be affected not only by rest/activity cycles, but also by pressure from peer groups. With humans, a larger number of factors are involved, such as personality, lifestyle, motivation, feeding routines, and the light/dark cycle, so that it is difficult to determine whether or not social cues are as important as auxiliary time cues for entraining human circadian rhythms.

A highly important function of the circadian system is to activate behavior and internal metabolic adjustments so that they are in time with events in the external environment. Consequently, there is a phase relationship between time-givers (exogenous cues) and each rhythm. This relationship can be either the difference between the timing of each (hours), or the difference between reference points of the two. The latter, for example, could be the difference between the first appearance of light and the beginning of activity. In other words, light might first appear at 6:00 A.M., but an individual's activity rhythm might not be stimulated until 9:00 A.M. As M. C. Moore-Ede and his associates have explained:

> When the natural period of the rhythm is shorter than that of the time cue, the entrained rhythm "phase-leads" as the time cue and hence, the phase relationship, is positive. In contrast, when the rhythm has a longer free-running period than the entraining time cue, then the entrained rhythm "phase lags" with reference to the time cue, and the phase relationship is negative.[1]

Similarly, the steady-state phase in which both the time cue and the rhythm are in close relationship to each other, occurs when "...the period of the Zeitgeber (time-giver) and the natural period of the pacemaker driving the rhythm...."[1] are in accord. (See Figure 13-1.)

When time cues are shifted by several hours, as in rotating shifts, rhythms desynchronize, some more rapidly than others. Reentrainment of the rhythms after a change in time-givers can be important for physiological reasons, performance and safety. It appears that reentrainment takes longer when time-givers are phase advanced and occurs more rapidly when they are phase delayed. This is an important concept in regard to shift work.

CIRCADIAN RHYTHMS IN SHIFT WORKERS

The working of the biological clock has particular implications for shift workers. W. P. Colquhoun and J. Rutenfranz have stated that:

> The circadian system represents a temporal order which is mediated by the mutual coupling of a variety of oscillators and by the synchronizing effects of Zeitgebers. It is likely that the well-being of man depends partly on the maintenance of this order, and that repeated or long-lasting disturbances to it will have harmful effects. Due to differences in the "inertia" of the various oscillators, the circadian system is temporarily out of order whenever it is phase shifted; and is more or less permanently out of order when under the influence of conflicting Zeitgebers

(time cues). One of the aims in designing shift work schedules is to minimize the possible effects of both these kinds of disorder.[2]

In an earlier work, Rutenfranz concluded, "Perhaps the most important physiological problem regarding shift work, particularly shift work which includes night work, is the problem of the reentrainment or resynchronization of physiological functions after a phase shift of working and sleeping times."[3]

The conclusions of both investigators have been supported by findings from studies of animals and humans under controlled environmental conditions. In shift work, the wake/sleep cycle is changed while clock-time and social stimuli remain the same. In other words, the worker may have to sleep at a time that is different from that of most of his cohorts and, consequently, is prevented from socialization with family and friends. The individual is unaware of the time cues that influence most of us.

A change in work time for one night apparently does not desynchronize circadian rhythms, but shifts of one to three weeks do. For example, the temperature rhythm, which is also associated with performance capability, is not affected by a single-night shift, but was desynchronized over the longer period, with the minimum temperature shifting to the new sleep period after seven or more days. As indicated earlier, the various physiological functions reentrain at different rates. With transmeridian flight, in general, a full day of recovery is required for each hour of time-zone change before circadian systems reentrain. This finding would suggest that when a work shift is rotated, one week would be needed to reentrain the circadian system and that some functions would reentrain at different rates.

Many occupational health and safety professionals are primarily concerned with the physiological aspects of rotating shifts, but most workers are not. They are more concerned with the psychosocial effects of shift work—they cannot spend enough time with their spouses, children, and friends. In a word, they are "out of sync" with the rest of their world. Consequently, health professionals must take into account the workers' feelings, as well as physiological responses, in designing or approving any rotating work schedule.

Shift work and employee health and safety

Although a number of physiological functions may be disturbed by change in the working time, the two that are most affected are the sleep/wake cycle and the gastrointestinal physiology. Sleep disturbance and altered eating patterns, then, are the primary worker complaints related to rotating shift work.

Sleep/wake cycles. The most important circadian rhythm in both humans and animals and the one that they are most aware of is the sleep/wake cycle. When this cycle is changed, daytime sleepiness or nighttime awakeness calls sharp attention to changes in the cycle, while important changes in endocrine and renal rhythms may go unperceived. The sleep/wake cycle is not to be confused with the rest/activity cycle because rest is not the same as sleep, and it does not provide the same rejuvenation. Activity is not the same as wakefulness either because people continue to be active during sleep.

In humans, the normal sleep/wake cycle approximates 8 hours of sleep and 16 hours of wakefulness, depending on individual needs. Animals, on the other hand, do not have such precise sleep/wake cycles, but will sleep and wake periodically throughout the 24-hour period of the dark/light cycle. With aging, humans spend more hours awake and less asleep. The newborn infant sleeps two-thirds of the 24-hour period in the beginning and slowly increases the period of wakefulness, but often takes a full two years to adapt to the concept of sleeping at night.

The quantity and quality of sleep in adult humans can be disturbed by changing the sleep time from night to day. The quantity of sleep of workers on various shifts is indicated by the finding that workers on morning shifts average 7.5 hours of sleep per 24-hour period; afternoon shift workers average 8.5 hours; and night shift workers average only 4 to 6 hours. Moreover, 50 percent of day shift workers sleep 7 to 8 hours, but only 15 percent of night shift workers sleep that long. Similarly, the mean length of sleep during the night is 7.55 hours; whereas for night workers, the mean length of sleep during the day is only 5.56 hours. Workers whose sleep period begins at the end of the night sleep even less (5.45 hours). The quantity of sleep that night shift workers achieve is considered insufficient.

Shift workers suffer sleep problems not only because the time of day when they must sleep is not conducive to sleeping, but also because of environmental conditions that are not conducive to sleep—there is more noise from traffic, children playing, dogs barking, and the like. Workers who must sleep in the day report a 40-60 percent greater incidence of sleep disturbance than those sleeping in the nocturnal hours, and, also, workers who live on well-traveled roads, near airports, or in homes that are not well insulated against noise suffer greater sleep disturbances.

Night shift work also affects the quality of sleep. Sleep occurs in five stages that follow successively—Stages I, II, III, IV, and REM (rapid eye movement). Through the use of electroencephalograms and other brain observation techniques, it has been determined that the sleep period is divided as follows: Stage I = 4.4-5.4%, Stage II = 48.7-51.2%, Stage III = 11.2-13.2%, Stage IV = 22.9-24.1%, and REM sleep = 22.9-24.1%.

The length of REM sleep, which is the most productive, does not decrease with aging, particularly during the normal working years. From the mid-30s on, however, Stage I sleep increases, while Stage IV decreases. If sleep disturbances occur during Stage II or the REM stage, sleep quality is substantially impaired.

Sleep disturbance is the major complaint of rotating shift workers. A study of such workers in a processed potash production plant who were phase advanced every seven days to an earlier (day to night to swing) shift, indicated that 81 percent of the workers required two to four days after each change for their sleep habits to adapt and 26 percent were unable to adapt at all. The investigators concluded that a rotating shift that does not take circadian rhythms into account will result in sleep disturbances, which, of course, can lead to lowered job performance. To test their conclusion, these investigators divided the workers into two groups and rescheduled both so that they were rotated periodically to a later (day to swing to night) shift rather than to an earlier one. One group was rotated every seven days and the other every three weeks. The workers were surveyed nine months after the change. Previous complaints that the schedule was rotated too frequently dropped from 90 percent of the workers to 20 percent in the three-week group. In addition to showing greater satisfaction in the scheduling, the workers exhibited improved health, greater productivity, and decreased turnover. Thus, phase delayed rotation (to a later shift rather than an earlier one) and the longer period for circadian reentrainment appeared to be more satisfying to the workers and more productive for the company.[4]

Decreases in either the quantity or quality of sleep can have a significant impact on the well-being of employees, and severe decrements can affect work performance and safety awareness. Consequently, employees who exhibit such health problems as lowered resistance, decreased work performance, or evidence of poor safety awareness, should be questioned closely about their sleeping habits, the length of time they sleep, whether or not they awake refreshed, and the environment in which they sleep. Occupational health and safety professionals should pay particular attention to the sleep conditions and habits of night shift workers and rotating shift workers. If they exhibit severe sleep problems, these workers may need to be transferred to another shift.

Disturbed eating patterns. Since all humans and animals are awake when they obtain, prepare and eat their food, it is evident that the feeding cycle is generally correlated with the sleep/wake and rest/activity cycle, although less affected by the latter. For example, night shift workers appear to have a higher incidence of gastrointestinal problems than day workers.[5] The frequency of gastrointestinal problems and appetite disturbances in shift workers, particularly those on night shifts—35-75 percent higher than other workers—suggests that eating habits play an important role. Thus, not only can the feeding cycle be disturbed, but other factors may play a part, for example, the difficulty of obtaining a hot, nutritious meal very late at night and the inability of having contact with family members and friends while eating. Sleep disturbances may also play a part in these problems. When employees exhibit gastrointestinal and appetite disturbances, they should be questioned about other factors that could have an effect. The personality of the worker, family problems, social difficulties, stress on the job, smoking, and excessive coffee drinking could also be causative. Occupational health and safety personnel should be especially alert to the foregoing, since serious cases may need to be transferred to other work shifts.

Other health problems. Some medical conditions, such as diabetes mellitus and epilepsy, have been found to be rhythmic in nature and consequently can be exacerbated by shift work. In diabetes mellitus patients, observers have noted periodic changes in rhythms, particularly in the excretion of citric acid and beta-hydroxbutyric acid, as well as production of ammonia in the kidney. These rhythms can also affect the time when insulin should be administered and the effectiveness of the drug once given. Not only must diabetic workers have medication at the proper time, but they must also have food to balance insulin intake. However, the rhythm of drug effectiveness may not correlate with the time of the shift work; and food may not be readily available on off-shifts. Consequently, medical professionals should thoroughly assess each diabetic worker before permitting him or her to be assigned to a different shift, particularly a night shift.

Epileptic seizures also follow a rhythm concerning the time of day when they occur. The incidence of seizures apparently rises steadily from 3:00 A.M. to a peak at 6:00 or 7:00 A.M. which is the highest seizure period. The second highest seizure period occurs between 10:00 and 12:00 P.M. Sleep disturbance may be a factor since many epileptics tend to have seizures during their normal sleep hours. Therefore, great care should be taken before allowing epileptic individuals to be transferred to other shifts. This is required not only to protect the epileptic, but also to protect the company from charges of discrimination if an epileptic worker is not permitted to fulfill career accomplishment.

Other medical problems may also be rhythmic in nature, but additional research is required to determine the extent of these problems and the effect that changing circadian rhythms will have.[6] Consequently, occupational health and safety professionals will need to be aware of the latest findings pertaining to disease and their circadian rhythms. They will be well advised to study each new finding carefully and balance it against the physical conditions of their work crews before transferring any employees to a different shift.

Psychological aspects of shift work

Shift work may also result in psychological problems for workers, and these problems can be as difficult to cope with as physiological problems. A broad spectrum of conditions and situations, ranging from disturbed family life to lack of social contact and even to sexual dysfunction, has been reported. Shift workers have fewer problems with social contact if they socialize primarily with others who work shifts. Otherwise they can feel isolated from family, friends, and the community because of their different hours.

Shift workers complain more frequently of sex-related problems, and also disturbed relationships with friends. But some appreciate the extra leisure time during the day to spend with family or friends and to conduct other affairs. Although they are unaware of it, some of this time must be accounted for by the reduced amount of sleep that they experience. Nonetheless, many workers resent their inability to attend theater or sports events, union meetings, religious services, and other activities that are scheduled for daytime employees.

Performance and safety in shift work

Job performance and safety should be primary concerns of occupational health and safety professionals, particularly in approving assignment of individuals to rotating work shifts. Although evidence to date is sparse on whether or not performance (including safety awareness) is degraded by shift work, it appears that there may be a connection. It is well known that human performance follows circadian rhythms, particularly the temperature rhythm. If there is a connection, the same type of work undertaken by the same individual at different times of the 24-hour period would result in a different output. Evaluation of decrements in work performance by time of day is difficult because of the effects of worker personality, life style, and motivation. With regard to worker personality traits, one group of investigators concluded that performance differed between morning- and evening-oriented individuals.[7] Their findings indicated that "morning people" performed better in the early hours and less well later, while performance of "evening-oriented" individuals was poor in the morning hours, but improved as the day went on. This factor may be useful in assigning individuals to shift work, even though it has not been entirely substantiated.

Knowledge about the safety awareness as it concerns shift workers remains inconclusive. Although circadian variations in accident occurrences have been studied, the findings often have been contradictory. For example, one study indicated that there were a greater number of accidents in the mining and steel industries in the morning and afternoon, but that accidents occurring during the night shift were more serious.[8] Other investigators have concluded that accidents peak during daytime hours at 10:00 A.M. and 4:00 P.M., while still others have claimed that the peaks occur at 9:30 and 11:45 A.M. and 2:00 P.M.

Despite the contradictory nature of the findings, the circadian pattern of accidents must take into account the effects of changes in the circadian sleep/wake cycle. Workers on rotating shifts where sleep has been disturbed have reported falling asleep during their work periods. If an operator of power machines or equipment falls asleep for only seconds, the results can be devastating.

When considering employees for rotating shifts, circadian rhythms, particularly the sleep/wake rhythm, must be taken into account because they may affect both performance and safety. Moreover, scheduling systems should be evaluated to determine whether or not they are supportive of the circadian system.

The optimal work shift schedule

Companies have been using three methods to solve the rotational shifting problem.

- The first has been to not rotate workers, but to keep employees on an assigned shift continuously. This approach has an inherent disadvantage as night shifts are difficult to staff on a permanent basis. Furthermore, environmental synchronizers influence the circadian rhythms during daytime-oriented weekends so that the rhythms are in a constant state of reentrainment.

- The second approach, which is preferred by many European and Asian companies, is to rotate the workers rapidly from one shift to the next to avoid disturbing the circadian rhythms. This approach demands a great deal of effort and cooperation from workers.

- Finally, the third approach is to select carefully individuals for rotational shifting, choosing only those who show the best tolerance for working or sleeping under adverse conditions. The disadvantages here are that considerable testing is required and that, once suitable individuals are identified, the available worker population could be limited.

Criteria. On the basis of the present knowledge of the effects of changing circadian rhythms, occupational health and safety professionals can use the following criteria in evaluating the adequacy of their company's scheduling approach:

• Single night shifts are preferred over consecutive, repetitive night shifts, because the shorter exposure does not alter circadian rhythms significantly.

• Each period on a night shift should be followed by at least 24 hours of free time to offset the harmful effects of sleep deprivation. This period should also be applied to early morning workers if their sleep routines are disturbed by their shift hours.

• The length of a shift should depend on the type of work to be performed. Shifts involving light or uncomplicated tasks can be longer than those requiring heavy physical work or difficult mental efforts.

• Older workers should not be placed on longer shifts (10 or 12 hours) because of the added stress they experience.

• The length of the rotating shift cycle should not be too long—for example, four weeks is better than forty weeks on the same shift.

• Shifts should be rotated regularly so that workers can plan their family and social lives.

• When rotating shifts, it is preferable to rotate workers forward to a later shift than to an earlier one.

• Workers on permanent night shifts should be given as many free weekends as possible so that they can participate in some family and social functions.

Selection of shift workers and health monitoring

Selection of workers for shift work should be a careful process because twenty percent of workers cannot tolerate this type of work. Unfortunately, there are few selection criteria since many health and performance effects of shift work remain to be identified.

Poor risks. However, research to date clearly indicates that certain types of individuals are poor risks for shift work. The following types of workers are not suitable for shift work:

• New employees under 25 years of age, for psychosocial reasons, particularly, if they are living alone.

• Employees over 50 years of age, unless they have previously adapted to shift work, in which case they should be permitted to undertake such work voluntarily.

• Workers with a history of gastrointestinal disorders because the gastrointestinal system may be disturbed by psychosocial problems, disruption of meal times, and the inability to obtain hot, nutritious meals during shift hours.

- Diabetics and individuals with thyrotoxicosis because of the need for regular food intake in the first case, and the need for proper timing of drug therapy in both cases. These can be difficult to manage during shift hours.

- Epileptic workers because sleep deprivation and desynchronization of circadian rhythms can result in seizures. Preferably, these workers should not be assigned to shift work at all, but if they are, they should be medically evaluated very carefully.

- Workers with sleep disturbance problems or those who have shown symptoms of emotional instability because circadian desynchronsis can exacerbate these conditions.

- Employees who live in noisy environments because noise aggravates sleep disturbances originating from shift work.

Occupational health and safety personnel will need to give more attention to shift workers. Companies, for the sake of the health and performance of their employees, should insist on periodic evaluation and examination of these workers. It is suggested that all potential shift workers be given medical exams before being assigned to shift work and, once assigned to a shift, should be examined six months thereafter, and then regularly every two years. Those over 50 years of age, however, should be examined more frequently.

ALTERNATIVE WORK SCHEDULING

As indicated earlier, alternative work scheduling has arisen because of growing consideration for the worker's well-being and job satisfaction and because of increasing transportation difficulties in heavily commercialized and industrialized areas. Under this system of work scheduling, there are two alternatives: flex time in which employees work the required five days a week but choose their own working hours—(e.g., 6:00 A.M. to 3:00 P.M.), and the compressed work week in which employees work their required 40 hours in fewer days. Workers like these arrangements because they allow for more leisure time.

Flex time. Studies so far have indicated that flex time does not involve any deleterious health effects. However, this method of scheduling has not been in practice very long, and few studies have been conducted on its health, performance, and safety effects. Consequently, until more information is available, it must be considered that flex time systems do not create circadian problems.

Compressed work week may be a different story. Although much study has been addressed to worker satisfaction with working fewer days a week for longer hours a day, little research has been accomplished on health and performance effects. Most compressed work weeks are 4/40, which means that workers work four 10-hour days each week. One study of this type of scheduling indicated that while workers exhibited greater fatigue, productivity was not decreased.[9] However, considerable study will be needed before it can be determined whether or not the compressed work week affects health, performance, and safety awareness. Despite the limitations of knowledge in this area, occupational health and safety professionals would be well advised to observe employees working on compressed work weeks more closely. This approach may not only alert management to any potential difficulties in this type of scheduling, but also could provide much needed on-the-job data pertaining to this type of scheduling.

CONCLUSIONS

To date, study findings on shift work—and other alternative scheduling—have been relatively inconclusive in terms of health, performance, and safety awareness.

Disruption of the sleep/wake rhythms appears to be the most serious problem, and the consequences of this desynchronization are only beginning to be clearly understood. It has only recently been determined, for example, that the normal quantity and quality of sleep result from a proper phase relationship between circadian pacemakers. When this relationship is disrupted by shift scheduling, sleep disturbances will follow. Moreover, the failures in homeostatic (stable-state) regulatory mechanisms that have been observed in research are also associated with a desynchronized circadian system. Shifts of the light/dark phase have been seen to result in a 5-20 percent decrease in longevity in insects and mammals. Although the effects of such shifts in humans are unknown, available data indicate that rotational shift workers are more prone to sleep and digestive disorders than others and that some individuals cannot tolerate shift work for any length of time.

Occupational health and safety personnel face a large task in monitoring and evaluating the work scheduling of their companies. Since much remains to be learned about the circadian effects of shift work, flex time, and compressed work week, they will need to keep abreast of the latest findings reported in the literature concerning all approaches to work scheduling.

REFERENCES

1. Moore-Ede, M. C., F. M. Sulzman, and C. A. Fuller. *The Clocks That Time Us*. Cambridge, Mass.: Harvard University Press, 1982.
2. Colquhoun, W. P., and J. Rutenfranz, eds. *Studies of Shiftwork*. London: Taylor & Francis Ltd., 1980.
3. Rutenfranz, J., *et al.* "Re-entrainment of Circadian Rhythms After Phase-Shifts of the Zeitgeber." *Chronobiologia* 2:23-78, 1975.
4. Czeisler, C.A., *et al.* "Rotating Shift Work Schedules That Disrupt Sleep Are Improved by Applying Circadian Principles." *Science* 217:460-463, 1982.
5. LaDou, J. "Health Effects of Shift Work." *West. J. Med.* 137(6):525-530, 1982.
6. Moore-Ede, M.C., C.A. Czeisler, and G.S. Richardson. "Circadian Timekeeping in Health and Disease." *New Engl. J. Med.* 309(8):469-476, 530-536, 1983.
7. Horne, J. A., *et al.* "Circadian Performance Differences Between Morning and Evening 'Types.' " *Ergonomics* 23(1):29-36, 1980.
8. Craig, A., *et al.* "Diurnal Variation in Vigilance Efficiency." *Ergonomics* 24(8):641-651, 1981.
9. Volle, M., *et al.* "Compressed Work Week: Psychophysiological and Physiological Repercussions." *Ergonomics* 22(9):1001-1010, 1979.

Chapter 14

Plant Inspection And Worker Protection

by Arthur R. Reich

PART I—PLANT INSPECTION

The success of a safety and health program depends on identifying hazards and taking remedial action; thus, periodic inspections of the plant are a necessity. Some state OSHA programs have come to realize the importance of safety and health inspections by specifying them in their regulations. An example of this type of regulation was recently put into California law for enforcement by state OSHA inspectors. The legal section is reproduced below in its entirety.

TITLE 8—GENERAL INDUSTRY SAFETY ORDERS 3204
(Register 81, No. 12—3—2—81) (p. 342.1)

3203. **Accident Prevention Program.**

(a) Effective October 1, 1977, every employer shall inaugurate and maintain an accident prevention program which shall include, but not be limited to the following:

(1) A training program designed to instruct employees in general safe work practices and specific instructions with respect to hazards unique to the employee's job assignment.

(2) Scheduled periodic inspections to identify and correct unsafe conditions and work practices which may be found.

This and similar regulations have begun to address behavioral aspects of health and safety. Inspection and control procedures have as their objectives achieving a safe work environment and controlling the unsafe behavior of people. It is obvious that even the best designed industrial installation can become hazardous if maintenance of facilities and equipment is inadequate. Management and staff attitudes and competencies are also factors. (See Chapter 4, "Safety and Health Management.")

All experienced safety and health professionals have seen design problems, defeated safety devices, and inappropriate behavior that resulted in injury, illness, and death. Industrial hygienists will tell the story of the poorly designed or improperly maintained local exhaust ventilation system. Safety engineers will tell stories of the improperly grounded electrical equipment or inappropriate personal protective equipment. Health physicists tell tales of design and behavioral problems in nuclear equipment and facilities. An effective plant inspection program, it is hoped, will identify most of these hazards before accidents and personal injuries occur.

A safety inspection is also one of the best methods by which management can demonstrate its sincerity regarding accident prevention. Other benefits that are shared by both management and employees can be increased work efficiency and improved public relations. Employees prefer working in a safe plant, and the reputation and community confidence in a company are affected by reports of hazardous episodes and accidents. Inspections can also be extremely valuable in highlighting unsafe areas, operations, or procedures that should be emphasized in management and staff training.

PLANT INSPECTION PROGRAM DEVELOPMENT

Many elements make up an effective plant safety inspection program. In general, the following questions must be asked by management and staff when developing a plant safety inspection program:

- What is the purpose and goal of the inspection program?
- What is to be inspected, and what are the inspection standards?
- Who will do the inspections?
- How often will the inspections be performed?
- What reports and records will be kept?
- How will inspection reports and records be communicated?
- How will corrective action take place?

Safety inspections should not be limited to finding unsafe physical conditions, but should also try to detect unsafe practices. Inspections of manufacturing processes are also important functions of quality control and preventive maintenance. Inspections should be conducted to determine whether an establishment or plant is operating satisfactorily, not to

OCCUPATIONAL HEALTH AND ENVIRONMENTAL CONTROL (29 CFR 1910.1000,.94)

Is management aware of the hazards caused by various chemicals used in the establishment? ☐ ☐

Is employee exposure to these chemicals kept within the acceptable levels? ☐ ☐

Are eye wash fountains and safety showers provided in areas where chemicals, such as caustics, are handled? ☐ ☐

Are all containers such as vats and storage tanks labeled as to their contents? ☐ ☐

Are employees required to wear personal protective equipment (gloves, eye protection, or respirators) when handling hazardous materials? ☐ ☐

If internal combustion engines are used, is carbon monoxide kept within acceptable limits? ☐ ☐

OCCUPATIONAL NOISE EXPOSURE (29 CFR 1910.95)

If a noise problem exists, have plans to reduce noise levels by engineering methods been formulated (e.g., enclosure, maintenance, or different methods of processing)? ☐ ☐

If engineering controls cannot reduce the noise to safe levels, have administrative controls, such as limiting worker-exposure in a given area, been implemented? ☐ ☐

Do all employees in high-noise areas wear hearing protection? ☐ ☐

Are annual noise surveys made to re-evaluate problem areas? ☐ ☐

HAZARDOUS MATERIALS

FLAMMABLE AND COMBUSTIBLE LIQUIDS (29 CFR 1910.106)

Are all connections on drums and combustible liquid piping vapor and liquid tight? ☐ ☐

Are flammable liquids kept in closed containers when not in use (e.g., parts cleaning tanks or pans)? ☐ ☐

Are all spills of flammable or combustible liquids cleaned up promptly? ☐ ☐

Are bulk drums of flammable liquids grounded and bonded to containers during dispensing? ☐ ☐

Are gasoline and flammable liquids stored in approved containers? ☐ ☐

Are LP-gas storage tanks guarded to prevent damage from vehicles? ☐ ☐

Are "NO SMOKING" signs posted on LP-gas tanks? ☐ ☐

Are storage cabinets for flammable and combustible liquids labeled "FLAMMABLE — KEEP FIRE AWAY"? ☐ ☐

Are all paints, lacquers, or thinners, kept for more than 30 days stored in approved metal or wooden cabinets or in storage rooms? ☐ ☐

PERSONAL PROTECTIVE EQUIPMENT (29 CFR 1910.132-.137)

Is personal protective equipment provided, used, and maintained wherever it is necessary? ☐ ☐

Is employee-owned personal protective equipment, such as gloves and protective shoes, adequate and properly maintained? ☐ ☐

Figure 14-1. Portion of a standard health and safety inspection form.

find out how many things are wrong.

An effective inspection program can only be developed after a thorough analysis of plant operations and work practices. A strong case can be made for inspecting for damage control as well as injury-accident prevention. Heinrich[1] observed this relationship more than 50 years ago and concluded that there were about 300 noninjury accidents to every one that resulted in a disabling injury. Considerable property damage was also observed in his discussion of the relationship. Location of failures that have an effect on the plant operation in terms of safety, reliability, maintainability, and the like is obviously desirable.

After agreeing on the purpose or goals of the inspection program, then decide what is to be inspected and what inspection standards are to be used. A company that develops an inspection program tailored to its own needs will have better staff involvement and support than with a ready-made inspection approach. Published inspection forms (a portion of one is shown in Figure 14-1) have a certain value in giving structure to the inspection approach and documentation, but it should always be understood that they are usually developed for particular industries, hazards, or equipment. Publications of the National Institute of Occupational Safety and Health (NIOSH) and many other published ref-

PERSONAL PROTECTIVE EQUIPMENT AND CLOTHING

Personal protective equipment for eyes, face, head, and extremities; protective clothing; respiratory devices; protective shields and barriers; and fall protection equipment shall be used wherever it is required by reason of hazardous processes, environment, chemicals, or physical agents encountered in a manner capable of causing injury or illness through absorption, inhalation, or physical contact.

All personal protective equipment must be properly designed and sufficiently well constructed to provide the protection for which it is intended. It must be properly fitted and maintained in a sanitary and reliable condition.

Read all of the questions in this section; discuss the questions with those persons at your establishment who know the most about the areas covered; and, while conducting a survey, complete this section with their help.

Eye Protection

1. Is wearing approved safety glasses required at all times in the following hazardous locations:

    a. Process areas? Yes ( ) No ( )
    b. Maintenance shops? Yes ( ) No ( )
    c. Shipping areas? Yes ( ) No ( )
    d. Plant roadways and walkways? Yes ( ) No ( )
    e. Mobile equipment operating areas? Yes ( ) No ( )
    f. Mine pits? Yes ( ) No ( )
    g. Laboratories? Yes ( ) No ( )
    h. All work activities or areas which are potentially harmful to the eyes? Yes ( ) No ( )

2. Are employees who use corrective lenses required to use approved industrial safety glasses or to wear protective goggles over their glasses? Yes ( ) No ( )

3. Is the wearing of metal frame glasses prohibited while performing electrical work? Yes ( ) No ( )

4. When opening pressurized valves or when exposed to sources of sulfuric acid, phosphoric acid, ammonia, or steam, are a full-face shield or tight fitting goggles required to be worn? Yes ( ) No ( )

Figure 14-2. One page of a five-page personal protective equipment and clothing inspection form.

From NIOSH Publication 80-129, *Self-Evaluation Instrument—Phosphate Fertilizer Industry.*

erences are available for guidance. (See Additional Readings at the end of this chapter.)

It is obviously important to meet legal code requirements for health and safety, but one should not lose sight of the overall purpose of the inspection—to identify and prevent potential accidents. Concentrating on physical conditions or specification standards brings you only part of the distance toward accident prevention. Enforcing behavioral standards is just as important in achieving accident prevention. Personal protective equipment in industry is a case in point. This type of internal safety program is only successful to the level that employees are properly trained and accept the program. Figure 14-2 presents the kind of standards used when making a personal protective equipment inspection to evaluate the adequacy of employee training, safety management decisions, and employee acceptance. By far the largest majority of standards used in this type of inspection addresses behavioral control.

In developing company inspection standards, it is important to evaluate accident records first. This is where management would like to see progress. A particular type of hazardous work area may require a specialized inspection or a particular type of equipment may require specialized inspections to remain functional and to prevent injury. Examples of forms used for these kinds of inspections are given in Figures 14-3 through 14-5. These examples of inspection reports can help guide management in the development of its internal inspection program.

There are two functional classes of inspections: those made by organizations external to the company and those made by company staff, such as the health and safety professionals, supervisors, and paid consultants.

Regulatory agencies in recent years have developed inspection standards and, in some instances, consulting units will help in making standards decisions. Take advantage of this source of free consultation. Inappropriate measures instituted to "meet code" may become costly; examples of this type are usually related to determining when it is best to use personal protective equipment and when it is best to use engineering and administrative controls.

In addition to regulatory agency consultants, many insurance companies provide health and safety consultations, educational materials, and plant inspections. The insurance company representative can be a valuable resource because the company may be insuring other companies whose processes are similar to yours. There are also private consultants who can help you.

The question almost certainly will arise as to who does the inspecting. There are several possibilities: the health and safety department representatives, the health and safety committee, a specialist in the

LABORATORY SAFETY CHECKLIST

DO YOUR WALKING AND WORKING SURFACES MEET THESE REQUIREMENTS?	OSHA Section	Yes	No	Action Taken
1. Working areas, passageways, storerooms, service rooms in clean, orderly, and sanitary condition.	1910.22a1			
2. Workroom floors clean and dry. Drainage maintained and dry standing places provided.	1910.22a2			
3. Floors free from protruding nails, splinters, holes and loose boards.	1910.22a3			
4. Permanent aisles and passageways marked.	1910.22b2			
DO YOUR EXITS MEET THESE REQUIREMENTS?				
1. Sufficient exits for leaving in case of fire or other emergency, with alternate escape means provided.	1910.36b1			
2. Exits unobstructed and unlocked.	1910.36b4, 1910.36d1			
3. Exits marked and readily visible. Non-exits clearly marked.	1910.36b5, 1910.36b6			
4. Fire alarms provided where necesssary.	1910.36b7			
5. More than one exit provided, sufficiently far apart.	1910.36b8			
6. Fire detection equipment, smoke alarms, sprinkler	1910.36d2			

Figure 14-3. Portion of the first page of a seven-page chemistry laboratory safety checklist. Note how pertinent OSHA requirement is referenced. Checklist continues with portions covering storing and handling of hazardous materials, personal protective equipment, signs, fire and electrical protection.

Lab Safety Supply Co., Janesville, Wis.

operation affected, the line supervisor, the plant maintenance group, or other knowledgeable persons. Part of the answer is based on the fact that, in most cases, there should be more than one type of inspection.

Depending on the size of the plant, type, or circumstances involved in each case, safety inspections by several persons might be necessary. Recently, a great deal of the legal thrust of OSHA has been directed at the employee's immediate supervisor. It is felt that greater day-to-day accident prevention and control can be exerted at the supervisory level.

Safety inspections conducted by the supervisor can be formal or informal. A supervisor should check daily to see that facilities and equipment are in proper condition prior to operations. The supervisors can also observe employees to determine whether or not they are conducting themselves properly. Written reports may or may not be necessary.

Almost always, more than one person should make safety and health inspections, and no one individual or group should have exclusive responsibility. A strong argument for the multiple inspection approach is that one person can become so familiar with an operation or area that the ability to discern even obvious hazards can become blurred. In some instances, this inability may, in reality, be the inspector's own sense of priorities (budgetary and professional).

In certain circumstances, the inspector should be an expert. For example, an industrial hygiene or a fire inspection requires a great deal of specialized training, experience, and ability. Plant inspections by experts usually involve detailed, in-depth surveys or measurements. It should also be recognized that specialized inspections may require the use of measuring equipment that requires some degree of experience. It is often useful to perform specialized expert inspections during the design and initial operating phases of a newly built unit or plant. A good example of a formal inspection of the fire protection organization of an industrial plant is given in the National Fire Protection Association booklet, "Conducting Fire Inspections—A Guidebook for Field Use."[13]

Some companies allow inspections by the safety and health committee. By seeing actual conditions and work methods, members of the committee can conduct more knowlegeable discussions in their meetings. In some companies, when inspections are made by a committee composed of workers or supervisors, the members inspect each other's areas of operation. An inspection of this type may gain greater support because employees are participating actively in the company's accident prevention program.

The important point is that inspections should be coordinated and directed by a responsible manager who can and will provide the supervision necessary to ensure effectiveness. Inspections should not become routine functions. When an organization has safety and health professionals, this responsibility is usually assigned to them. Literally, everything in the establishment should be inspected at some time by someone. The competent safety/health professional surveys the entire establishment in detail, finds out just what is covered by inspection and to what

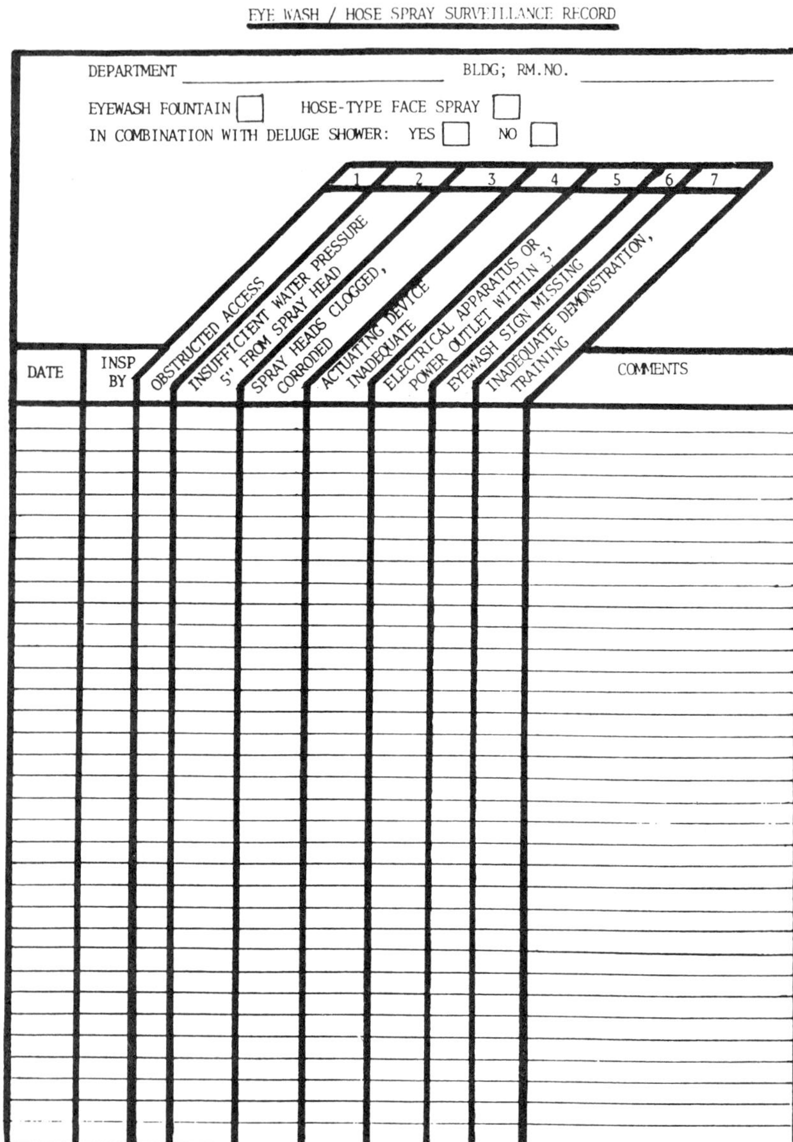

Figure 14-4. Emergency equipment surveillance record to be used with procedure shown in Figure 14-5.

University of California, San Francisco.

extent, and reaches a decision as to what should be included in a systematic inspection program.

EYEWASH/HOSE SPRAY SURVEILLANCE PROCEDURE

Surveillance Responsibility

The responsibility for inspection and testing of eyewash fountains and hose-type face spray devices lies with the Department or Institute in charge of the space in which they are located.

Surveillance Frequency

Eyewash fountains and hose-type face spray devices should be inspected and tested at least monthly. The Eyewash/Hose Spray Surveillance Record should be used to keep a written record of the results. Inspection tags on the devices are encouraged, but the written record must be maintained by Departments.

Inspection Criteria

(1) Obstructed Access: The device must be easily accessible to the user. Mark deficient if access to the device is obstructed by plumbing, chemicals, glassware, overhanging cabinets, etc.

(2) Insufficient Water Pressure: Water pressure must be sufficient to develop a stream from the delivery head of the device. The water pressure should not be too great because it can drive the chemical deep into the eye.

(3) Spray Heads Clogged, Corroded: The perforated holes in the delivery heads of the device should be checked for clogging due to rust or corrosion.

(4) Actuating Device Inadequate: The device must have a stay-open valve. The device, once activated, should remain on until deliberately deactivated. The user should be able to have both hands free after actuating.

(5) Electrical Apparatus/Power Outlet: No sources of electricity should be located within 3 feet on either side of the device or under it to prevent electric shock.

(6) Eyewash Sign: Devices should have highly visible signs next to them.

(7) Inadequate Demonstration, Training: Employees should know how the device operates. They should know about the necessity of flushing exposed areas for at least 15 minutes.

Figure 14-5 Surveillance procedure explains how to use the record form shown in the previous figure.

University of California, San Francisco.

Although there is no standard sequence for the most productive safety inspection, it is best to follow a process or operation from the beginning, step by step, through to the finished product. The safety/health professional should acquire and digest all pertinent information available on the operations and equipment to be inspected. The inspection program should be worked out in sufficient detail to make sure that everything of any importance is properly covered. Coordinating safety

S = Satisfactory
U = Unsatisfactory

INSPECTION CHECK LIST
FIRE HAZARDS &
FACILITIES AND EQUIPMENT

DATE: ______

DEPT. ______ SHOP ______ W.C. ______

SHIFT ______ INSPECTOR(S) ______

FIRE HAZARDS FACILITIES & EQUIPMENT	S	U	COMMENTS
1. FLAMMABLE LIQUIDS - Containers a) Contents Identified			
b) Marked Class I, II or III Label (pint or less)			
- Dispensing a) From Safety Cans (Class I & II liquids in excess of 1 pt.)			
b) Drum Pumps used for Class I liquids			
c) Drums of Class I liquids properly grounded			
- Storage a) In Cabinets when not in use			
b) Cabinets free of ordinary combustibles			
c) No smoking within 10 feet			
2. WASTE DISPOSAL - Proper disposition of oily rags			
- No overflowing trash containers			

Figure 14-6. Portion of first page of a four-page facilities and equipment checklist.
United Airlines.

inspection criteria with production inspection criteria is the most economical method in the long run. (See Figure 14-6). Published material on typical hazards of the industry or equipment, as well as a consideration of the plant accident experience, should be integrated into the inspection procedure.

The timing (frequency) of inspections is another critical point in developing an effective inspection program. There are a number of factors to consider. One obvious consideration is the budget. Safety and health items should be submitted to management as consistently as any other budget item. Another factor is the nature of the potential hazard. Frequency of inspection may be weekly, monthly, or annually, according to accident and illness experience and previous inspection results.

Furthermore, coordination between production, inspection and maintenance schedules can be important because many times corrections can be pointed out to those in charge of these activities, resulting in immediate solutions. Inspection according to a proper sequence is easier, saves time, is more efficient, and identifies needs more readily. Scheduling inspections should be well thought out in advance if a company expects to avoid meaningless activity, often referred to as "paper program syndrome."

Recordkeeping and review

In determining what and how many inspection reports and records should be kept, it is necessary to consider legal requirements and liability considerations. Some safety and health regulations require recordkeeping as a form of accountability. These recordkeeping requirements include such documents as fire extinguishing-equipment inspection records, local exhaust system testing records, water quality testing records, and hazardous waste removal records. Some recordkeeping is not obligatory, but is considered a state-of-the-art requirement in a particular industry. For example, deluge shower and eye wash equipment testing records are considered necessary in the chemical industry.

The advent of computers triggered industry recordkeeping, internal statistical studies, and data management to become automated. (See Chapter 11, "Information Management Systems.") Recordkeeping requirements have gained impetus as the link between medical, environmental, employee, and safety data base record systems has become stronger. Microcomputers can provide a low-cost means of eliminating many of the burdens associated with storing and retrieving safety records while facilitating the preparation of accurate statistical reports. Recent publications by Garber[2] point to this exciting development.

Every inspection program must define how inspection results will be handled, filed, classified, and followed up. Rarely does a plant or area inspection produce a complete result in and of itself; more often, it establishes what should be done. In most cases, further action, assistance, reinspection, or correction of some sort is necessary. The results of the inspection should be communicated to the supervisor of the area, if not at the time, certainly as soon thereafter as possible. Since supervisors are the key persons in safety and health, it is important to communicate the results and, if possible, recommendations. For example, if a machine guard appears to be necessary, it is important to know why one is not already in place. The mere recording of the hazard represents communication. It may be that the supervisor recognizes the hazard, but needs assistance in process change and design as well as budget support. Whoever inspects should make every effort to communicate at the supervisory level as well as at the management level.

The plant safety and health committee should review inspection results. Committees generally are charged with developing educational programs and safety procedures. They establish priorities for safety and health needs and correction, and communicate them to management. Sometimes corrections require considerable expenditures. This situation can usually be dealt with by implementing interim programs, such as the use of personal protective equipment. Corrections approved by management should be assigned to the appropriate department for action through regular channels. In some cases, it may be a function of maintenance or engineering, a training function, or a combination.

Clearly, communication breakdowns result in safety attitude breakdowns followed by employee morale problems, government regulatory inspections, and sometimes, ultimately what we are trying to avoid—accidents and illnesses.

PART II—WORKER PROTECTION

The remainder of this chapter is concerned with four areas of worker protection that often receive insufficient attention in many worker protection programs because the subjects receive little emphasis in occupational health and safety degree curricula or, for some reason, are considered too technical. They are: *Electrical safety, cross-connection control, ventilation,* and *emergency preparedness.* A basic understanding of these principles should improve hazard-recognition abilities during plant inspections.

ELECTRICAL SAFETY

OSHA cites electrical violations more frequently than any other violation. The first set of OSHA's electrical safety standards that became effective April 16, 1981, are called "Design Safety Standards for Electrical Systems" (Subpart S, Part 1910 of Title 29, CFR). They cover specific, mandatory rules on "all electrical equipment and installations used to provide electric power and light for employee work places." As with OSHA's previous electrical standard, it was derived from *National Electrical Code*®, primarily a design standard. The new standard applies to any installations made since 1981. A retroactive clause is presented in the standard regarding requirements of electrical installations made prior to that date. The new electrical standard was carefully developed to cover only those parts of any electrical system that an employee would use or come in contact with. The requirements are set down more in terms of performance than in terms of specification.

The California OSHA program reports that on the average 28 workers are killed and 559 workers disabled annually in that state alone because of occupational accidents involving electrical shock.[4] Many states match or exceed this unenviable statistic. It can be assumed that a great many electrical safety problems result from poor hazard recognition by employees, design people, and inspectors. In a study performed by the California Division of Industrial Safety in 1972, of 1,090 electrical injuries,[5] it was possible to determine what the workers were doing at the time of the accident, and unsafe acts contributed to seven out of ten of the injuries.

The most frequently cited human errors are:

- Failure to de-energize equipment during repairs or inspection.
- Assuming an unsafe position near energized equipment.
- Use of defective or unsafe tools or equipment.
- Unsafe acts by others.

Electrical terms and units

Everyone should have a basic understanding of how electricity works. It is frequently easier to explain how electricity behaves rather than to define what it is. Electricity can be described in terms of pressure (volts), volume (coulombs), flow (amperes), and resistance (ohms). Materials are generally referred to as conductors or insulators. Conductors refer to materials that allow electricity to pass readily through them. Insulators refer to a category of materials that do not allow electricity to pass readily through them. Insulators are used to help prevent humans from accidentally contacting the electrical current.

There are two types of power-generated electricity: alternating current (AC) and direct current (DC). These terms refer to the way electricity moves in a wire or circuit. Alternating current changes direction at some rate and therefore has a frequency that is measured in cycles per second or hertz. Most electrical current used to power tools and small equipment is 60 Hz (hertz).

Electrical shock

Electrical shock can only occur when an electrical source, external to

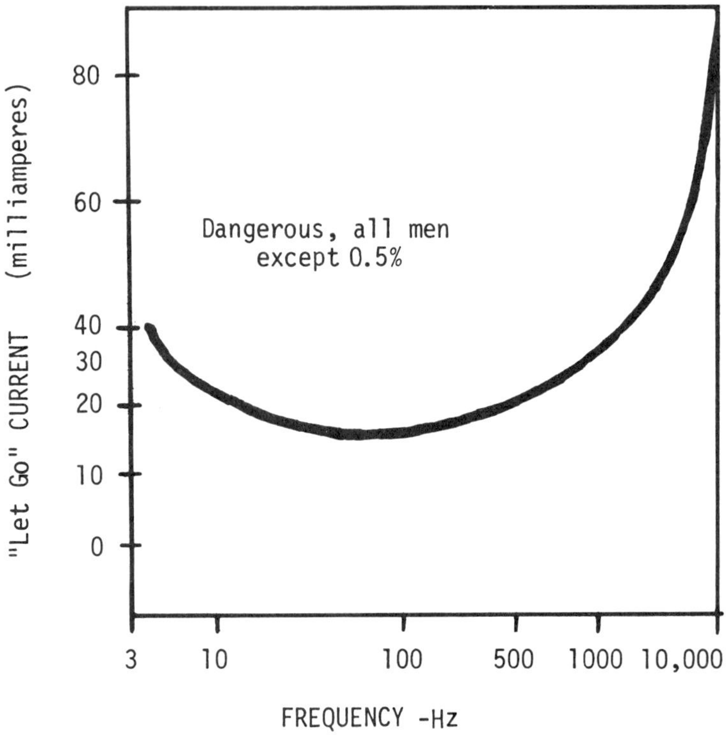

Figure 14-7. "Let go" current versus frequency.

Adapted from "Shock Hazards of Electrical Currents," *J. Eng. Inst.*, Canada, Nov. 1959.

the body, can use the human body as part of its circuit. Three specific conditions must exist simultaneously for an electrical shock to occur: (*a*) two conductive surfaces must be available to contact the body; (*b*) some portion of the body must be in contact with the conductive surfaces; and (*c*) electrical currents must be available to flow from one conductive surface to the other.

Between 1940 and 1950, experiments were performed on engineering students and other volunteers to study the sensitivity of the body to electrical currents of different frequencies. Curves were generated by having the test subject hold a metal rod in each hand (see Figure 14-7). The rods had been connected through wires to an electric generator. The experimenter would select a specific frequency and then turn up the current intensity until the test subject could not let go of the metal rods. As the curve indicates, the test subjects could not let go at relatively small currents (15-20 milliamperes) over the frequency range, of about 10-100 Hz. Greater current intensities, below and above this frequency range, were required to prevent the test subjects from "letting go."

A point to be remembered is that the human body is most sensitive at 60 Hz, which just happens to be the frequency that operates most electrical equipment (Table 14-A).

TABLE 14-A
PHYSIOLOGICAL RESPONSE OF THE ADULT BODY TO 60 Hz CURRENTS

Milliamperes	Response
0.70	"Tingle" level, average adult female.
1.0	"Tingle" level, average adult male.
1-8	Shock, not painful, no loss of muscular control.
8-15	Shock, painful, no loss of control of arm muscles.
15-20	Shock, painful, loss of control of arm muscles.
20-50	Shock, painful, severe chest muscle contraction, difficult breathing.
100-300	Ventricular fibrillation.
Over 300	Burns, severe muscle contractions, no ventricular fibrillation.
	Comparative Currents:
830 (AC)	Current through a 100 watt, 115 VAC incandescent lamp.
20,000 (AC)	Trips a typical circuit breaker in the home.
60,000 (DC)	Defibrillator output, 400 W-seconds.

Protective devices

Both fuses and circuit breakers are overcurrent devices used to prevent damage to wiring and equipment and to prevent fires caused by an excessive flow of current. The current rating required to activate these devices is so high (15 to 30 amperes) that they cannot protect humans against shock. When the current flow exceeds the rated levels, the fuse melts or the circuit breaker is tripped. In either case, the circuit is broken and current can no longer flow.

Grounding. Systems are grounded to protect the wiring and equipment against excessive flow of current. The equipment is grounded to protect the operator against shock.

There are two types of grounding:

• System grounding in which one of the current-carrying wires is connected to ground.

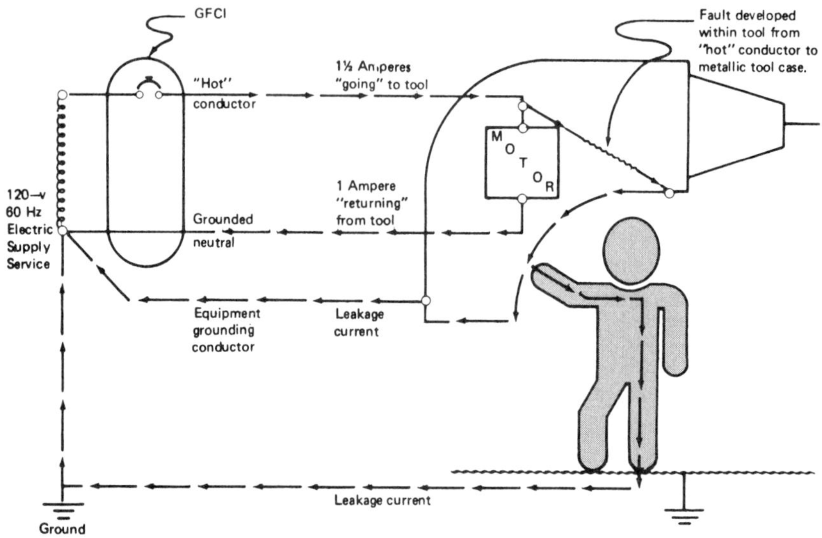

Figure 14-8. Ground-fault circuit interrupter monitors the difference in current flowing into the "hot" and out of the grounded neutral conductors. The difference (1/2 ampere in this case) will flow back through any available path, such as the equipment grounding conductor, and through a person holding the tool if that person is in contact with a grounded object.

State of California, Department of Industrial Relations, Division of Occupational Safety and Health.

• Equipment grounding in which the metal frame of the equipment is grounded.

If the ungrounded frame of a hand tool becomes energized, the quickest and easiest path of the current to take to ground may be through the worker. Depending on the conditions, the effects of the shock will range from slight to fatal. If the frame of the tool is grounded, the connection to ground provides an easier pathway for the current to follow. This does not completely protect the worker; a certain amount of current will flow through the worker. But the chances of serious injury or death are reduced because most of the current will follow the grounding wires.

The ground-fault circuit interrupter (GFCI) is a fast-acting circuit breaker that senses small imbalances in the circuit caused by current leakage to ground and, in a fraction of a second, shuts off the electricity. The GFCI continually matches the amount of current going to an electrical device against the amount of current returning from the device along the electrical path. Whenever the amount "going" differs from the amount "returning" by approximately 5 milliamps, the GFCI interrupts the electric power within as little as 1/40 of a second (see Figure 14-8).

However the GFCI will not protect the employee from line-to-line contact hazards (such as a person holding two "hot" wires or a hot and a neutral wire in each hand). It does provide protection against the most common form of electrical shock hazard—the ground fault. It also provides protection against fires, overheating, and destruction of insulation on wiring.

Double insulation. Two factors should be investigated when buying a double insulated or three-wire grounded tool. First, some insulated electrical tools have a single layer of functional insulation and generally have metal cases. Double insulation can be provided by encasing the tool in a nonconductive material that is also shatterproof. In such a tool, the switch is also nonconductive so that no metal part comes in contact with the operator. This protects the user from electrical insulation failure.

Second, large tools may need the rigid design provided by a metal case. When such tools are double insulated, an internal layer of insulation isolates all electrical components from the metal housing. The insulation around the electrical components is in addition to the functional insulation found in such conventional tools as electric drills.

Ideally, double insulated tools are designed with a nonconductive handle, an insulated armature shaft, and a completely insulated motor. The extra insulation should be physically separated from the functional insulation so that the deteriorating effects of temperature, contaminants, and wear will not affect both sets of insulation at the same time. A double insulated (or all-insulated) tool does not need a third (grounding) wire.

Electrical system testing

Companies should periodically inspect for electrical hazards. The nature of the electrical system and the types of equipment will determine the frequency and types of measurements required. This type of inspection usually requires a tagging program (see Figure 14-9).

To prevent electrical shocks, periodic inspections and testing must make certain that the ground is intact, along its entire path, from the electrical equipment to the earth (ground). Ground tests on portable equipment may be performed by using a simple ohmmeter and measuring the ground resistance between the metal equipment case and the power cord plug pin. The grounds must also be intact at the electrical receptacles. The receptacle grounds and wiring can be simultaneously tested using a testing device usually called a polarity checker.

Visual inspections of wall receptacles and equipment are desirable before testing for ground. The receptacle should be scrutinized for looseness in the wall and any obvious signs of damage. Equipment should be

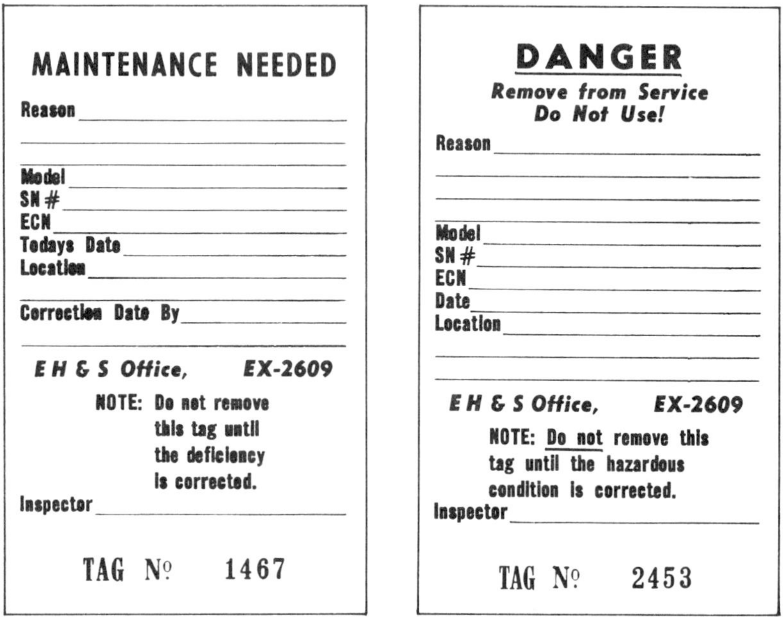

MAINTENANCE NEEDED

Reason ______

Model ______
SN # ______
ECN ______
Todays Date ______
Location ______

Correction Date By ______

E H & S Office, EX-2609

NOTE: Do not remove this tag until the deficiency is corrected.

Inspector ______

TAG No 1467

DANGER

Remove from Service
Do Not Use!

Reason ______

Model ______
SN # ______
ECN ______
Date ______
Location ______

E H & S Office, EX-2609

NOTE: Do not remove this tag until the hazardous condition is corrected.

Inspector ______

TAG No 2453

Figure 14-9. Two self-sticking safety tags for use by an inspector on electrical tools and equipment. Use of tags should ensure timely maintenance.

University of California, San Francisco

inspected to make sure that a ground pin is intact in the plug, that strain reliefs at each end of the cord are effective, and whether the body of the equipment shows any sign of damage, such as rattles, dents, or broken parts, or signs of fluids spilled into the equipment.

By making sure that electrical grounding is intact, a high degree of protection against electric shock, fire, and destruction of equipment will be achieved.

CROSS-CONNECTIONS OF WATER SYSTEMS

A cross-connection is any link or channel between a drinking water system and a source of contaminated water, or water of questionable or unsafe quality that enables the contaminating substance to enter the drinking water system.

The problem of cross-connection is a serious one in industrialized societies. As more and larger industrial complexes are being built, reports of contaminated drinking water and the spread of waterborne illnesses have increased. Outbreaks caused by cross-connections are probably more prevalent than disease statistics now indicate.

Between 1946 and 1970, cross-connections caused more waterborne disease outbreaks in the United States than any other reported factor.[6] From 1971 to 1974, they caused 32 percent of all confirmed waterborne disease outbreaks involving municipal water systems.[7] In 1975 and 1976, they were responsible for 47 percent of the disease outbreaks in municipal water systems.[8]

Causes

Cross-connections and the backflow of contaminated substances into the drinking water systems can occur in either of two ways—by back-siphonage or back pressure.

Back-siphonage occurs when pressure is reduced in the drinking water supply line. This reduction causes a negative or subatmospheric pressure to exist at that site in the water system which allows substances to siphon into the potable water supply. The principal causes of back-siphonage in cross-connections are:

- Undersized water supply piping
- Water line repairs or breaks at a location lower than a water service point
- High water usage, such as when water is used in fighting fires or when a water main is being flushed, reducing the water system pressure
- Reduction of water system pressure if the supply line is on the suction side of an on-line booster pump when the pump is activated.

An example of a serious back-siphonage condition would be a garden hose that is connected to a building sillcock when a liquid pesticide is being applied from a hose spray. If the water is siphoned back into the building system, a lethal contamination could occur.

Back pressure. Backflow caused by back pressure occurs when the pressure of the contaminated substance exceeds the pressure of the potable water supply, causing the contaminated substance to flow into potable water. This occurs when pressurized feed systems, such as boilers, create pressure in the potable water systems. Pumps are often indirectly connected to potable water systems, because the water is frequently used as a seal-flushing agent. If the pump is handling toxic fluid, the pressure of the pump may force the fluid back into the potable water pipes.

Almost every water-using facility has actual or potential cross-connections. Therefore, a cross-connection control program should be established and have as its objectives:

- *Protecting* the public water supply by isolating within the facility contaminants or pollution that may occur because of some undiscovered or unauthorized cross-connection on the premises.

- *Safeguarding* the drinking water supply within the facility where plumbing defects or cross-connections may endanger it.
- *Eliminating* cross-connections between drinking water systems and other sources of water that jeopardize the safety of the drinking water supply.
- *Preventing* future cross-connections.

Cross-connections or backflow conditions are often difficult to detect. However, investigations are warranted whenever there are complaints of bad-tasting or discolored water, or where the water meter is recording backward or water pressure is low.

Cross-connection control

Although difficult to detect, cross-connections can be controlled or eliminated. The control of backflow and back-siphonage requires either the removal of the cross-connections or the installation of a proper cross-connection control device or backflow eliminator.

For example, the link between the potable water and the contaminated substance can be eliminated by a vertical airgap. By definition, the airgap is the unobstructed vertical distance through the free atmosphere between the lowest opening from any pipe or faucet supplying water to a tank, plumbing fixture, or other device and the flood-level rim of the receptacle. Although airgaps will eliminate cross-connections, from an industrial point of view, they are often impractical to use and can be easily bypassed.

Before making any corrections, be sure to check the regulations of the administrative authority having jurisdiction.

Backflow or cross-connection control devices consist of:

- Atmosphere vacuum breakers
- Pressure vacuum breakers
- Double check valve assemblies
- Reduced pressure principle devices.

An atmosphere vacuum breaker (AVB) is a device that has a moving element inside, which during flow prevents water from spilling from the device and during cessation of flow, drops down to provide a vent opening. This device should not remain under pressure for long periods, and it cannot have any shut-off valve downstream (see Figure 14-10). They are used, for example, in irrigation installations and hose bibs.

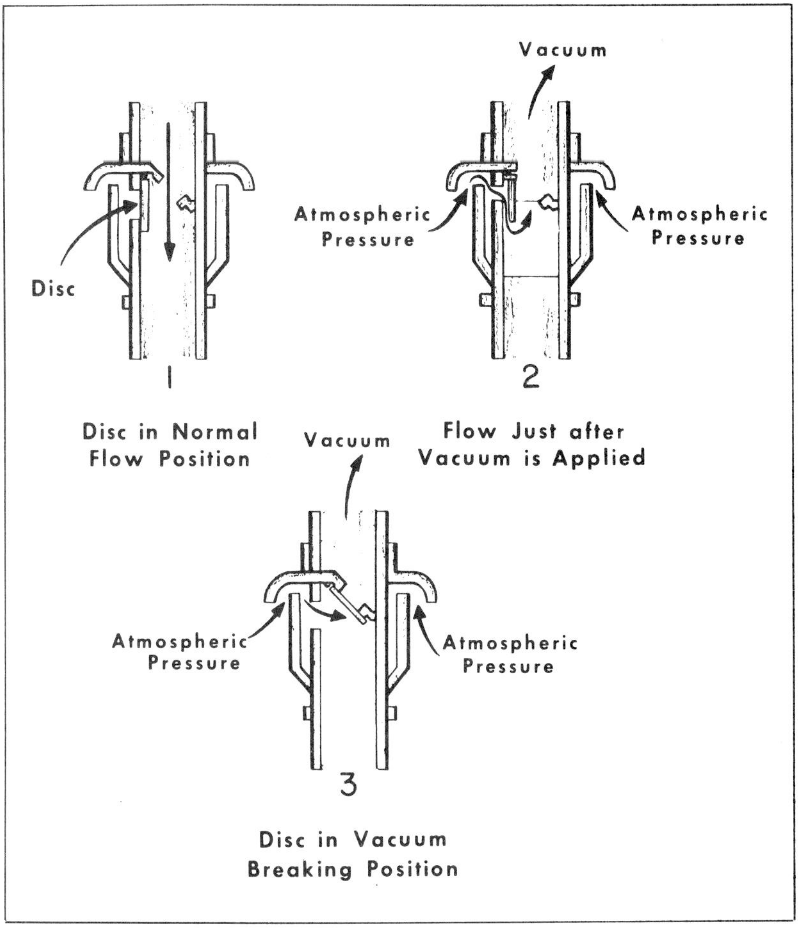

Figure 14-10. Operation of one type of an atmospheric vacuum breaker.

From U.S. Environmental Protection Agency, Water Supply Division, Publication EPA-430/9-73-002, *Cross-Connection Control Manual.*

The pressure vacuum breaker (PVB) is a device that contains within a single body, a single loaded check valve and a loaded air opening valve that open to admit air whenever the pressure within the body of the device approaches atmospheric pressure. The body of the device has two tight-closing shut-off valves and it is fitted with test cocks, appropriately placed for testing the device (see Figure 14-11).

The double-check valve (DCV) assembly is composed of two single, independently acting check valves. The unit also has two tightly closing shut-off valves located at each end of the device and four test cocks for the testing of the check valves (see Figure 14-12).

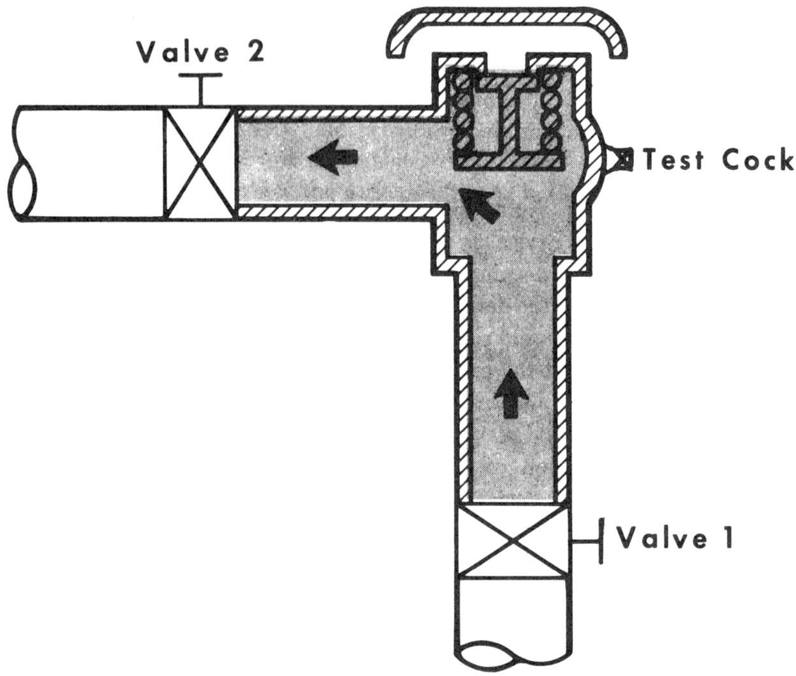

Figure 14-11. Operation of a pressure vacuum breaker.

From U.S. Environmental Protection Agency, Water Supply Division, Publication EPA-430/9-73-002, *Cross-Connection Control Manual.*

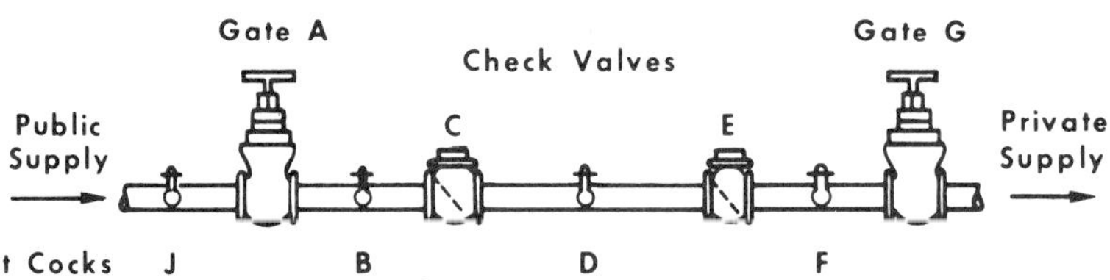

Figure 14-12. The double check-/double gate valve assembly should include test cocks as shown here. The identifying letters refer to the methods for testing the check valves that are explained in the reference source.

From U.S. Environmental Protection Agency, Water Supply Division, Publication EPA-430/9-73-002, *Cross-Connection Control Manual.*

The reduced pressure principle device (RPPD) consists of two independently operating check valves, together with an automatically operating pressure differential relief valve located between the two check valves. The first check valve reduces the supply pressure at a predetermined amount so that during normal flow the pressure between the two check valves will be lower than the supply pressure. If either check valve

Figure 14-13a. Reduced pressure principle device.

University of California, San Francisco.

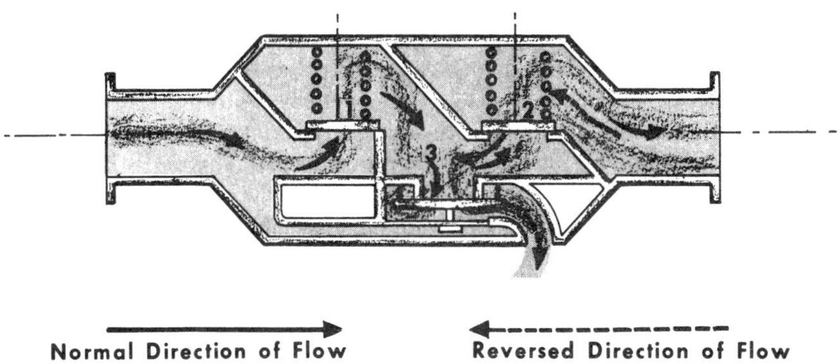

Figure 14-13b. Operation of a reduced pressure principle device.

From U.S. Environmental Protection Agency, Water Supply Division, Publication EPA-430/9-73-002, *Cross-Connection Control Manual.*

leaks, the relief valve will discharge into the atmosphere. This will maintain the pressure in the zone between the two check valves lower than the supply pressure (see Figure 14-13).

Manufacturers of backflow prevention devices and cross-connection control devices test their products before they are sold. However, once installed, these devices are frequently forgotten or, more than likely, neglected as far as maintenance and periodic inspections are concerned. It is possible that a check flow valve that has been in service for several years could fail and allow water to backflow through the unit. The only way to prevent such a failure is to purchase check valves fitted with test cocks and conduct periodic inspections to determine that the checks are operating properly.

The selection of cross-connection control measures depends on the degree of the hazard. Many states have enacted regulations addressing this problem in the form of design standards, approved devices, and testing performed by certified persons. State sanitary engineering agencies and local environmental health departments should be consulted. Many examples of cross-connection control solutions are published by the State of California, "Manual of Cross-Connection Control Procedures and Practices."[14]

In addition to the selection of appropriate cross-connection devices, some industries install an industrial water system that is separate from the drinking water system. This solution should include a pipe identification system to distinguish the two systems.

PLANT VENTILATION

Plant ventilation is a major environmental control system for protection of employees. Standard methods of ventilation engineering control became legal requirements with the passage of the OSHAct. All members of the occupational health and safety team should have some knowledge of ventilation principles. For those desiring a better understanding of the use of ventilation in comfort and contamination control, the author suggests two excellent publications on the subject: the American Conference of Governmental Industrial Hygienists—*Industrial Ventilation Manual*[9] and the *Handbook of Ventilation for Contamination Control.*[10] The treatment of the subject of ventilation here is in general principles. Technical and design decisions are the purview of the engineer and industrial hygienist. But, in any case, all members of the occupational health and safety team can profit from a grasp of ventilation fundamentals, discussed next.

Ventilation principles

A first and basic principle of ventilation is that air blown from a small opening retains its directional effect for a considerable distance beyond the plane of the opening (see Figure 14-14). However, if the flow of air through the same opening were reversed so that it operated as an exhaust, the air flow would become almost completely nondirectional and its range of influence would be reduced. For this reason, local exhaust

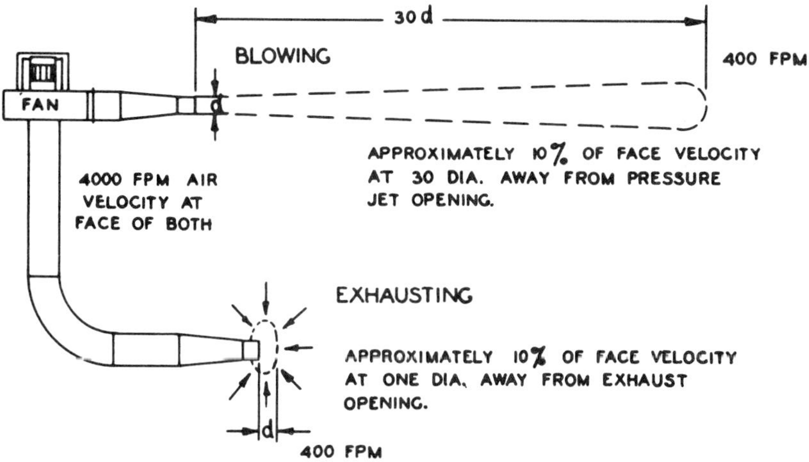

Figure 14-14. Air flow characteristics of blowing and exhausting are greatly different.

From American Conference of Governmental Industrial Hygienists, Lansing Mich., *Industrial Ventilation.*

ventilation must be in the immediate vicinity of the emission source. This principle is also applied in situations where it is desirable to "throw" air or create directional air movement through the worker's breathing zone or work station toward a contaminant removal system.

A second principle of air movement by a ventilation system is that whenever air is exhausted from a building, supply or outside air must be introduced to take its place. The incoming air may enter by infiltration through random cracks around windows and doors or be delivered through supply air ducts. This required air is referred to as make-up air. If the work area make-up air is insufficient, the exhaust systems can become air-starved. Common symptoms of this situation are doors that become hard to move and high-velocity air drafts through windows or other openings in the work area. In most cases, the volume of make-up air should equal the total volume of air removed by the exhaust system, or the result will be a decrease in the designed exhaust flow rate from the work area.

In some situations, the design engineer will try to direct air from area to area by increasing or decreasing the supply and exhaust volumes in spaces within a building. For example, an excess supply of air in a room will help make a clean room process area to be under positive pressure with respect to external areas. The reverse situation (negative pressure) is needed to control and remove odors and contaminants from a confined work area.

It is important to recognize the recent demands of energy conserva-

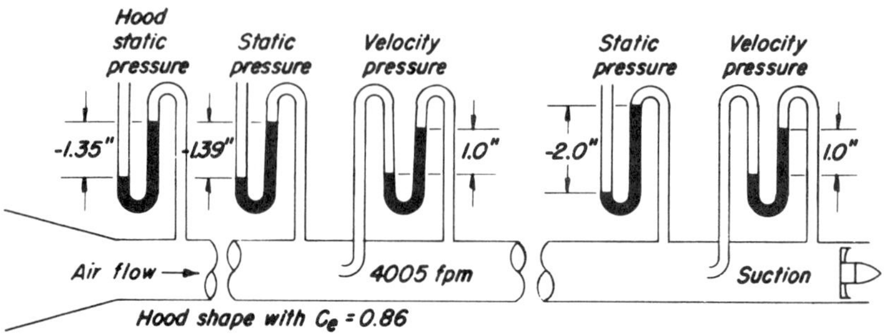

Figure 14-15. Typical static and velocity pressures are shown in a long section of exhaust duct equipped with a defined hood.

From American Conference of Governmental Industrial Hygienists, Lansing Mich., *Industrial Ventilation.*

tion made on ventilation systems. Uncontaminated, heated exhaust air has been reused to heat parking garages, outside mechanical rooms, and the like. A recent document by NIOSH[11] has opened up the question of the reuse of contaminated exhaust air by the application of filter technologies. There is no reason not to consider and apply these approaches. It will require vigilance on the part of the occupational health and safety team to make sure that energy conservation needs and requirements are coordinated with those of occupational health.

Ventilation terms

Air can be placed in motion naturally or by a mechanical system such as a fan. The problem with natural means such as the wind or convection currents is that they are unpredictable and unacceptable when unhealthy contaminants are to be controlled.

Air movement in mechanical ventilation systems is described in terms of pressure, velocity, and volume. Air in motion through a duct exerts a pressure called velocity pressure, which is analogous to kinetic energy. It exists only when air is in motion, acts in the direction of air flow, and is always positive. Static pressure produces initial air velocity, overcomes the resistance in a system caused by friction of the air against the duct walls, and overcomes turbulence caused by a change in the direction or air velocity. Static pressure is analogous to potential energy and exists even when there is no air in motion. It acts equally in all directions and either tends to collapse the walls of the duct upstream from the fan (negative in sign) or tends to expand the walls of the duct on the downstream side (positive in sign). The driving force for air flow is a pressure difference. This pressure is required to start and maintain flow and is called total pressure. Static pressure (SP), velocity pressure (VP), and total pressure (TP) are all interrelated. This phenomenon was stated by Bernoulli:

$$SP + VP = TP$$

Figure 14-15 shows the relationship between static, velocity, and total pressures at different points in a duct system. The total pressure represents the amount of energy in the system. The pressures are interconvertible—for example, if the velocity is reduced at some point in a duct system, a portion of kinetic energy (velocity pressure) is converted into potential energy (static pressure). It should be understood that this conversion is always less than 100 percent efficient because of a net loss of total pressure resulting from turbulence and shock.

The results of turbulence, friction, acceleration, and hood-entry losses on air flowing through ductwork lead to a pressure drop in the system. Bernoulli's theorem can be restated as:

$$\underset{\text{(upstream)}}{SP_1 + VP_1} = \underset{\text{(downstream)}}{SP_2 + VP_2} + \text{losses}$$

The reader now begins to see the importance of pressure measurements at various locations in a ventilation system. In some instances, it is advisable to fix such measuring equipment permanently.

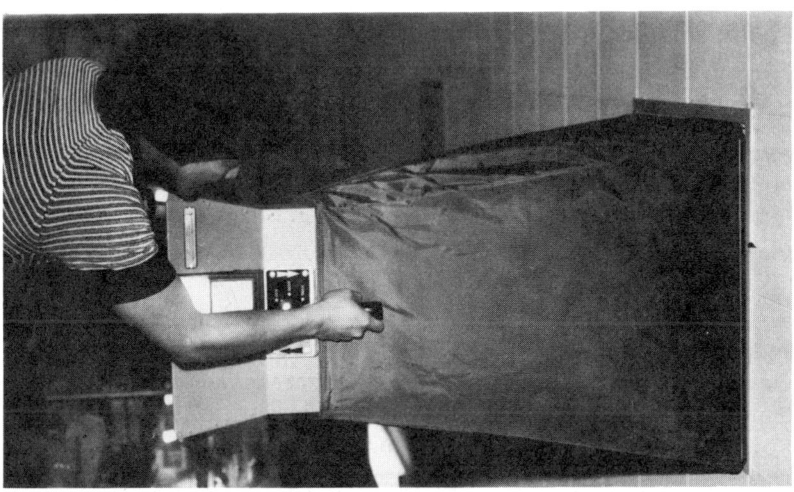

Figure 14-16. Technician measures volumetric flow rate with a flow hood.

University of California, San Francisco.

Air velocity is expressed in units of feet per minute (fpm) while velocity pressure is usually expressed in units of inches of water. A useful equation to show the relationship between velocity and velocity pressure is:

$$V = 4005 \sqrt{VP} \text{ fpm}$$

Note: Equation assumes standard air density of 0.075 lb/cu ft

Air volume flow rate is expressed in cubic feet per minute (cfm). This measurement is usually taken at the intake or discharge point of the system. The use of a flow hood which gives direct measurement in cfm is shown in Figure 14-16. This measurement is usually made on building ventilation systems (usually referred to as a general or comfort ventilation system). It is generally used to calculate what is called an air change rate. The air change rate is expressed in air changes per hour within a space. The air change rate is calculated by the following equation:

$$ACR = \frac{VFR \times 60 \text{ min/hr}}{RV}$$

where

ACR = Air change rate (changes per hour)
VFR = Volumetric flow rate (cfm)
RV = Room volume (cu ft)

The air change rate is used frequently in discussions of ventilation requirements. Unfortunately, it is often used improperly. The term is useful when applied to offices, schools, and similar spaces where the purpose of ventilation is simply to control odors, temperature, or humidity. Ventilation requirements based on room volume alone have no validity in calculating requirements for contaminant control. Calculations of this type require information on the rate of generation of the contaminant.

Dilution ventilation

Dilution ventilation is the class of ventilation systems that controls contaminants by diluting contaminants with clean air. It is expressed in cfm. The important point to remember is that you are *diluting* the contaminant but *not removing* it at the rate a local exhaust system would provide. In general, dilution ventilation is not as satisfactory as local exhaust systems. There are, however, occasional circumstances where the design of a local exhaust system is impractical. The following criteria should be used when considering the selection of dilution ventilation to control contaminants:

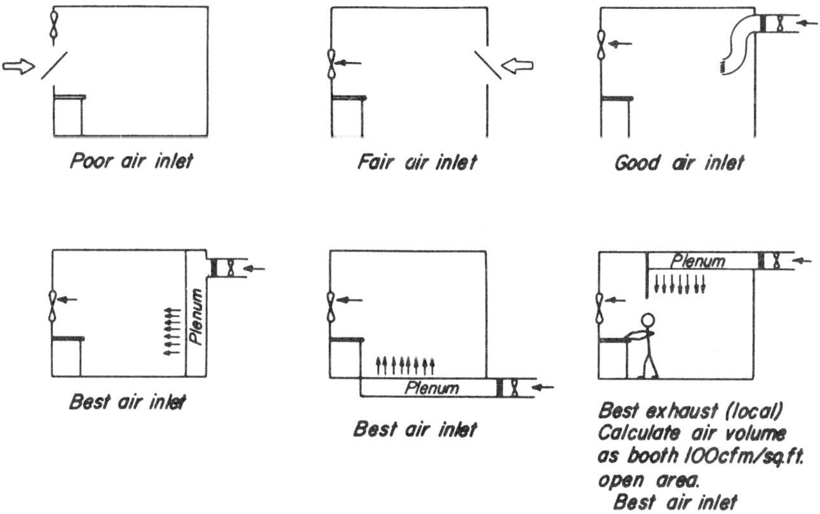

Figure 14-17. Comparison of fan and inlet locations for dilution ventilation.

From American Conference of Governmental Industrial Hygienists, Lansing Mich., *Industrial Ventilation.*

• The quantity of contaminant generated must not be too great or the air volume necessary for dilution is impractical.

• Workers must be far enough away from the contaminant evolution or the evolution must be sufficiently low so that they will not have excessive exposures. Sufficient dilution must occur. Fan location is important (see Figure 14-17).

• The toxicity of the contaminant must be low.

• The evolution of the contaminants must be reasonably uniform.

To provide the reader with some appreciation of the quantities of air needed to dilute a contaminant, it may be useful to calculate a problem. An equation used in calculating dilution ventilation is:

$$\text{Cubic feet of air/pint evaporated} = \frac{(403)\ (\text{sp gr liquid})\ (K)\ (10^6)}{(\text{MW of liquid})\ (\text{TLV})}$$

K = A dimensionless safety factor normally ranges from 3 to 10 based on such things as uniformity of evolution, number of exposed workers and so forth.

Assuming that two pints of methyl alcohol are being released each hour and a *K* value of 6 is selected, the following dilution rate is arrived at:

$$\text{Dilution rate of methyl alcohol} = \frac{(403)\,(0.792)\,(6)\,(10^6)\,(2)}{(32.04)\,(200)\,(60)} = 9950 \text{ cfm}$$

A local exhaust system might require less than 2000 cfm for control of this vapor-type evolution. Dilution ventilation is usually not recommended for fumes or dust (both are tiny particles that are suspended in air), because the velocity and rate of evolution are usually very high, resulting in locally high concentrations.

Local exhaust ventilation

A local exhaust system to control contaminants is usually the most desirable approach because of the following:

- Less air needs to be exhausted to control the contaminant resulting in energy conservation because less conditioned make-up air is required and a smaller fan will be used.

- The contaminant is usually captured before it enters the worker's breathing zone.

- If community air pollution control of the contaminant is required, the contaminant is more highly concentrated in the conveying airstream.

- The local exhaust system is more easily monitored, because it is isolated from the general ventilation system.

Before any exhaust system is installed, however, every effort should be made to control the contaminant by substitution of materials, revision of the process, or perhaps isolation of the process. In discussions of local exhaust systems, additional terms are used to describe air movement (see Figure 14-18)—they are capture, face, slot, plenum, and duct velocities. These control velocities are measured in feet per minute, and it is particularly important that they are measured as they are described. For example, a requirement that a chemical fume hood have an average face velocity of 150 fpm when working with carcinogens implies that measurements will be taken in the opening plane of the hood. The important thing to remember is that the goal of the design is to achieve these control and capture velocities. Health and safety personnel should avoid a common misunderstanding that exhaust hoods have some magical power to reach out and draw the contaminated air into them.

A common mistake made by inexperienced people is the misunderstanding of the basic equation for airflow into hoods:

$$Q = VA$$

where

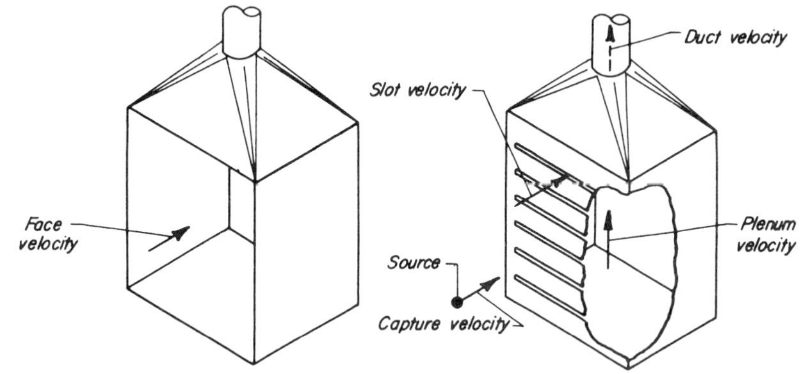

Capture Velocity – *Air velocity at any point in front of the hood or at the hood opening necessary to overcome opposing air currents and to capture the contaminated air at that point by causing it to flow into the hood.*

Face Velocity – *Air velocity at the hood opening.*

Slot Velocity – *Air velocity through the openings in a slot-type hood, fpm. It is used primarily as a means of obtaining uniform air distribution across the face of the hood.*

Plenum Velocity – *Air velocity in the plenum, fpm. For good air distribution with slot-types of hoods, the maximum plenum velocity should be 1/2 of the Slot Velocity or less.*

Duct Velocity – *Air velocity through the duct cross section, fpm. When solid material is present in the air stream, the duct velocity must be equal to the Minimum Design Duct Velocity.*

Minimum Design Duct Velocity – *Minimum air velocity required to move the particulates in the air stream, fpm.*

Figure 14-18. Principles of exhaust hoods and descriptions of the various velocities.

From American Conference of Governmental Industrial Hygienists, Lansing Mich., *Industrial Ventilation.*

Q = Volumetric flow rate (cfm)
V = Air velocity (fpm)
A = Area of opening (sq ft)

To meet OSHA requirements, a person may be tempted to reduce the face area of the hood opening to increase the face velocity. But the capture velocity depends on the total air flow entering the hood. A high face velocity or slot velocity is not the only important factor. Total air flow (Q) is reduced because of the increased hood resistance. The capture of air contaminants depends on mass air movement (Q). Always keep in mind that while OSHA has standards for local exhaust design, there are also performance standards for contamination control—that is to say,

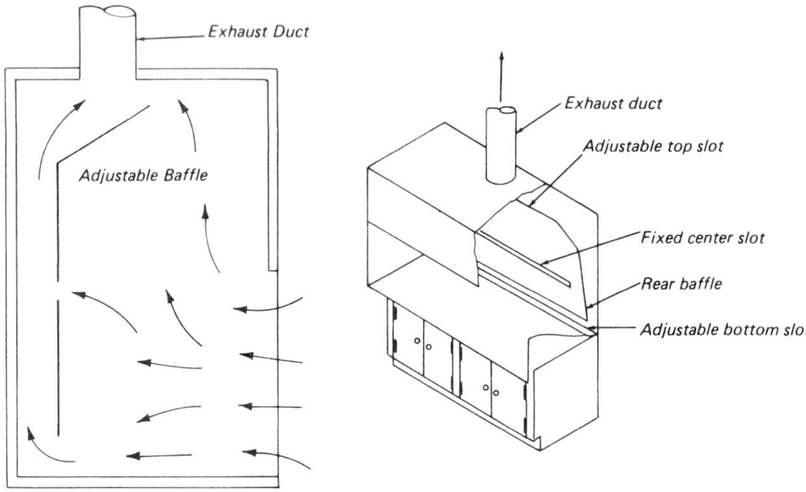

Entry loss = 0.5 duct VP
Duct velocity = 2000 fpm minimum

Figure 14-19. Laboratory hood with distribution panels.

From NIOSH Publication 76-162, *Recommended Industrial Ventilation Guidelines.*

the object of all ventilation controls must be to lower the exposure.

Hood design principles

There are principles of good local exhaust system design that are not followed in some installations, thus rendering the system inefficient or useless. Problems can develop after design because of plant structural or process changes and because of lack of employee understanding of the system. The design principles that must be observed are:

- Locate the hood as close as possible to the source of the contaminant, so that the initial velocity of the contaminant will throw it into the hood opening.
- Locate the hood so that the contaminant is moved away from the breathing zone of the worker.
- Locate the hood so that it will not interfere with the worker's task.
- Air flow velocities into hoods should be distributed equally (see Figure 14-19).
- Enclose the source of contamination insofar as possible to reduce the volumetric air flow (cfm) needed and to prevent cross drafts from disrupting control and capture air velocities.

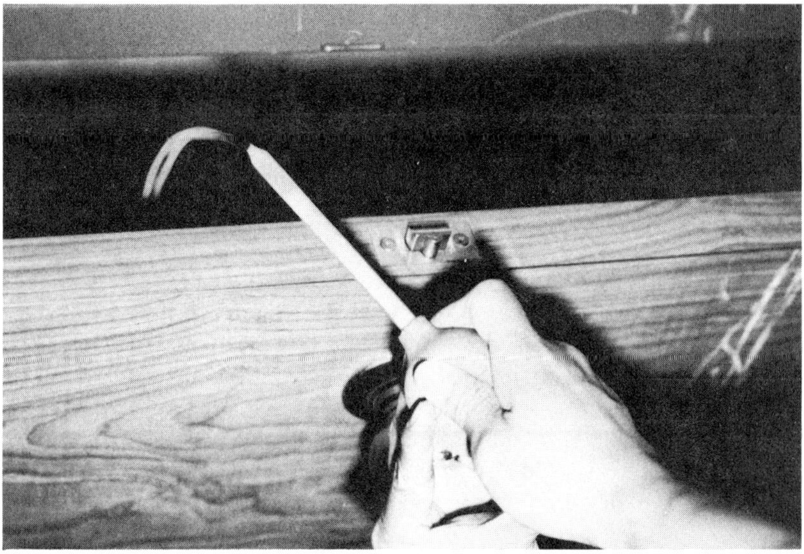

Figure 14-20. Smoke tube test showing air flow direction through a doorway.
University of California, San Francisco.

The use of smoke tubes during a plant walk-through can provide a great deal of information on how well the system is adhering to these basic principles (see Figure 14-20).

There is one last concept that occupational health and safety professionals should be aware of before leaving the problems of hood design. It is an explanation of the heavier-than-air argument. This argument arises when the specific gravity of a chemical is reported to be heavier than air and therefore is thought to settle to the floor. As a result, some ventilation designs locate the exhaust hood near the floor. To resolve this argument, one must show by calculation the *effective* specific gravity of the liquid. An example of such a calculation is presented here:

The effective specific gravity (ESG) of a 10,000 ppm perchoroethylene-air mixture:

SG of air	= 1
SG of C_2Cl_4 vapor (air = 1)	= 5.8
10,000 ppm	= 1 part C_2Cl_4 to 99 parts air
0.01×5.8	= 0.058
0.99×1.0	= 0.990
	1.048 ESG

This means the mixture would have a tendency to move downward at a rate of 105/100 for the 100 ppm mixture or 10,005/10,000 in the case of a 10,000 ppm mixture and *not* by a ratio of 5.8/1 as is frequently believed.[9] It should be recognized that almost all occupational health chemical standards are expressed in concentrations less than 1000 ppm. This implies that their effective specific gravity would be essentially the same as air and follow the air movement of the work area.

Ventilation testing

Testing of the ventilation system can be performed by any member of the occupational health and safety team or the maintenance staff, assuming that the individual performing the evaluation knows about ventilation control and design flow rates and velocities. This information can be obtained from the plant engineer or industrial hygienist. Employee training programs should also stress the importance of the system and its relationship to air contaminant concentration levels.

EMERGENCY PLANNING

No company can fulfill its responsibility to workers if it does not prepare for emergencies. Emergency planning is important for security, production continuity, damage control, and prevention of injury or death. Emergencies can develop externally from natural disasters, or internally from fires, chemical spills, structural mishaps, and employee injury. Since each company and its organizational structure are different, the program of planning for and participating in emergency actions should be a coordinated effort, custom-made for each plant.

Planning for emergencies requires thinking about the worst possible situation. Planning should start with a compilation of the potential hazards that exist in a plant or establishment and their relationship to the environment. In preparing the plan, such primary hazards as fire, explosion, electrical shock, and sudden chemical exposure should be listed. The magnitudes of these events should also be considered. Safeguards to prevent these disasters should be specified, and locations of fire sprinkler systems, electrical emergency backup, air and liquid monitoring systems, fire hose connections, and emergency medical equipment should be evaluated for adequacy.

It is also useful at this point to evaluate available community resources that will assist in dealing with emergencies. The following questions might be asked:

- What is the response time and capabilities of the local fire department?

• What is the response time of the local ambulance service?

• What are the capabilities of local medical care providers?

• What are the response time and capabilities of the local police department?

• What are the structure and response time of the local government disaster plan?

• How long will it take to repair equipment and resume production?

At this point, top management should consider the extent to which it expects action. Some companies rely heavily on outside community resources, while others develop comprehensive in-plant emergency response personnel and equipment.

A chain of command should be established so that there is no confusion as to who has authority for making or approving all decisions. There must be a definite assignment of responsibility. Actions to be taken by designated staff or departments at the outset of the emergency must be established. All employees should receive instructions commensurate with their work, because each employee is capable of providing some assistance, if he knows well in advance what is expected. The response of a company to an emergency will be reflected in the actions of its employees.

Loss control team

Many companies establish what is considered a loss control team (LCT). The LCT is designed to respond to general and localized emergencies to facilitate personnel evacuation and safety, shut down building services and utilities, as needed, work with responding civil authorities, protect and salvage company property, and evaluate areas for safety prior to reentry. The following is a well-developed LCT organization suggested by Ed Sawicki, consulting safety engineer.[12] It is not suggested that every organization adopt this program; rather, it is presented to demonstrate thoroughness of preparation.

LOSS CONTROL TEAM COMPOSITION

The LCT is generally comprised of selected personnel from plant maintenance, line maintenance, security, safety, and others as needed, using the minimum number required to efficiently complete the task.

SECURITY—Since security generally receives the first report, it will assume the responsibility of notification of all affected parties and outside agencies, initiation of general evacuation if needed, escorting outside agencies, crowd control, first aid, and guiding emergency vehicles to the affected site.

PLANT MAINTENANCE—will assume the responsibilities for shutdown, as needed, of utilities (electrical, natural gas), ventilation system, major chemical systems such as liquid hydrogen, and oxygen, and, if necessary, fire hose and sprinkler system after it has been activated.

LINE AND PRODUCTION MAINTENANCE—has the responsibility of shutting down essential manufacturing equipment and minor chemical systems (compressed gas cylinders).

SAFETY—will augment security and serve as advisor to the LCT during an emergency. Its primary responsibility will be to help minimize injury and loss of life. Safety will reevaluate the area prior to allowing reentry.

OTHERS—It may be necessary to augment the LCT by using persons with special expertise in the affected area(s).

Loss control team structure. The team leader, of course, has responsibility for the entire loss control team. This person evaluates the scope and nature of each emergency; uses information and advice from the assistant leader, sub-leaders, and others as appropriate; makes decisions on extent of shutdown and evacuation (if not already done); handles search and rescue attempts, coordinating with civil authorities; makes decisions on disposition of general personnel; and works with the safety department in evaluating property prior to reentry.

Assistant LCT leaders assist in carrying out the above responsibilities. They assume the leader's role if that person is absent or incapacitated. LCT sub-leaders carry out the leader's and the assistant leader's orders, using team members under their command. LCT members carry out the leader's and sub-leaders' instructions, using the required equipment. Emergency vehicle guides are individuals who, when needed, are immediately dispatched by the LCT leader to a key point(s) on company property (i.e., property entrance) to steer emergency vehicles toward the site of the emergency.

Organizing the LCT. Each building should have an LCT for each shift, whether it be a solitary facility or a part of a multibuilding complex. The leaders and subleaders, usually on the day shift, begin writing comprehensive shutdown procedures for that facility, which will be followed by all shifts. Final selection of regular team members will be made after the shutdown procedures are written, since the number of people required for each team depends on the portion of the shutdown procedure accorded to it.

Facility shutdown procedure. The facility shutdown procedure is written in coordination with the facilities maintenance and production maintenance departments (see Figure 14-21). It should contain complete information on a swift, general shutdown of all critical equipment and utilities in the case of an extreme emergency. This will include shutting down main electrical, natural gas, water, chemical tanks, and critical machinery.

Since the LCT is primarily geared for major emergencies, its shutdown procedures usually do not include localized shutdown information, for example, turning off individual breaker boxes or power to a particular machine. However, it would be wise for maintenance groups to record and retain this information. Critical machinery is defined as any machine, instrument, or device that, if not shut down in a prescribed manner, prior to general electrical shutdown, would result in an unacceptable loss of information or property. Computers are critical, for example.

The shutdown procedures are to be written in such a clear, concise manner that a person unfamiliar with the equipment or area will be able to easily find and shut down that item. It is often helpful to have critical valves or panels marked by distinctive symbols or colors to enhance identification.

The LCT leader will be responsible for initiating a quarterly review of the shutdown procedures to assure their validity.

Evacuation assembly points

The writing of the shutdown procedures should be accomplished by the designation of assembly points for general personnel during an evacuation, to provide a method of accounting for all employees, and to keep them out of danger. Locations for general assembly points should meet the following criteria:

- At least 300 feet away from the building to protect against explosion hazards.
- A large enough area to accommodate the number of people required without overcrowding.
- Out of the way of all likely paths of emergency vehicles.
- Not in the direct line of sight of a particularly hazardous area, if possible. Sites can be chosen so that a building lies between the hazardous area and the assembly point.
- Situated so that personnel can be evacuated without violating above criteria.

LOSS CONTROL TEAM LEADER'S EMERGENCY CHECKLIST

PRIORITY ITEMS

Building Evacuated (or portions thereof as needed) with personnel reporting to Evacuation Assembly point	YES___	NO___
Civil Authorities (fire dept., police, ambulance, etc.) notifed (if needed)	YES___	NO___
Main Electrical Panels	ON___	OFF___
Natural Gas	ON___	OFF___
Liquid Oxygen Tank	ON___	OFF___
Liquid Hydrogen Tank	ON___	OFF___
Fab & T.D. Lab Gas Pad	ON___	OFF___
Emergency Generator	ON___	OFF___
Post Indicator Valve (PIV)	ON___	OFF___
Liquid Nitrogen Tanks	ON___	OFF___

SECONDARY ITEMS

Fab Shutdown	DONE___	NOT DONE___
T.D. Lab Shutdown	DONE___	NOT DONE___
Test Area Shutdown (prior to power shutdown if possible)	DONE	NOT DONE
Digitizing Area Shutdown (prior to power shutdown if possible)	DONE___	NOT DONE___
Scrubber Exhaust #1 & 2 (Fab)	ON___	OFF___
Scrubber Exhaust #3 (T.D. Labs)	ON___	OFF___
D.I. Water System	ON___	OFF___
D.I. Water R.O. (Reclaim) Unit	ON___	OFF___
City Water	ON___	OFF___

TERTIARY ITEMS

Acid Storage	CHECKED___	NOT CHECKED___
Solvent Room	CHECKED___	NOT CHECKED___
Fire Riser (drain valve)	OPEN___	CLOSED___
Sodium Hydroxide Tank	ON___	OFF___
Elevator	ON___	OFF___

Figure 14-21. LCT leader's emergency checklist (semiconductor industry).

From E. Sawicki (Reference 12).

In the case of multi-building sites, a separate assembly area for each building's personnel is needed that corresponds to the above criteria. These areas can be adjacent to one another so long as they are somewhat separated. All personnel should be familiar with the location of their assembly point. Evacuation maps showing assembly points and the safest routes of travel from each exit should be given to all supervisory personnel. Employees *must be instructed* to use emergency exits and practice sessions will help. In many emergencies, many employees, unfortunately, will use normal exits to leave the building.

Location for the LCT assembly points should meet the following criteria:

- Close to the building but far enough away to protect from most hazards, also close enough to facilitate rapid assembly and response. However, LCT assembly points should not be near a hazardous area, such as a chemical processing unit, unless some substantial structure lies between to afford protection.
- In the case of multibuilding complexes, all LCTs meet at a centrally located site to ensure proper communication. Since most buildings in a complex are interconnected, it is most important that each LCT know exactly what others are doing to avoid conflict.
- LCT emergency equipment should be stored nearby.

It is important to keep the team size down to the minimum number needed to do the job effectively. For each selected member, there should also be an alternate member, who is not part of any other team, who can assume the duties of the primary member should he or she be absent or incapacitated. The subleader should select the most capable primary member of his team to be the alternate subteam leader. When the final teams are chosen, the LCT leaders should conduct reviews of the teams with the subleaders to cover personnel changes and to make sure that replacements are made.

LCT training

All members of the LCT should be familiar with the shutdown procedures, especially the subteam members. It is the responsibility of the subteam leaders to drill their crews on their various shutdown tasks until they become proficient. The LCT leaders should make sure that the subteams receive that training. When time allows, the subteams of each LCT should learn each other's shutdown procedures to achieve maximum flexibility.

The teams should also receive instruction in the use of emergency equipment, i.e., self-contained breathing apparatus, acid suits, gas detection instruments, and so forth.

In time, all LCT members should be certified in basic first aid procedures and cardiopulmonary resuscitation (CPR).

Equipment

Adjacent to the LCT assembly point should be the emergency equipment storage area. This may be a simple box or a shed or trailer as desired for that facility. It should be a locked outdoor unit and contain enough equipment to supply the team (or teams in a complex) using it. Emergency equipment should be locked with keys issued only to the LCT leaders, security, and safety. The equipment should be checked and maintained monthly and should consist of an appropriate number of some or all of the following items (depending on the facility):

- 30-minute pressure/demand self-contained breathing apparatus
- Spare air bottles
- Bullhorns
- First aid supplies
- Safety belts, lanyards, life lines, and harnesses
- Acid bubble suit(s)
- Disposable acid suits
- Disposable coveralls
- Hard hats
- Fire axes
- Keys, to be used for entering locked areas requiring shutdown
- Shutdown procedures—enough copies for all LCT members
- Air sampler and appropriate detection tubes
- Flammable gas detectors
- Hydraulic jack
- Explosion-proof portable flashlights, lanterns, and batteries
- Walkie-talkies (optional) for communication
- Salvage tarp—heavy-duty plastic sheets or rolls used to cover up and protect equipment.

LCT activation and reporting procedure

The LCT is activated by one of two methods: the general evacuation alarm (automatic activation) or announcement over the public address system (verbal call activation).

Automatic activation procedure. The LCT is automatically activated when the general alarm for the building is sounded (usually a sustained, dual-tone, high-pitched signal over the public address system). The alarm is augmented by frequent pages from the security post stating that an emergency exists and that *all* personnel are to leave the building *immediately* and report to their designated assembly point.

All persons, including LCT members, are to leave the building at once without stopping unnecessarily to turn off machines, go to lockers, or the like, since an evacuation alarm means that danger exists and it would be unsafe to remain in the building.

LCT members then report to their assembly point, using the routes of exit prescribed in the evacuation maps. Once at the assembly point, subteam members report to their leaders for instruction.

Verbal call activation. The LCT can also be activated by means of a paging system. Such an announcement, without the evacuation alarm, usually indicates a less serious emergency. If such an announcement is made, teams should report immediately to the location given over the public address system. Once there, the subteam member reports to his leader for instructions.

LCT drills and practice evacuations. The LCTs should run shutdown drills every three months. The drill should consist of assembling the entire team on each shift, without the sound of any alarms, walking through a complete mock shutdown, and practicing the use of emergency equipment.

Practice evacuations of each building are held annually and should be coordinated with the production managers so that work loads can be adjusted. Additionally, skeleton crews in manufacturing areas can be allowed to remain in the building to perform critical production tasks to make sure lost production time is kept to a minimum.

After each drill and practice evacuation, a critique session should be held by the LCT leaders and subteam leaders to identify problems and deficiencies encountered and to initiate corrective action.

General evacuation. All employees should be instructed on what to do when the building alarm sounds and take the appropriate actions. A sample checklist for the LCT to use in emergencies is shown earlier in Figure 14-21.

SUMMARY

The reduction of accidents and injuries cannot be accomplished without the combined efforts of the entire company. Motivating, educating, and training employees to recognize, evaluate, and react properly to those elements in the plant or establishment that have hazard potential is a continuing process. This process must become an integrated part of equipment and process design, maintenance, and employee training.

The more knowledge an occupational health and safety professional has of specific types of hazards that may exist in the plant or establishment operation, the closer he or she comes to being an expert. The natural consequence of this greater ability to perceive and recognize potential accident causes is greater acceptance of safety and health objectives within an organization.

ADDITIONAL READINGS

Industrial Fire Hazards Handbook, 1st ed. Quincy, Mass: National Fire Protection Association, 1979.

Inspection Manual, 5th ed. Quincy, Mass.: National Fire Protection Association, 1982.

Cross Connection Control Manual, U.S. Environmental Protection Agency, Water Supply Div., Pub. No. EPA-430/9-73-002, 1973.

Naidich, A. *Do-It-Yourself Safety Inspection Manual.* Man and Manager, Inc., 1977.

Inspection Checklists. Houston, Texas: International Safety Academy, 1982.

"Self-evaluation Instrument—Phosphate Fertilizer Industry," DHHS (NIOSH) Pub. No. 80-129.

Thumann, A. *Electrical Design, Safety and Energy Conservation.* Atlanta, Ga.: Fairmont Press, 1978.

"Recommended Industrial Ventilation Guidelines," DHHS (NIOSH) Pub. No. 76-162.

REFERENCES

1. Heinrich, H.W. *Industrial Accident Prevention,* 4th ed. New York: McGraw-Hill, 1959.
2. Garber, B. T. "A Microcomputer Approach to Safety Record Keeping." *Occup. Health and Safety* Jan. 1983, pp. 39-41.

3. Garber, B. T. "Industrial Hygiene Applications of Microcomputers—On Air Sampling Data Base Management Systems." *Am. Ind. Hyg. Assoc. J.* 42 (1981):8 pp. 625-29.

4. "Electrical Safety," Bull. S550, Sacramento: CAL-OSHA Communications, August 1982.

5. "Electrical Work Injuries in California, 1972." Sacramento: California Division of Labor Statistics and Research, December 1973.

6. Craun, C. F., and L. M. McCabe "Review of the Causes of Waterborne Disease Outbreaks. *J. Am. Water Works Assoc.* 65 (1973):1 p. 74.

7. "Waterborne Disease Outbreaks in the U.S. 1971-1974." *J. Am. Water Works Assoc.* 68, (1976):8 p. 420.

8. Craun, C. F., and R. A. Gun. "Outbreaks of Waterborne Disease in the United States: 1975-1976." *J. Am. Water Works Assoc.* 78(1979):8 p. 422.

9. *Industrial Ventilation—A Manual of Recommended Practice.* 17th, ed. Lansing, Mich.: American Conference of Governmental Industrial Hygienists, Committee on Industrial Ventilation, 1982.

10. McDermott, H. T. *Handbook of Ventilation for Contaminant Control.* Ann Arbor, Mich.: Ann Arbor Science Publishers, 1976.

11. "Evaluation of Air Cleaning and Monitoring Equipment Used in Recirculation Systems," DHHS (NIOSH) Pub. No. 81-113.

12. Sawicki, E. "Emergency Response," in *Symposium Material for Fire and Safety Hazards in the Semiconductor Industry.* Menlo Park, Calif., 1981.

13. "Conducting Fire Inspections—A Guidebook for Field Use." Quincy, Mass.: National Fire Protection Association, 1982.

14. "Manual of Cross-Connection Control Procedures and Practices." Sacramento: State of California Department of Health Services, Sanitary Engineering Section, July 1981.

Chapter 15

Information Sources for Health and Safety Professionals

by Barbara Cohrssen

Knowing where to go to get relevant and up-to-date information on health and safety subjects is essential for effective decision making. Since the information in the field of occupational health and safety is expanding rapidly, methods of literature searching or finding health and safety reference data in typical sources of information are provided in this chapter. Then a list of recommended publications is presented. Because of the dynamic aspect of this professional field, many of the publications listed may be out of date shortly. However, knowing where to find information and the basic techniques for searching the literature will enable readers to constantly update their references.

Ask these questions first. Knowing "Who needs the information and why?" affects the level of information sought. For example, does the manager need it to make a decision on protective equipment or new engineering controls to recommend, or have new epidemiologic data been published that may affect a number of workers? Does the employee need the information to work more safely, or does the supervisor need it to provide a healthier or safer environment?

"How urgently is it needed?" is another question to ask. Is the information needed for an emergency that demands a quick response, such as a spill or an acute exposure, or is the information needed to plan training, to determine the scope of monitoring a program, or to know to what extent the legal requirements of a standard have changed? Certainly, how urgently the information is needed affects how much information to obtain. Are only toxicologic or health data needed, or are the chemical properties of a substance important to know? It is important to know who will use the information, why it is needed, and how it will be used. Answers to these questions affect which information sources or references will be used.

"How new is the information?" Another aspect that affects which information sources to use is the currency of the information. New knowledge about the health effects of a chemical, or new legal requirements can change the concentrations to which an employee may be exposed, the company's policies on medical surveillance, or the protection and training of the employees. The physical properties of a chemical or the operation of a piece of equipment, however, may not change so that older sources of literature or information can provide the necessary data. Depending on who needs the information, how urgently, and in how much depth, the reader can determine where to go for the information.

Sources of Information

The health and safety professional has a choice of two major sources of information. One source of data that is not always considered is corporate- or company-generated information. The other more commonly used source is the published literature by members of the professional community, manufacturers of equipment or chemicals outside the company, and consultants.

Within the company. Information on chemicals and equipment used within a company that might present potential health hazards to employees can be obtained from the engineering department in the form of chemical or product flow sheets showing how and where various chemicals are used within a plant. Engineering also may have equipment layouts showing potential health or safety hazards, as well as engineering controls in use. The procurement department or purchasing office of a company usually knows the specifications of the chemicals and equipment it buys. Additional information from manufacturers or suppliers can be obtained from purchasing departments. Specifications for the procurement of new equipment or products and/or operating manuals or data sheets may indicate the physical properties or potential hazards of chemicals or equipment that have been purchased.

The maintenance department can also be a source of information about equipment and potential hazards since maintenance workers are aware of where repairs have been made and, perhaps, are even aware of what they were exposed to while making the repairs. These employees know the inner workings of a company or a plant and can provide much information to managers.

Published information. The other choice the health and safety professional has is to go to external sources of information. These external sources are in two different forms. *(a) A static or in-print information form* which consists of encyclopedias, directories, or textbooks. The currency of the information presented in these publications depends on the time it took to print them. Usually encyclopedias and handbooks take the longest time so that if the most recent data on health effects of a particular chemical are desired, an encyclopedia would be the least likely place to go. However, if information about the physical properties of a particular chemical is wanted, a handbook would be a good source, since this information does not change. *(b) More dynamic sources* include journals, newsletters, newspapers, and consultants. Sometimes the reliability of the data presented in these sources should be questioned. Therefore, they should be considered as a means for learning about recently discovered information, or the basis for alerting a person about future legislation or requirements.

A relatively quick way to find out about what literature exists on a particular subject, or about the most recently published information is to use one of the many computer-based abstracting or indexing services. These services can provide up-to-the-minute information on the toxicity of a chemical, retrospective or state-of-the-art information, or sometimes data about the availability of training films and materials on a particular subject.

Other sources of "dynamic information" can be government publications, such as the National Institute for Occupational Safety and Health (NIOSH) publications, standards or legislation that provide information on the most recent health or safety data developed, as well as recent court decisions and new regulations that may affect an individual workplace. Manufacturers and suppliers of chemicals and equipment are also a source of recent or current information. Some chemical companies provide catalogs about the chemicals they manufacture. These catalogs can include chemical handling precautions, disposal methods, special properties, and chemical identification numbers. All manufacturers are now required to prepare and supply material safety data sheets. These data sheets present information on the chemical and physical properties of a chemical product, including health effects, first aid measures, spill precautions, and the chemical formula. Since they are prepared by manufacturers, the data sheets may not have been reviewed by peer professionals and the scientific reliability of the information can be questionable. (The telephone book Yellow Pages, under "Chemicals," is an excellent source for finding the names of chemical suppliers.)

Organizations, societies, and associations in the field of health and safety can also provide information about their specific branch of health and safety. This resource category usually publishes information, sponsors conferences, and can provide the names of consultants actively working in the field who can provide up-to-date information for the health and safety professional. These dynamic sources, if needed in an emergency, are not always accessible, which can be critical during an emergency. It is better to have several published references on hand, and for the health and safety professional to know which reference to use first and why, so that the kind of information available in each reference can be anticipated and valuable time saved.

The following types of resources provide information about the current knowledge of a subject: (*a*) computer-based information, indexing or abstracting systems; (*b*) recently promulgated state or federal standards; (*c*) literature reviews, such as criteria documents and journal articles; and (*d*) encyclopedia articles. For quick, factual information, the types of resources or references to go to would be: (*a*) reference texts, (*b*) handbooks, (*c*) journal articles, (*d*) material safety data sheets, and (*e*) consultants.

PRINT-BASED INFORMATION SOURCES

No single book or reference source can provide all of the information that a health and safety manager needs. The reader should know the types of information resources available and the kinds of information that can be obtained from them. Therefore, it is necessary to know what reference sources are available, but not to be overwhelmed by them. The best resource is a resourceful manager. The information sources available are as follows (particularly useful sources are prefixed with a ★). Note that some sources may appear under more than one category in order to ensure completeness of that category.

Professional and specialized organizations

These organizations publish reports, textbooks, and guidelines and provide information about consultants in the field of occupational health and safety.

American Chemical Society (ACS)
1155 Sixteenth St. N.W.
Washington, D.C. 20036

American Conference of Governmental Industrial Hygienists (ACGIH)
6500 Glenway Ave., Bldg. D-5
Cincinnati, Ohio 45211

American Industrial Hygiene Association (AIHA)
475 Wolf Ledges Parkway
Akron, Ohio 44311

American Society of Safety Engineers (ASSE)
1800 E. Oakton St.
Des Plaines, Ill. 60018

National Safety Council (NSC)
444 North Michigan Ave.
Chicago, Ill. 60611

Government agencies

Federal agencies, together with various state agencies, can provide excellent information and resource materials.

Environmental Protection Agency (EPA)
401 M Street, S.W.
Washington, D.C. 20460

National Institute for Occupational Safety and Health (NIOSH)
4676 Taft Parkway
Cincinnati, Ohio 45226

Occupational Safety and Health Administration (OSHA)
U.S. Department of Labor
200 Constitution Ave. N.W.
Washington, D.C. 20210

Directories

Wasserman, Paul, ed. *Health Organizations of the United States, Canada, and the World: A Directory of Voluntary Associations, Professional Societies, and Other Groups Concerned with Health and Related Fields,* 5th Ed. Detroit: Gale Research Co., 1981.

Provides a list of such organizations, arranged alphabetically, with information on membership as well as fees, and includes the names of the officers and an index for locating an organization in a specific subject field.

U.S. Bureau of Labor Statistics. *Directory of National Unions and Employee Associations, 1979,* Bul. 2079. Washington: U.S. Government Printing Office, 1980.

Lists national unions and professional and state employee associations.

NIOSH. *Occupational Safety and Health Directory.* DHHS (NIOSH) Pub. No. 80-124. Cincinnati: NIOSH, 1980. Order from NTIS No: PB-81-171-035/A08 (NTIS is explained on the next page.)

Lists federal, state, and local government agencies responsible for worker health and safety.

Grupenhoff, John T., ed. *National Health Directory.* Rockville, Md.: Aspen Systems Corp., 1982.

Lists sources of information on health programs and legislation. Also names congressmen, congressional health committees, the executive branches of government, and the various state agencies.

International Molders and Allied Workers Union. *Health and Safety Resource Directory.* Cincinnati, 1980.

Contains a list of lung associations, workers' clinics, universities that present health and safety programs, public interest groups, and related organizations.

Public Media Center. *The Hazard Connection: Occupational Safety and Health Resources Guide.* San Francisco, 1981.

Lists organizations in California concerned with occupational health hazards, gives names of publications and of legislative and media contacts interested in occupational health and safety.

Tchobanoghlous, George, Robert Smith, and Ronald Crites. *Wastewater Management: A Guide to Information Sources.* Detroit: Gale Research Co., 1976.

Includes references and sources of information on wastewater management: treatment, collection, effluent reuse, and disposal sludge treatment and disposal.

U.S. Environmental Protection Agency; Office of Pesticides and Toxic Substances. *Chemical Information Resources Handbook, Toxic Integration Information Series.* Washington, January 1981.

Handbook describes most of the more important chemical information resources available through government and private organizations. Most of these resources are computer-based information systems.

Akey, Denise S., ed. *Encyclopedia of Associations,* 19 ed. Detroit: Gale Research Co., 1985.

Comprehensive resource for information concerning nonprofit American Organizations.

Young, Margaret L., and Harold C. Young. *Special Directory of Special* Libraries and Information Centers, 5th ed. Vol. 3, "Health Sciences Libraries." Detroit: Gale Research Co., 1979.

Bibliographies

A number of publications listed here are bibliographies or present extensive bibliographies in the occupational health and safety field.

Hyslop, Frances, and W. M. Gafafer. *Bibliography of Occupational Health 1909-1953.* Public Health Service Pub. No. 300. Washington: U.S. Government Printing Office, 1954.

A comprehensive listing of publications on occupational health covering chemistry, dentistry, dermatology, education, and environmental factors.

★Key, M. M., *et al. Occupational Diseases: A Guide to Their Recognition,* rev. ed., DHHS (NIOSH) Pub. No. 77-181. Washington, U.S. Department of Health, Education and Welfare, June 1977. Order from NTIS No: PB-83-129-528/A99.

Contains an extensive occupational health bibliography.

★*NIOSH Publications Catalog.* 6th ed. DHHS (NIOSH) Pub. No. 84-118. Cincinnati: 1984. National Institute for Occupational Safety and Health. Order from Publications Dissemination DSDTT, NIOSH.

A catalog of all NIOSH publications. An excellent resource for locating health hazard evaluations, contract reports, and criteria documents.

NIOSH publications are available from several sources:

Publications Dissemination DSDTT, NIOSH
4676 Columbia Parkway
Cincinnati, Ohio 45226

National Technical Information Service (NTIS)
5285 Port Royal Road
Springfield, Va. 22161
(Order desk telephone is 703/487-4650)

Superintendent of Documents
U.S. Government Printing Office (GPO)
Washington, D.C. 20402

Olishifski, Julian B., ed. *Fundamentals of Industrial Hygiene,* 2nd ed. Chicago: National Safety Council, 1979.

Contains an entire chapter of listings: associations and organizations, emergency services, government agencies, as well as an extensive bibliography.

Peck, Theodore P. *Occupational Safety and Health: A Guide to Information Sources.* Detroit: Gale Research Co., 1974.

Covers the major sources in occupational safety and health: government and private organizations, training opportunities, research institutions, and books and other materials in this field.

U.S. Department of Labor, Occupational Safety and Health Administration. *A bibliography.* Springfield, Va.: NTIS, 1974.

An annotated bibliography of published sources from 1950-70. Includes bibliographies and catalogs, films and other audio visual materials.

Toxic Substances Sourcebook, Series 2. New York: Environment Information Center, 1980.

Lists NIOSH literature, federal government agencies, data bases, periodicals, selected U.S. chemical companies. Has a bibliography.

Wexler, Phillip. *Information Resources in Toxicology.* New York: Elsevier Science Publishing Co., 1982.

A selective and annotated bibliography.

Process information

The following are sources that can provide information on the chemical or physical hazards that may exist during a plant operation or process.

Burgess, William A. *Recognition of Health Hazards in Industry: A Review of Materials and Processes.* New York: John Wiley & Sons, 1981.

Reviews unit operations encountered in different industrial settings: the materials used, potential exposures, and methods of control. Reviews selected production facilities and cites types of exposures and some control techniques (mostly for industries with large numbers of workers). Describes equipment used as well as hazards.

Cralley, Lester V., and Lewis J. Cralley. *Industrial Hygiene Aspects of Plant Operations,* vol. 1, "Process Flows"; Vol. 2, "Unit Operations and Product Fabrication"; vol. 3, "Equipment Selection, Layout and Building Design." New York: MacMillan Publishing Co., 1982, 1984, 1985.

Chapters cover the hazards resulting from the production of various materials and products, and the equipment to control the hazards.

Kirk, R. E., and D. E. Othmer. *Encyclopedia of Chemical Technology.* New York: John Wiley & Sons, 1968.

An all-inclusive and extensive resource for chemical processes and technology in 22 volumes.

★Shreve, R. H. *Chemical Process Industries,* 3rd ed. New York: McGraw-Hill Book Co., 1967.

Discusses the entire range of industries engaged in the conversion of raw materials into products.

Health and safety references

GENERAL

Allen, R. W., *et al. Industrial Hygiene.* Englewood Cliffs, N.J.: Prentice-Hall, 1976.

Provides information on safety and medical programs for both large and small companies.

American Industrial Hygiene Association. *Hygienic Guides.* Akron, Ohio.

Provides information on permissible exposure levels, significant chemical and physical properties, industrial hygiene practices, first aid procedures, and medical management recommendations for approximately 200 different chemicals. (Various dates.)

Ashford, Nicholas. *Crisis in the Workplace.* Cambridge, Mass.: MIT Press, 1976.

Analysis of technical, legal, political and economic problems in occupational health and safety as of 1975. Bibliography includes descriptions of foreign activities.

★Brief, R. S. *Basic Industrial Hygiene: A Training Manual.* Linden, N.J.: Exxon Corp., 1975. (Available from the AIHA.)

Designed as a training course.

★*Workplace Injury and Illness Prevention Guide,* CS-1 rev. San Francisco: Cal/OSHA Consultation Service, 1982.

Purpose of the booklet is to assist employers in meeting California OSHA requirements. Included are a workplace program for illness and injury prevention and suggested checklists for safety and health inspections for different work environments and equipment.

Hopf, Peter. *Designer's Guide to OSHA*. New York: McGraw Hill Book Co., 1974.

Handbook for architects or engineers to use in designing buildings.

Hricko, Andrea, and Melanie Brunt. *Working for Your Life: A Woman's Guide to Job Health Hazards*. Berkeley, Calif.: Labor Occupational Health Program, 1976.

Written so that nontechnical persons can understand the health hazards associated with their work.

International Labour Organization; *Encyclopedia of Occupational Health and Safety*, 3rd ed. Geneva, Switzerland, 1983.

Good overall reference in two volumes.

★Key, M. M., *et al. Occupational Diseases: A Guide to Their Recognition*, rev. ed., DHHS (NIOSH) Pub. No. 77-181. Washington: U.S. Department of Health, Education and Welfare, June 1977. Order from NTIS No: PB-83-129-528/A99.

Lists occupations and the chemicals that workers may be exposed to in these occupations. The text covers routes of entry and modes of action; chemical, physical, and biological hazards; sources of consultation and reference aids. Occupational diseases are discussed in terms of occupational health hazards as a means to recognition of disease.

McCann, Michael. *Artist Beware*. New York: Watson-Guptill Publications, 1979.

Although the book emphasizes health hazards to artists, the presentation of the information is of great help to all in the health field. Part I gives general hazards and precautions. Part II gives hazards and precautions in specific artistic activities.

Fire Protection Handbook, 15th ed. Quincy, Mass.: National Fire Protection Association, 1981.

Contains information on characteristics and behavior of fires, storage practices, hazards, public protection, and special fire suppression agents and systems, and process fire hazards. Encompasses information in industrial fire hazards handbook.

The Industrial Environment, Its Evaluation and Control, 3rd ed, DHHS/NIOSH Pub. No. 74-117. NIOSH, Rockville, Md.: 1973. Order from GPO No: 017-001-00396-4.

A textbook covering a broad range of subjects. (Currently being revised.)

Olishifski, Julian B., ed. *Fundamentals of Industrial Hygiene,* 2nd ed. Chicago: National Safety Council. 1979.

Provides practicable information for those persons who either have an interest in or responsibility for the recognition, evaluation, and control of occupational health hazards.

★*Patty's Industrial Hygiene and Toxicology,* vol. I *General Principles,* vol. IIA *Toxicology,* vol. IIB *Toxicology,* vol. IIC *Toxicology,* vol. III *Theory and Rationale.* New York: John Wiley & Sons, 1978 to 1982.

A basic reference text in industrial hygiene which has been updated and expanded since it was first published in the 1960s.

Peterson, Jack. *Industrial Health.* Englewood Cliffs, N.J.: Prentice Hall, 1977.

Schilling, R. S. F. *Modern Trends in Occupational Health.* London, England: Butterworth, 1960.

Written prior to the OSHAct, it shows the need for an occupational health team. A broad brush of all subjects for management involved in health and safety of employees.

Schilling, R. S. F. *Occupational Health Practice,* 2nd ed. London, England: Butterworth, 1981.

Details the functions of the physician, industrial hygienist, and nurse in prevention of occupational disease for lay people.

Seeger, Nancy. "Alternatives for the Artist: Guides to the Safe Uses of Materials." Chicago: The School of the Art Institute of Chicago, Health Hazards in the Arts Program, 1984.

Describes hazards of and precautions to take with materials used in the arts.

Stellman, J. M. *Women's Work, Women's Health.* New York: Pantheon Books, 1977.

Appendix I lists health hazards in selected women's occupations; Appendix II lists health effects of selected occupational health hazards; tables list reproductive hazards to offspring, adult males, and adult females.

Stellman, J. M., and S. M. Daum. *Work is Dangerous to Your Health.* New York: Vintage Books, 1973.

Written in nontechnical language, book discusses health hazards that exist both at work and in the general environment.

CHEMICAL AND PHYSICAL PROPERTIES OF CHEMICALS

Aldrich Catalog/Handbook of Fine Chemicals, 1982-1983. Milwaukee, Wis.: Aldrich Chemical Co.

A catalog containing over 12,000 chemicals with data on CAS numbers, literature references, handling precautions, disposal methods, and physical properties, RTEC number, cross reference to Sax's *Dangerous Properties of Industrial Materials,* 5th ed.

Gosselin, Robert E., *et al. Clinical Toxicology of Commercial Products,* 5th ed. Baltimore: Williams & Wilkins, 1984.

Marzulli, F. N., and H. I. Maibach, eds. *Dermato-Toxicology,* 2nd ed. New York: Hemisphere Publishing Corp., 1983.

A reference book on local and systemic effects of chemicals following skin exposure.

Rose, Arthur, and Elizabeth Rose. *The Condensed Chemical Dictionary,* 7th ed. New York: Reinhold Publishing, 1966.

Alphabetical list of chemical and product names. Provides information on chemical properties, composition, and manufacturer.

Sittig, Marshall: *Priority Toxic Pollutants: Health Impacts and Allowable Limits.* Park Ridge, N.J.: Noyes Data Corp., 1980.

Information on EPA's 65 priority toxic pollutants is provided: occurrence, physical properties, chemical properties, uses, tax effects, groups at risk, basis for health criteria, and references.

Weast, R. C. *Handbook of Chemistry and Physics.* Boca Raton, Fla.: CRC Press, yearly.

A ready-reference book of chemical and physical data. A standard handbook.

Another fine source for information about chemicals is the Material Safety Data Sheets (MSDSs) prepared by the chemical manufacturers. Names and addresses of chemical manufacturers can be obtained from the *Directory of Chemical Producers of the U.S.A.* (published annually by SRI International, Ravenswood Ave., Menlo Park, Calif. 94025.)

The telephone book Yellow Pages, under "Chemicals" can provide the names, numbers, and addresses of other chemical manufacturers or suppliers; and publications such as *Chemical Marketing Reporter* (formerly *Oil, Paint and Drug Reporter*), Schnell Publishing Company, 100 Church Street, New York, N.Y. 10007, can also provide the names and addresses of chemical manufacturers.

Sampling And Analysis

American Conference of Governmental Industrial Hygienists. *Air Sampling Instruments for Evaluation of Atmospheric Contaminants,* 6th ed. Cincinnati, Ohio, 1982.

Describes uses, principles, physical and performance data, operating and maintenance instructions, and commercial sources for air sampling equipment.

American Industrial Hygiene Association. *Combustible Gas Detectors: A Manual of Recommended Practice.* Akron, Ohio, 1982.

Summarizes the techniques of correct usage of direct-reading combustible gas indicators.

American Industrial Hygiene Association. *Direct-Reading Colorimetric Tubes: A Manual of Recommended Practice,* Akron, Ohio, 1980.

Summarizes the techniques for use of colorimetric detector tubes to enable broader and more precise usage.

American Public Health Association. *Methods of Air Sampling and Analysis,* 2nd. ed., Washington, 1972.

Presents 136 methods of sampling and analyzing ambient and workplace air and tells general precautions and techniques to take when calibrating instruments and when sampling and analyzing chemicals.

Crable, John V., ed. *NIOSH Manual of Analytical Methods,* 2nd ed. vols 1-7. Washington: U.S. Dept. of Health, Education, and Welfare, 1977.

Contains 337 sampling and analytical methods for use in industrial hygiene environmental monitoring. Some of the procedures are useful for measuring toxic constituents in blood, urine, or body tissues; a majority of the methods are intended for use in personal air exposure monitoring.

Jacobs, Morris B. *The Analytical Toxicology of Industrial Inorganic Poisons.* New York: Interscience, 1967.

A classic reference in analytical toxicology. Many of the techniques have been outdated by the availability of new equipment.

Toxicology

The field of occupational diseases or toxicology is one of the fastest growing in health and safety. Therefore, only a select few are listed, as a guide.

Adams, Robert M. *Occupational Skin Disease.* New York: Grune & Stratton, 1983.

The definitive textbook of occupational dermatology.

American College of Obstetricians and Gynecologists. *Comprehensive Bibliography on Pregnancy and Work.* DHHS (NIOSH) Pub. No. 78-132. Rockville, Md.: NIOSH, 1977. Order from GPO No: 017-033-00288-0.

American College of Obstetricians and Gynecologists. *Guidelines on Pregnancy and Work,* DHHS (NIOSH), Pub. No. 78-118. Rockville, Md.: NIOSH, 1977. Order from GPO No: 017-033-00279-7.

Guidelines to protect pregnant workers.

American Conference of Governmental Industrial Hygienists. "TLVs® Threshold Limit Values for Chemical Substances and Physical Agents in the Work Environment and Biological Exposure Indices with Intended Changes." Cincinnati, yearly.

Lists air-borne concentrations of substances to which nearly all workers may be exposed without adverse effect, based on available information. Some of the concentrations are lowered as additional toxicologic data become known. (The address of the ACGIH is 6500 Glenway Ave., Bldg. D5, Cincinnati, Ohio 45211.)

Anderson, Kim, and Ronald Scott. *Fundamentals of Industrial Toxicology.* Ann Arbor, Mich.: Ann Arbor Science Publishers, 1981.

Basic concepts of industrial toxicology are presented. Presents a good bibliography of sources of information on industrial toxicology: books, journals, guides, reviews, and standards.

Casarett, L. J., and J. Doull. *Toxicology, the Basic Science of Poisons,* 2nd ed. New York: Macmillan, 1980.

Designed primarily as a textbook for courses in toxicology, this book presents a perspective on the subject for the generalist or manager.

Council on Environment Quality. *Chemical Hazards to Human Reproduction.* Washington: U.S. Government Printing Office, 1981.

Contains a compilation of human and animal evidence for adverse reproductive effects of chemicals and chemical processes.

Daugaard, J. *Symptoms and Signs in Occupational Disease: A Practical Guide.* Copenhagen, Denmark: Munksgaard (available from Year Book Medical Publishers, Chicago), 1978.

A quick guide by symptoms or signs of occupational disease and also arranged by chemical.

★Dreisbach, Robert H. *Handbook of Poisoning: Prevention, Diagnosis and Treatment,* 11th ed. Los Altos, Calif.: Lange Medical Publications, 1983.

A concise summary of the diagnosis and treatment of poisoning. Poisons are organized into industrial, agricultural, household, medical, and natural hazards.

Fishbein, Lawrence. *Potential Industrial Carcinogens and Mutagens.* New York: Elsevier Scientific Publishing Co., 1979.

Brings together relevant facts about carcinogenic and/or mutagenic organic chemicals and related compounds—arranged by structural category. Lists uses of the chemicals.

Gosselin, Robert E., *et al. Clinical Toxicology of Commercial Products,* 5th ed. Baltimore: Williams & Wilkins, 1984.

Assists the physician in dealing quickly and effectively with acute chemical poisonings arising from misuse of commercial products. Best part is the trade name index. Contains: (*a*) first aid and general emergency treatment; (*b*) ingredients index; (*c*) therapeutics index; (*d*) supportive treatment; (*e*) trade names index (lists chemicals found in the product); (*f*) general formulations; (*g*) manufacturer's index (addresses and phone numbers).

★Hamilton, Alice, and Harriet Hardy. *Industrial Toxicology* (revised by Asher J. Finkel). Boston: John Wright, Publishing Sciences Group, 1983.

Combines industrial hygiene practice with industrial toxicology. Good overview and background regarding occupational exposures. Previous editions are also excellent.

Hunt, Vilma R. *Work and the Health of Women.* Boca Raton, Fla.: CRC Press, 1979.

A general review; also discusses influence of environmental agents on male reproductive failure.

Hunter, Donald. *The Disease of Occupations,* 6th ed. London, England: Stoughton, 1978.

General review; describes chemical composition of the materials handled in various occupations. Detailed descriptions of industrial processes and expected working conditions.

Kaye Sidney: *Handbook of Emergency Toxicology: A Guide for the*

Identification, Diagnosis and Treatment of Poisoning, 4th ed. Springfield, Ill.: Charles C Thomas, 1980.

Lists signs and symptoms of exposure, the name of the chemical under it, screening exams, health effects of exposure. (More detailed than Proctor & Hughes; see next page).

Kusnetz, Stanley, and Marilyn K. Hutchinson, eds. *A Guide to the Work-Relatedness of Disease,* rev. ed., DHHS (NIOSH), Pub. No. 79–116 Washington: U.S. Government Printing Office, 1979. Order from NTIS No: PB-298-561/A12.

Presents a method for assembling and evaluating evidence that may be relevant in determining work-relatedness of disease, 14 agents are discussed.

Lauwerys, Robert R. *Industrial Chemical Exposure: Guidelines for Biological Monitoring.* Davis, Calif.: Biomedical Publications, 1983.

★Maibach, H. I., and G. A. Gellin. *Occupational and Industrial Dermatology.* Chicago: Year Book Medical Publishers, 1982.

A primer and intermediate text in occupational dermatology designed for use in recognizing, managing, and treating work-related skin diseases.

★NIOSH/OSHA. *Occupational Health Guidelines for Chemical Hazards,* DHHS (NIOSH), Pub. No. 81-123. Washington: U.S. Government Printing Office, 1981. Order from GPO No: 017-033-00337-8.

Summarizes information on permissible exposure limits, chemical and physical properties, and health hazards. It provides recommendations for medical surveillance respiratory protection and personal protection and sanitation practices for approximately 380 chemicals.

NIOSH/OSHA. "Pocket Guide to Chemical Hazards," DHHS (NIOSH) Pub. No. 78-210. Washington: U.S. Government Printing Office, 1978. Order from GPO No: 017-033-00342-4.

Presents in tabular form, information and recommendations relating to permissible exposure limits, chemical and physical properties, health hazard information, respiratory protection and personal protection and sanitation practices for 380 specific chemicals.

NIOSH. *Registry of Toxic Effects of Chemical Substances 1981-82,* DHHS (NIOSH) Pub. No. 83-107. Washington: U.S. Government Printing Office, 1983. Order from NTIS No: PB-81-154-478/A99.

An exhaustive listing of known toxic substances that may exist in the environment or that are manufactured. Also available on microfiche and on-line computer.

Peto, Richard, and Marvin Schneiderman. *Quantification of Occupational Cancer,* Banbury Report No. 9. Cold Spring Harbor, N.Y.: Cold Spring Harbor Laboratory, 1981.

Well-edited collection of papers on above subject. Could be useful in doing risk assessment.

Physicians' Desk Reference, Oradell, N.J.: Medical Economics Co., yearly.

Makes available essential information on major pharmaceutical products. Provides product information (including photograph), diagnostic product information. Information is presented alphabetically by company name, then by product name. Can find who a manufacturer is by using product name index.

Parkes, Raymond W. *Occupational Lung Disorders,* 2nd ed. London, England: Butterworth, 1982.

Detailed and comprehensive text related to lung diseases.

★Proctor, N. H., and J. P. Hughes. *Chemical Hazards in the Workplace.* Philadelphia: J. B. Lippincott, 1978.

Intended for the physician. Describes the potential health effects of 386 chemicals. Presents fundamentals of toxicology, clinical signs of overexposure, medical measures most successful in the early detection of adverse effects, and accepted methods of treatment.

Rom, William N. *Environmental and Occupational Medicine.* Boston: Little Brown, 1983.

The definitive resource on the history, causes, prevention, and treatment of work-related diseases.

Sax, N. I. *Dangerous Properties of Industrial Materials,* 6th ed. New York: Van Nostrand Reinhold, 1984.

Provides a single alphabetical source of concise hazard information on about 15,000 common industrial and laboratory materials. Does not cite source of toxicologic data or reason for hazard rating.

Shaw, Charles R. *Prevention of Occupational Cancer.* Boca Raton, Fla.: CRC Press, 1981.

An overview on the subject. Well organized and well written.

Sittig, Marshall. *Handbook of Toxic and Hazardous Chemicals and Car-*

cinagens, 2nd ed. Park Ridge, N.J.: Noyes Publications, 1985.

Presents, alphabetically, chemical, health and safety, and suggested disposal methods for approximately 600 chemicals. Includes many substances regulated by OSHA, priority toxic water pollutants of EPA and many EPA hazardous chemicals.

★ *Toxic and Hazardous Industrial Chemicals Safety Manual: for Handling and Disposal with Toxicity and Hazard Data.* Tokyo, Japan: The International Technical Information Institute, 1976.

Practical data on more than 700 chemicals: chemical name and synonym, uses, properties, criteria for hazard, criteria for toxicity, handling and storage, emergency treatment and measures, spills and leakage, disposal and waste treatment.

U.S. Coast Guard, Cargo and Hazardous Materials Division. *Chemical Data Guide for Bulk Shipment by Water,* Stock No. 5012-0068. Washington: U.S. Government Printing Office.

Information on bulk chemical shipment.

Zenz, Carl. *Occupational Medicine: Principles and Practical Applications,* 2nd ed. Chicago: Year Book Medical Publishers, 1986.

This book discusses various aspects of toxicology and medicine such as epidemiology in occupational health, physical and chemical environments at work, and psychosocial considerations. Health effects of selected chemicals are discussed.

Poison Control Centers associated with hospitals or county health departments, 911 telephone services, and other local emergency services are frequently good sources of toxicologic information. (See the discussion under "Computer-Based Information Systems" later in this chapter.)

Engineering/Air Pollution Control

★American Conference of Governmental Industrial Hygienists, Committee on Ventilation. *Industrial Ventilation.* Lansing, Mich., (yearly).

The "standard" used for designing industrial fume and dust control systems. A comprehensive text.

American Industrial Hygiene Association. *Air Pollution Manual,* 2 vols., Part I, "Evaluation," 2nd ed. Part II, "Control Equipment." Akron, Ohio, 1972.

Part I has the information needed to evaluate an air pollution problem. Part II describes the equipment used for control of air pollution sources.

Committee on Indoor Pollutants. *Indoor Pollutants.* Washington: National Academy Press, 1981.

An overview of indoor air pollution, what characterizes it, potential health effects, and methods to reduce it.

Environmental Protection Agency. *Air Pollution Engineering Manual,* 2nd ed. Pub. No. AP-40. Washington: U.S. Government Printing Office, 1973.

The manual deals with the control of air pollution at specific sources. These sources include metallurgical, mechanical, incineration, combustion, chemical and organic-solvent-emitting processes.

McDermott, H. J. *Handbook of Ventilation for Contaminant Control,* 2nd ed. Stoneham, Mass.: Butterworth Publishers, 1985.

A practical and comprehensive book covering ventilation including OSHA ventilation standards, evaluation of employee exposures, and methods for solving ventilation problems.

Stern, Arthur C. *Air Pollution,* 3rd. ed. (5 vols.) Vol. 1, "Air Pollutants, Their Transformation and Transport." Vol. 2, "The Effects of Air Pollution." Vol. 3, "Measuring, Monitoring and Surveillance of Air Pollutions." Vol. 4, "Engineering Control of Air Pollution." Vol. 5, "Air Quality Management." New York: Academic Press, 1977.

Almost everything you ever wanted to know about air pollution.

NOISE

Alliance of American Insurers. "Background for Loss of Hearing Claims," 2nd ed. Schaumburg, Ill.: Alliance of American Insurers, 1981.

Provides information about the worker's compensation claims aspect of occupational noise exposure and hearing impairment.

★American Industrial Hygiene Association. *Industrial Noise Manual,* 3rd ed. Akron, Ohio, 1975.

Physics of sound, instruments and techniques of sound measurement, noise surveys, effects of noise on man, hearing impairment, and other noise-related topics are discussed.

American Industrial Hygiene Association. *Noise Monograph.* Akron,

Ohio, 1981.

Covers topics such as physiological effects, measurement, exposure, conservation, and control.

Bragdon, Clifford R. *Noise Pollution: A Guide to Information Sources,* Detroit: Gale Research Co., 1979.

Sources of information on: physiological effects, behavioral effects, abatement; architectural and engineering abatement; government and administration; acoustics; education. References: primary and secondary periodicals and indexes.

Burns, W., and D. W. Robinson. *Hearing and Noise in Industry.* London, England: Her Majesty's Stationery Office, 1970.

Reference for noise standard. Nonspecific to Hearing Conservation Amendment, but gives helpful ideas on how to implement a program in general.

Cheremisinoff, Paul N., and Peter P. Cheremisinoff. *Industrial Noise Control Handbook.* Ann Arbor, Mich.: Ann Arbor Science Publishers, 1977.

Provides information for engineer and planner on how to reduce noise.

National Institute for Occupational Safety and Health. *Industrial Noise Control Manual,* rev. ed., DHHS (NIOSH) Pub. No. 79-117. Washington: U.S. Government Printing Office, 1978. Order from NTIS No: PB-297-534/A16.

Contains basic information on understanding, measuring, and controlling noise and a number of case studies on how noise problems were solved.

Olishifski, J. B., and E. R. Harford. *Industrial Noise and Hearing Conservation,* Chicago: National Safety Council, 1975.

Provides information on how to set up a hearing conservation program.

Peterson, Arnold P. G. *Handbook of Noise Measurement,* 9th ed. Concord, Mass.: GenRad, Inc., 1980.

Comprehensive, but slightly difficult to read.

U.S. Dept. of Labor, Occupational Safety and Health Administration. "Noise Control: A Guide for Workers and Employers." OSHA Pub. No. 3048, 1980. (Available from OSHA offices.)

Concepts involved in noise control, effects of noise on human health, principles to be applied for reducing noise, and an explanation of the legal requirements for noise control that employers must follow. For workers and employers interested in reducing workplace noise.

Radiation

★Cember, Herman. *Introduction to Health and Physics.* Oxford, England: Pergamon Press, 1969.

Main purpose of the book is to present information for attaining technical competency in health physics. Reviews physical principles, atomic and nuclear structure, radioactivity, interaction of radiation with matter, radiation dosimetry, biological effects of radiation, health physics instrumentation, protection methods, and evaluation of protective methods.

Moss, C. E., *et al. Biological Effects of Infrared Radiation,* DHHS (NIOSH) Pub. No. 82-109. Cincinnati, Ohio: NIOSH, 1982.

Presents information on types of IR sources, modes of biological interaction, existing exposure standards and control measures.

Reese, D. J. *Health Physics: Principles of Radiation Protection.* Cambridge, Mass.: MIT Press, 1967.

Written for people with no prior knowledge of radiation protection in various fields.

★Shapiro, Jacob. *Radiation Protection: A Guide for Scientists and Physicians.* Cambridge, Mass.: Harvard University Press, 1972.

Provides the radiation user with information needed to protect self and others. Presents information on principles of radiation protection, radiation dose calculations, radiation measurements, practical aspects of the use of radionuclides, and ionizing radiation and public health.

U.S. Bureau of Radiological Health and Environmental Control Administration Training Institute. *Radiological Health Handbook,* rev. ed. Washington: U.S. Government Printing Office, 1970.

A basic reference and resource guide in radiological health. Presents physical, chemical, and mathematical data; radioisotope, decay, and radioassay data; and radiation protection data.

Wade, James E., and G. E. Cunningham. *Radiation Monitoring: A Programmed Instruction Book,* EDM-123. Oak Ridge, Tenn.: U.S. Atomic Energy Commission.

Self-instruction on radiation monitoring.

Additional information on ionizing radiation can be obtained from the following:

Health Physics Society
4720 Montgomery Lane, Suite 506
Bethesda, Md. 20014

The primary objective is to develop scientific knowledge and a practical means whereby mankind and the environment can be protected against the harmful effects of radiation. The society is an affiliate of the International Radiation Protection Association and publishes *Health Physics.*

National Council on Radiation Protection and Measurements
NCRP Publications
P.O. Box 30175
Washington, D.C. 20014

Publishes guides on radiation protection and measurements. Congressionally chartered; a nonprofit organization.

Legal/Regulatory Concerns

Bacon, Lawrence S. *Bargaining for Job Safety and Health.* New York: The MIT Press, 1980.

Technical, economic, political, and administrative problems that must be overcome if job hazards are to be controlled. Evaluate performance of OSHA in addressing problems with suggestions. Collective bargaining approach to solving problems.

Morris Davis, ed. *Labor Occupational Health Program: A Worker's Guide to Documenting Health and Safety Problems.* Berkeley: Regents of the University of California, 1978.

A guide to questions workers should ask even though employers have responsibility for health and safety. Checklists for walk-throughs are provided; sample of various recordkeeping forms are provided.

LaDou, Joseph. *Occupational Health Law: A Guide for Industry.* New York: Marcel Dekker, 1981.

To help health and safety professionals avoid legal problems.

⋆NIOSH: *Standards, Interpretations and Audit Criteria for Performance of Occupational Health Programs.* Division of Training and

Manpower Development. (Available from the American Industrial Hygiene Assoc.) University of Cincinnati, Sept. 1979.

Provides checklists for evaluating occupational health programs: administration, medical, nursing, hygiene, and health physics. They include: policies, staff, facilities, equipment, education, recordkeeping, and emergency preparedness. A good checklist for self-evaluation of the disciplines of an occupational health team.

Nothstein, Gary Z. *The Law of Occupational Safety and Health.* New York: Free Press, 1981.

Provides answers to the whys, hows, and wherefores of the occupational safety and health law; presents a good understanding of the law in a systematic manner.

Occupational Health in Canada: Current Status. Ottawa: Minister of National Health and Welfare, 1977.

Provides a summary of federal and provincial programs, roles, and workers' compensation system. Good charts on roles and responsibilities of federal and territorial governments.

Peterson, Dan. *The OSHA Compliance Manual,* rev. ed. New York: McGraw-Hill Book Co., 1978.

A good reference on how to comply, records to keep, checklists, how to correct violations, how to become familiar with OSHA.

Toxic Substance Control Source Book. Rockville, Md.: Aspen Systems Corp., 1978.

Summary of TSCA, and EPA's implementation plan. Compilation of information on the TSCA, the laws, nature, and status of toxic and hazardous substances subject to TSCA, and economic impact on industry, identifies and classifies recommended management techniques for control of toxic and hazardous substances posing risk to worker health.

U.S. Department of Labor, OSHA: *Training Requirements in OSHA Standards,* OSHA 2254. Washington, 1979.

A compilation of OSHA's training-related requirements for 29CFR 1910, 1926, and 1928.

The *Federal Register,* published daily by the U.S. government, is the source of all notices of proposed regulations or promulgated regulations. The *Federal Register* may be ordered from Superintendent of Documents, U.S. Government Printing Office, Washington, D.C. 20402.

Several weekly publications, which are commercially available, review the *Federal Register,* comment on how the regulations may affect employers, and reprint, in an organized, updated manner, the regulations from the *Federal Register.* These publications include:

The Occupational Safety and Health Reporter

The Chemical Regulations Reporter

The Environment Reporter

They are published by the Bureau of National Affairs, Inc., 1231 25th Street, N.W., Washington, D.C. 20037.

General Management

★American Nurses Association, Occupational Health Nurses Section. *Guide for the Development of a Manual for an Employee Health Program,* New York, 1962.

Old, but has good basic ideas of how to set up a program and local/community resources that may be available.

Arthur D. Little, Inc. *Costs and Benefits of Occupational Health Nursing,* DHHS (NIOSH), Pub. No. 80-140. Cincinnati: NIOSH, 1980. Order from GPO No: 017-033-00368-8.

Develops basic information regarding costs and benefits of occupational safety and health programs and the needs for an occupational health nurse. Four case studies are examined.

Brown, Mary L. *Occupational Health Nursing.* New York: Springer Publishing Co., 1981.

Chissick, S. S., and R. Derricott. *Occupational Health and Safety Management.* New York: John Wiley & Sons, 1981.

Reviews policies, regulations, hazards, and administrative solutions to occupational safety and health management in Great Britain.

The Conference Board. *Industry Response to Health Risk,* Report No. 811. New York, 1981.

Examples of how large companies have developed health and environmental programs. Presented for the corporate manager to inform that level of management of programs that have been instituted.

Connolly, Walter B., Jr., and Donald R. Crowell, II. *A Practical Guide to the Occupational Safety and Health Act: Law, Principles and Practices,* vol. 1. New York: New York Law Journal Press, 1977.

A comprehensive review and analysis of the OSHAct including procedural and substantive aspects of citations, defenses, and recordkeeping requirements.

Heinrich, H. W., Don Petersen, and Nestor Roos. *Industrial Accident Prevention: a Safety Management Approach,* 5th ed. New York: McGraw-Hill Book Co., 1980.

Presents information on the basis and philosophy of accident prevention, accident prevention methods, law, risk management, and experience modification.

Hopkins, Bill, *et al. Behavior Management for Occupational Safety and Health.* Cincinnati: Division of Training and Manpower Development, U.S. Dept. of Health and Human Services, NIOSH, 1979.

Study of factors important in behavior management; discusses lessons and examples used to modify behavior to solve safety and health problems in business organizations.

Lee, Jane A. *The New Nurse in Industry: A Guide for the Newly Employed Occupational Health Nurse,* DHHS (NIOSH) Pub. No. 78-143. Cincinnati: NIOSH, 1978. Order from GPO No: 017-033-00295-9.

A guide to the duties and responsibilities for the nurse with no experience in occupational health as well as for plant managers and other employers to know what to look for when hiring an occupational health nurse. (Includes where to find sources of consultation and assistance.)

Metropolitan Life Insurance Co. *Correlated Activities in an Employee Health Program: A Guide for Management, the Physician and the Nurse.* About 1962.

Presents the activities of a medium-sized, company-operated employee health program and responsibilities of each group. Does not include industrial hygiene. Slightly out of date, but good.

Safety Program Practices in Record-Holding Plants, DHHS (NIOSH) Pub. No. 79-136. Morgantown, W.Va.: NIOSH, Div. of Safety Research, 1979. Order from GPO No: 017-033-00358-1.

A successful survey of five companies to understand safety programming and practices in industry.

U.S. Department of Labor, Bureau of Labor Statistics. *Occupational Injuries and Illnesses in the United States by Industry,* 1978. Washington: Superintendent of Documents, 1980.

Source of statistics by industry, by size of industry; the statistics come from OSHA, No. 200 forms.

TRAINING

National Safety Council. "1983-1985 College and University Safety Courses," 15th ed. Chicago, 1983.

Gives name and address; tells whether undergraduate or graduate level, non-degree, and/or summer courses offered.

U.S. Department of Labor, OSHA. *Training Requirements in OSHA Standards.* OSHA 2254. Washington, 1979.

Intended to serve as an aid for training employees.

U.S. Department of Labor, OSHA. *A Guide to Worker Education Materials in Occupational Safety and Health.* Washington, 1982.

Lists the resource materials produced or acquired by New Directions Grantees and state programs. Most of the materials can be used for training.

Urban Environment Conference. *A Guide to Worker Education Materials in Occupational Safety and Health.* Vol. I: "Materials produced by citizens' organizations and College and University Labor Education Programs." Vol. II: "Materials Produced by Labor Unions." Washington: U.S. Dept. of Labor, OSHA, 1980.

Lists materials that can be used as training aids.

Weis, William J., and Thomas C. Purcell. *Directory of Academic Programs in Occupational Safety and Health,* 1979. DHHS (NIOSH) Pub. No. 79-126. Cincinnati: NIOSH, 1979. Order from NTIS No: PB-80-149-503/A07.

Provides basic information concerning various academic programs being offered in the field of occupational safety and health: tuition fees, programs, addresses, and contacts.

Other sources of training materials include:

Worker's compensation insurance carriers.

Chemical suppliers, such as J. T. Baker, that provide training in chemical laboratory safety and spill cleanup.

Personal protective equipment suppliers, such as E.A.R. Corporation (hearing protectors).

Associations, such as the American Foundrymen's Society, American Iron and Steel Institute, and National Safety Council.

Occupational health resources centers such as the Women's Occupational Resource Center, School of Public Health, Columbia University, 60 Haven Ave., B-1, New York, N.Y. 10032.

A number of companies and organizations have training materials available, either for rent or purchase, including:

Aetna Life & Casualty Company
151 Farmington Avenue
Hartford, Ct. 06115

American Hospital Association
840 North Lake Shore Drive
Chicago, Ill. 60611

The Bureau of Business Practice
24 Rope Ferry Road
Waterford, Ct. 06385

Factory Mutual System
1151 Boston-Providence Turnpike
Norwood, Mass. 02062
(Fire films only)

Fisher Scientific Company
711 Forbes Avenue
Pittsburgh, Pa. 15219

General Electric Educational Films
Corporations Park
Building 705
Scotia, N.Y. 12302

Resource, Inc.
14007 North Dale Mabry
P.O. Box 271570
Tampa, Fla. 33688

Several states have requirements for hazardous substances information and training. This training in hazardous substances requires having material safety data sheets (MSDSs) readily available for each substance used in the workplace. Not only must the MSDSs be available, but the employees must be informed about the information in an MSDS, either in written form or through training programs. MSDSs can provide the following information about a chemical: physical data, chemical name and synonyms, fire and explosion hazard data, health hazard data, reactivity data, spill or leak procedures, and special protection and precautions information. See section on Chemicals for information on how to obtain Material Safety Data Sheets.

Personal Protective Equipment

Birkner, L. R. *Respiratory Protection Program: A Manual and Guidelines.* Akron, Ohio: American Industrial Hygiene Association, 1982.

Details selection, maintenance, and use of respirators.

NIOSH. *Certified Equipment List* as of June 1980, Supplement, Oct. 1, 1981, DHHS (NIOSH) Pub. No. 80-144 (Order from GPO No: 017-033-00374-2) and Pub. No. 82-106, (Order from GPO No: 017-033-00397-1) respectively. Washington: Superintendent of Documents.

These publications list products certified by NIOSH for personal protection and hazard measurement.

NIOSH. *A Guide to Industrial Respiratory Protection, 1976,* DHHS (NIOSH) Pub. No. 76-189. (Order from NTIS No: LA-667-IM/A09.)

A standard reference in respiratory protection.

Other sources of information on personal protective equipment are the manufacturers or vendors of this equipment. The telephone Yellow Pages under "Safety Equipment and Clothing" is a good reference for finding the names, addresses, and phone numbers of manufacturers, distributors, and sales offices.

Industrial Hygiene News, Chilton Company, Ames Division, Radnor, Pa. 19087, is a monthly publication that carries many ads for manufacturers who can provide information on personal protective equipment.

National Safety and Health News, published monthly by the National Safety Council, carries ads and informative articles on personal protective equipment. Its March issue carries a Safety and Health Buyers' Guide.

Biological

U.S. Department of Health and Human Services, National Institutes of Health, Division of Safety. *Laboratory Safety Publications and Training Aids: An Annotated Bibliography.* Bethesda, Md., December 1981.

Serves as a resource for persons interested in biomedical research laboratory health and safety. Subjects covered include aerosol control and containment, animal handling, biological safety, and disinfection and sterilization.

University of California—San Francisco, Office of Environmental Health and Safety. *UCSF Biosafety Procedure Guide.* San Francisco, 1977.

U.S. Department of Health and Human Services, National Institutes of Health. "Guidelines for Research Involving Recombinant DNA Molecules." *Federal Register,* v. 47, No. 167, August 27, 1982.

These guidelines include several bibliographies.

Health and safety newsletters and reports

AOMA Report. American Occupational Medical Association, 2340 S. Arlington Heights Rd., Arlington Heights, Ill. 60005.

Industrial Hygiene Digest. Industrial Health Foundation, Inc., 5231 Centre Ave., Pittsburgh, Pa. 15232.

Contains abstracts of recent articles and books on industrial medicine, chemical hazards, occupational health management, environment problems, industrial hygiene.

Industrial Hygiene News Report. Flournoy and Associates, 1845 West Morse Ave., Chicago, Ill. 60626. Monthly.

Occupational Health and Safety Letter. Gershon W. Fishbein, Publisher, 1097 National Press Building, Washington, D.C. 20004. Semi-monthly.

Occupational Safety and Health Reporter. Bureau of National Affairs, 1231 25th St., N.W. Washington, D.C. 20037. Weekly.

OSHA Up to Date. National Safety Council, 444 N. Michigan Ave., Chicago, Ill. 60611. Monthly.

Radiological Health Data and Reports. Environmental Protection Agency (formerly published by the Bureau of Radiological Health, Rockville, Md.), Washington, D.C. 20460. Monthly.

Health and safety journals and magazines

Articles on general principles of industrial hygiene sampling techniques and analysis, toxicology, and other occupational health and safety subjects can be found in the following journals and magazines.

American Industrial Hygiene Association Journal. American Industrial Hygiene Association, 475 Wolf Ledges Parkway, Akron, Ohio 44311. Monthly.

American Journal of Industrial Medicine. Alan R. Liss, Inc., 41 E. 11th St., New York, N.Y. 10003. Monthly.

American Journal of Public Health. Journal of the American Public Health Association, 1015 Fifteenth St., N.W., Washington, D.C. 20005. Monthly.

Analytical Chemistry. American Chemical Society, 1155 16th St., N.W., Washington, D.C. 20036. Monthly.

ASHRAE Journal. American Society of Heating, Refrigerating and Air-Conditioning Engineers, Inc., 345 East 47th St., New York, N.Y. 10017. Monthly.

British Journal of Industrial Medicine. British Medical Association, Tavistock Sq., London WCI H9JR, England. Quarterly.

Chemical and Engineering News. American Chemical Society, 1155 16th St., N.W., Washington, D.C. 20036. Weekly.

Environmental Science and Technology. American Chemical Society, 1155 16th Street, N.W., Washington, DC. 20036. Monthly.

Ergonomics. Ergonomics Society, International Ergonomics Association, Taylor & Francis Ltd., London, England. (Available from: Publications Expediting Inc., 200 Meacham Ave., Elmont, N.Y. 11003.) Bimonthly.

Excerpta Medica. Section 35: "Occupational Health and Industrial Medicine." Excerpta Medica Foundation, Nassau Bldg., 228 Alexander St., Princeton, N.J. 08540. Monthly.

Health Physics. Pergamon Press, Maxwell House, Fairview Park, Elmsford, N.Y. 10523. Sponsored by the Health Physics Society. Monthly.

Journal of Air Pollution Control Association. Air Pollution Control Association, 4400 Fifth Ave., Pittsburgh, Pa. 15213. Monthly.

Journal of Occupational Medicine. American Occupational Medical Association, 2340 S. Arlington Heights Rd., Arlington Heights, Ill. 60005. Monthly.

Journal of the Society of Occupational Medicine. John Wright & Sons Ltd., 823-825 Bath Rd., Bristol, BS4 5NU, England. Quarterly.

National Safety and Health News. National Safety Council, 444 N. Michigan Ave., Chicago, Ill. 60611. Monthly.

Occupational Hazards. Occupational Hazards, 614 Superior Ave., W., Cleveland, Ohio 44113. Monthly.

Occupational Health & Safety. Medical Publications, Inc., 4901 Bosque Blvd., Waco, Texas 76710.

Occupational Health Nursing. American Association of Occupational Health Nurses, 79 Madison Ave., New York, N.Y. 10016. Monthly.

Occupational Safety and Health Subscription Service. Standards, Interpretations, Regulations, and Procedures. U.S. Occupational Safety and Health Administration, Washington, D.C. 5 vols. (OSH 01 through OSH 05). Available from Government Printing Office, Washington, D.C. 20402. Catalog No. L35.6/3-1 through -5.

Professional Safety (formerly *American Society of Safety Engineers Journal).* ASSE, 1800 E. Oakton St., Des Plaines, Ill. 60018. Monthly.

Scandinavian Journal of Work, Environment & Health. Haartmaninkatu 1, SF-00290, Helsinki 29, Finland. Monthly.

Sound and Vibration. Acoustical Publications, Inc., 27101 E. Oviatt Rd., Bay Village, Ohio 44140. Monthly.

State of the Art Reviews: Occupational Medicine. Hanley & Belfus, Inc., 210 S. 13th St., Philadelphia, Pa. 19107. Quarterly.

COMPUTER-BASED INFORMATION SYSTEMS

Information published in the journals listed above, plus much more, can be rapidly and readily searched by using computer-based information systems. A list of computerized information services follows. Some of these information systems are based in the National Library of Medicine (NLM).

Books in Print.

Information source on U.S. book publishing; from R. R. Bowker Co., 1180 Avenue of the Americas, New York, N.Y. 10036.

International Dissertation Abstracts.

Doctoral dissertations from accredited universities (predominantly in the U.S.); from University Microfilm International, 300 North Zeeb Rd., Ann Arbor, Mich. 48106.

GPO Monthly Catalog.

The machine equivalent of the printed "Monthly Catalog of U.S. Government Publications," from U.S. Government Printing Office, Washington, D.C. 20402.

National Newspaper Index.

An index to *The Christian Science Monitor, The New York Times,* and *The Wall Street Journal;* from Information Access Corporation, 11 Davis Dr., Belmont, Calif. 94002.

National Technical Information Service.

Citations to government-sponsored research, mostly federal, but also contains some state and local reports; from National Technical Information Service, U.S. Dept. of Commerce, Springfield, Va. 22161.

Pharmaceutical News Index.

Indexing records for seven major drug industry newsletters; from Data Courier, Inc., 620 S. Fifth, Lexington, Ky. 40202.

TSCA Initial Inventory.

A nonbibliographic dictionary listing of chemical substances in commercial use in the United States as of June 1, 1979. It is not a list of toxic chemicals. From Dialog Information Retrieval Service, 3460 Hillview Avenue, Palo Alto, Calif. 94304.

Ulrich's International Periodicals and Irregular Serials.

Information on periodicals and serials; from R. R. Bowker Co., 1180 Avenue of the Americas, New York, N.Y. 10036, 1983.

National Library of Medicine. The following data banks can be reached through the National Library of Medicine, 8600 Rockville Pike, Bethesda, Md. 20814. For reference to the various data bases described next, call the information desk at (800) 638-8480 and ask for the desired database (for example, CHEMLINE).

AVLINE.

Audiovisual materials in the health sciences that have been evaluated by experts from health sciences professions.

BIOETHICSLINE.

Bibliographic information on questions of ethics and public policy arising in health care or biomedical research.

CANCERLIT.

Scope includes journal articles, proceedings of meetings, government reports, symposia reports, selected monographs, books and theses relating to all aspects of cancer; information from the National Cancer Institute.

CANCERPROJ.

Summaries of on-going cancer research projects provided by cancer scientists in many countries; information from the National Cancer Institute.

CHEMLINE.

On-line chemical dictionary containing CAS Registry Numbers, molecular formulas, CAS preferred chemical nomenclature, synonyms, ring information and structural fragments.

TDB.

Toxicology Data Bank contains facts and data from standard reference textbooks, handbooks, and critical documents; information from NIH's Study Section in Toxicology.

TOXLINE/TOXBACK.

Contains citations for human and animal toxicity studies, effects of environmental pollutants, adverse drug reactions, and analytical methodology.

RTECS.

Registry of Toxic Effects of Chemical Substances contains toxicity data for chemical substances; information from the National Institute for Occupational Safety and Health.

For additional information on how to query the RTECS file, contact:

> RTECS
> Toxicology Information Program
> Specialized Information Services
> National Library of Medicine
> 8600 Rockville Pike, Room C-6B
> Bethesda, Md. 20814
>
> Telephone: (301) 496-1131

NLM Regional Medical Libraries

Region I

New England Regional Medical Library Service (Conn., Mass., Maine, N.H., R.I., and Vt.)

Francis A. Countway Library of Medicine
Harvard University
10 Shattuck St.
Boston, Mass. 02115

Region II

New York and New Jersey Regional Medical Library
New York Academy of Medicine Library
2 East 103rd St.
New York, N.Y. 10029

Region III

Mideastern Regional Medical Library Service (Del. and Pa.)
Library of the College of Physicians
10 South 22nd St.
Philadelphia, Pa. 19103

Region IV

Mid-Atlantic Regional Medical Library (D.C., Md., N.C., Va., and W.Va.)

National Library of Medicine
8600 Rockville Pike
Bethesda, Md. 20814

Region V

Kentucky-Ohio-Michigan Regional Medical Library Program
Wayne State University
Shiffman Medical Library
4325 Brush St.
Detroit, Mich. 48201

Region VI

Southeastern Regional Medical Program (Ala., Fla., Ga., Mass., S.C., Tenn. and B.R.)

A.W. Calhoun Medical Library
Emory University
Atlanta, Ga. 30322

Region VII

Midwest Health Sciences Library Network (Iowa, Ill., Ind., Minn., N.Dak., and Wisc.)

John Crerar Library
33 West 33rd St.
Chicago, Ill. 60616

Region VIII

Mid-Continental Regional Medical Library Program (Colo., Kans., Mo., Ne., S.Dak., Utah, and WY.)

Library of Medicine
University of Nebraska Medical Center
Omaha, Neb. 68105

Region IX

South Central Regional Medical Library Program (Ark., La., N.Mex., Okla. and Texas)

University of Texas Health Science Center
5323 Harry Hines Blvd.
Dallas, Texas 75235

Region X

Pacific Northwest Regional Health Sciences Library (Alaska, Idaho, Mont., Oregon., and Wash.)

University of Washington Health Sciences Library
Seattle, Wash. 98195

Region XI

Pacific Southwest Regional Medical Library Service (Ariz., Calif., Hawaii, and Nev.)

Biomedical Library
Center for the Health Sciences
University of California at Los Angeles
Los Angeles, Calif. 90024

NOTE: Non-U.S. MEDLARS centers are located in Australia, Brazil, Canada, France, Japan, Mexico, South Africa, Sweden, the United Kingdom, and West Germany.

Part V

Legal Aspects of Occupational Health and Safety

Chapter 16

Governmental Impact on Occupational Health And Safety

by Charles R. Goerth

People entering the work force in the 1980s, either as workers or managers, accept as a fact of business life, governmental regulation of workplace safety and health. It would come as a surprise to some of the newcomers that the federal government's major involvement in safety and health only began as early as January 1969, when the first proposal for an occupational safety and health act was introduced into the House of Representatives. A Senate version was initiated later that same year.

Prior to that time, occupational safety and health had been a piecemeal concern of the federal government. Early labor-oriented legislation included provisions for safety. For example, the Walsh-Healey Public Contracts Act of 1936 set limited standards for working conditions in factories. It required contractors furnishing or manufacturing materials with a value of $10,000 or more to maintain safe and healthy workplaces. The standards for safety and health were vague. The Labor Management Relations Act of 1947 (Taft-Hartley Act) contained a provision allowing employees to walk off a job if it was "abnormally dangerous." Vagueness again.

Conditions in particular industries led to specialized legislation: Metal and Non-Metallic Mine Safety Act of 1966; McNamara-O'Hara Public Service Contract Act; Maritime Safety Act; Coal Mine Safety Act; Contract Work Hours and Safety Standards Act (also known as the Construction Safety Act). In 1969, the Coal Mine Health and Safety Act was enacted. It was the direct result of a coal mine explosion in 1968 that killed 79 people. The disaster triggered calls for action in Washington to stop what some viewed as the carnage in the workplace.

Dramatic statistics of the late 1960s gave added impetus to federal action. Testimony before House and Senate committees revealed the tragic stories of people dying as a result of industrial accidents. The loss in work days from disabling injuries and illness numbered in the hundreds of millions. One estimate held that workers lost \$1.5 billion annually in wages because of illness and disability. The impact on the economy was a staggering drop of \$8 billion in the Gross National Product. Even more serious was the contention that the number of deaths and injuries—and resulting losses, both personal and economic—were accelerating at an alarming rate. Further, it was feared that chemicals used in industry added a health impact to the impact of accidents on the job.

Spurring those calls for federal regulation was the demonstrated success of safety programs operated on a voluntary basis. National Safety Council (NSC) member-companies had significantly lower rates of disabling injuries per million workers. For example, in 1968, NSC members showed 5.3 disabling injuries per million worker-hours compared with the nonmember tally of 17.9 per million. That same year, after 10 years of operation of the Maritime Safety Act, the rate of injury among longshoremen and shipyard workers had dropped by 53 percent and 41 percent, respectively.

Almost two years after the first version was initiated in the House of Representatives, the Occupational Safety and Health Act of 1970 was signed into law. It created the Occupational Safety and Health Administration in the Department of Labor and a research organization, the National Institute of Occupational Safety and Health, in what is now known as the Department of Health and Human Services. The law went into effect in April 1971. It was the result of numerous compromises and, like most compromises, has inherent problems. Certain sections of the act are vague, others are redundant, and some are even contradictory.

A problem experienced in enforcing the law has been interpreting the intent of Congress. The many points of view—the bargaining, and the back-and-forth of the debate that characterized its legislative history—have made it impossible to tell precisely what the legislators had in mind. The result has been continuing controversy over the years, swirling around charges that OSHA inspectors are either nit-picking or ignoring dangerous conditions. Employers have complained that some standards are unnecessary or are counterproductive. They contend that safety and health requirements are unduly burdensome to business or fail to protect workers, particularly from health hazards. Others say that the federal government has exceeded its authority, on the one hand, or is failing to perform the functions assigned to it by Congress, on the other.

Presidential administrations have, of course, imparted their partisan stamps. Democratic administrations, influenced by labor unions and worker activists, have worked to strengthen the enforcement of existing regulations and to broaden and toughen the regulatory pace. Republican administrations have attempted to downplay OSHA's activities where they have been viewed as an excessive burden on business.

President Ronald Reagan took office in January 1981, highlighting the shifting philosophy toward the role of government intervention in the workplace. Reagan's oft-stated goal is to take government "off the backs" of the people and of business wherever possible. As a consequence, existing regulations of the various agencies were to be reexamined to determine their impact, particularly the economic impact, and their benefits. New regulations must pass through the powerful Office of Management and Budget for approval.

OSHA was targeted for drastic revision, if not elimination, during Reagan's 1980 presidential campaign. That did not happen, but changes did occur under the tenure of Assistant Secretary of Labor Thorne C. Auchter, a former Florida construction executive and the first agency head under the new Administration.

Principal changes have been:

- Targeting industries and companies with highest safety and health problems for inspection.

- Performing only brief surveys, if that, at companies with good records.

- Passing over companies with active safety and health programs, if their programs generally meet OSHA criteria for effectiveness.

- Reducing the number of follow-up inspections on the basis of statistics showing that 99 percent do not reveal any "failure-to-correct" violations.

- Reducing employer lawsuits by giving OSHA regional directors authority to negotiate settlements.

- Pressuring the states to assume more enforcement duties as part of the Reagan administration's philosophy of government.

OSHA claims that the 61,000 inspections that it made in the Reagan administration's first fiscal year were 3,000 more than in President Jimmy Carter's last year in office. These inspections were accomplished with the same number of field inspectors, 1,200.

Some labor unions have complained vigorously about the new agenda under Reagan. They interpret these changes in modes of enforcement and the decline in legal challenges as indicative of a relaxation of requirements for worker safety and health. But labor is divided in its position. Leaders of the construction industry unions—construction was a prime target of OSHA activity during the first two years of the Reagan administration—have stated privately their satisfaction with the results of OSHA targeting on their industry, because safety performance did improve.

Additional developments are shifting the direction and structure of OSHA and its regulatory and enforcement activity:

• *Judicial challenges.* The courts have ruled on crucial issues pertaining to the scope and detail of standards, looking at the issues of significant risk, best available evidence, medical surveillance, economic feasibility and judicial review of rulemaking.

• *Search warrants.* The Supreme Court has ruled on the legal issues arising out of OSHA inspections. NIOSH inspections and demands for data are also under scrutiny.

• *Private actions.* Employees are going to court in increasing numbers in connection with work-related injury and illness. The targets are manufacturers of products used in the workplace, labor unions, insurance companies, and employers themselves.

This chapter examines these trends more closely, exploring the day-to-day operations of OSHA, NIOSH, and the Occupational Safety and Health Review Commission (OSHRC). It considers the other agencies that have an impact on workplace safety and health. Little known to most workers—and managers—is the influence that other federal agencies have on occupational safety and health. These agencies range from the Environmental Protection Administration (EPA) to the Food and Drug Administration (FDA) to the Coast Guard. They operate under statutes authorizing regulation and enforcement of workplace safety and health within their areas of jurisdiction. Sometimes these activities are in concert with OSHA. At other times, there have been bitter battles over jurisdiction. Duplication and overlap occur, creating gray areas confusing to everyone. Those manufacturers falling under the jurisdiction of other agencies must deal with one or more regulatory bodies on matters of workplace safety and health.

OSHA's response has come by way of two standards: (*a*) the Access to Medical Records Standard—to give workers access to company records dealing with hazardous working conditions, their medical data, and information about their exposure to hazards; and (*b*) the Hazard Communication Standard—to provide workers with information about the hazardous substances with which they work on the job. Both standards originated within the framework of the principle that a worker has a right to know the hazards associated with his or her job. (Also see Chapter 10, "High-Risk Workers.")

The right-to-know issue has become a controversial topic. For example, by early 1985, 22 states* had enacted laws requiring that the names of hazardous substances used in the workplace be made available to workers and, in some cases, to the public. Proponents of the state legislation contended the local action was necessary because OSHA had failed to do anything about it.

The proposed Federal Product Liability Act places significant restrictions on the rights of workers injured on the job to obtain damages beyond those provided by workers' compensation. At the same time, the act imposes liability on employers and co-workers whose negligence has led to employee injury. The statute is yet another interpretation of the federal government's new role in mandative safety and health in the workplace.

THE REGULATORY AND ENFORCEMENT SPHERES OF THE FEDERAL AGENCIES

The principal vehicles of the federal government to regulate safety and health in the workplace are the Occupational Safety and Health Administration, the National Institute for Occupational Safety and Health, and the Occupational Safety and Health Review Commission. All were created by the Occupational Safety and Health Act of 1970. They will be discussed first; other involved federal agencies will be covered in the second part of this section.

Principal federal agencies

Employers in every state, the District of Columbia, Puerto Rico, and the Territories have two main obligations. (*a*) Employers must keep their workplaces free from hazards that cause or are likely to cause death or serious harm to their employees. (*b*) Employers must also comply with all regulations and standards promulgated by OSHA under its authority granted by the act.

OSHA standards have been developed in three ways: (*a*) by adoption of national consensus standards and existing federal safety and health standards, an approach used until 1973; (*b*) by modifying, revoking, or issuing new standards after a rulemaking procedure; and (*c*) by issuing emergency temporary standards. Emergency temporary standards may

---

*Alaska, California, Connecticut, Delaware, Florida, Illinois, Maine, Maryland, Massachusetts, Michigan, Minnesota, New Hampshire, New Jersey, New Mexico, New York, Oregon, Pennsylvania, Rhode Island, Utah, Washington, West Virginia, and Wisconsin.

be issued if employees are subject to grave danger as a result of exposure to substances that are known to be toxic or physically harmful. The action must be necessary to protect employees from immediate danger.

An employer may also obtain a temporary or permanent variance from a promulgated standard. A temporary variance may be granted if the employer is unable to comply with a particular standard because of lack of staff, facilities, or equipment. The employer is expected to take all possible steps to protect employees from the hazard, and must submit plans for achieving compliance as soon as possible. A permanent variance may also be issued if OSHA determines that the workplace is as safe and healthful as it would be by compliance with the established standard.

The OSHAct specifies that compliance officers be empowered to inspect workplaces. An OSHA inspector must present his or her credentials to the owner, operator, or agent in charge of a business establishment before proceeding. Both the employer and an employee representative have the right to accompany the inspector.

After the inspection, a conference takes place during which safety and health conditions and possible violations are discussed. The compliance officer cannot issue citations on the spot. He or she must go over the report with an OSHA area director. The area director decides whether or not to issue a citation and computes any penalties to be assessed. A citation, if issued, is mailed to the employer no later than six months after the alleged violation occurs. All citations must be in writing and describe in detail the violations and must refer to the pertinent standards and regulations.

The following penalties may be imposed:

De Minimis Notice	$0
Not Serious	$0–$1,000
Serious	$1–$1,000
Repeated	$0–$10,000
Willful	$0–$10,000
Failure to Abate Notice	$0–1,000 per day

Various factors are considered in selecting the penalty: the employer's good faith, the severity of the violation, past compliance of the employer, and the size of the company.

Penalties for nonserious violations cannot be assessed the first visit unless an inspection shows ten or more violations. Small employers who have requested on-site consultation and make efforts to correct violations are not subject to penalties for nonserious violations. Criminal sanctions are available for willful violations where death has occurred.

An employer, an employee, and/or an authorized employee representative (e.g., from the union) have 15 working days in which to file a Notice of Contest of the citation of a violation. Before a citation can be appealed to a higher level, do any one of these three things: contest the violation, the date set for the abatement (i.e., date for elimination of the violation), or a proposed penalty. Filing a Notice of Contest in good faith stops the abatement period. A hearing must then be scheduled. An employer can file a petition for modification of the abatement if it is not possible to comply within the time schedule.

Whenever a Notice of Contest is filed, it must go to the Occupational Safety and Health Review Commission (OSHRC). This independent, quasi-judicial agency reviews all citations of violations. The federal government, in the person of the Secretary of Labor or a representative, is called the complainant. The government has the burden of proving any violation. The employer is the respondent.

An administrative law judge presides over the hearing. The judge's decision affirms, modifies, or vacates the citation, the penalty, or the abatement date. The decision is reviewed automatically by the three-member commission. Parties dissatisfied with the judge's decision can petition for a review by the commission to reconsider the evidence. The commissioners have the power to reverse the preliminary finding of the judge. Such reviews must take place within 30 days, otherwise the judge's decision is final.

When the commission issues a final order, any party adversely affected may petition for review by the U.S. Court of Appeals, either in the judicial circuit in which the violation occurred, or in the District of Columbia branch of the U.S. Court of Appeals. The appeal must take place within 60 days of the final order.

The act makes it possible for the states to operate their own compliance plans. As of March 1985, the following 23 states and territories had submitted plans that were approved by OSHA:

Alaska	North Carolina
Arizona	Oregon
California	Puerto Rico
Hawaii	South Carolina
Indiana	Tennessee
Iowa	Utah
Kentucky	Vermont
Maryland	Virginia
Michigan	Washington
Minnesota	Wyoming
Nevada	Virgin Islands
New Mexico	

Why have fewer than half the states and territories taken steps to set up plans? Why did other states that had plans (Connecticut and Colorado) terminate them? The answer is money. A state plan that meets the basic requirement for federal approval—being "at least as effective" as the federal law—is expensive. Federal financial support for state plans has been declining over the years to the point where it is now just 50 percent. Many state legislatures are simply not willing to pay the remaining cost.

Companies covered by the act are those engaged in interstate commerce. (The federal government itself, also state and lower political subdivisions, are not covered by OSHA.) An employer is a company with one or more employees. Certain employers are exempt. Farm businesses with ten or fewer employees at any one time during the year do not have to comply. Members of a farmer's family are not considered to be employees.

The OSHAct requires that each federal agency establish, develop, and maintain a health and safety program that is consistent with OSHA standards although federal agencies are not subject to regulation. Beginning in 1980, federal employees obtained added protection through an executive order of the president. These agencies are obligated to comply with OSHA standards except where the Secretary of Labor approves alternatives. The executive order also requires that actions be taken to make workplaces free of recognized hazards causing or likely to cause death or serious physical harm. Federal employees are protected against discrimination for exercising such rights as making formal complaints about unsafe or unhealthy working conditions. Finally, the executive directive establishes occupational safety and health committees to monitor the agency's safety and health performance, and permits unannounced inspections by the Labor Department.

The OSHAct has a distinctive feature that differentiates it from other work-oriented legislation, for example, the National Labor Relations Act and Equal Employment Opportunity Act. It does not prohibit specific employment practices. Rather, it assigns the Secretary of Labor responsibility for developing detailed occupational safety and health standards. These standards generally do not list prescribed conduct, but rather set forth the requirements to maintain safe conditions or to adopt and make use of various practices, means, methods or operations necessary for reducing hazards in the workplace.

There are four broad categories of standards:

- Construction
- Agricultural
- General Industry
- Maritime and Longshoring

These standards can also be divided into classes based on methods of promulgation: (*a*) existing standards adopted into the program; (*b*) new standards; and (*c*); emergency temporary standards. The standards in the first group were existing federal standards and national consensus standards, thus providing immediate protection for workers.

To issue new standards, a complicated rule-making procedure is set forth in the statute. The Secretary of Labor has the responsibility for initiating the procedure, although interested parties can suggest new standards. The law gives the secretary the power to issue an emergency temporary standard if employees face "grave danger from exposure to substances or agents determined to be toxic or physically harmful or from new hazards." The rule-making procedure can be bypassed in such a situation, but the temporary standards can remain in effect only six months. Thereafter, a permanent standard must be promulgated. A challenge of any rule goes to the U.S. Court of Appeals, and a petition for review must be filed within 60 days after issuing of the standard.

The National Institute of Occupational Safety and Health (NIOSH) provides one source of standards. This agency was authorized under the statute to conduct studies and research in order to develop recommended safety and health standards. NIOSH is under the jurisdiction of the Department of Health and Human Services. NIOSH's role is not only to make recommendations for new standards, but also to update and revise standards that have already been issued. Its work can be initiated by the Secretary of Health and Human Services, the NIOSH director, or an employer. OSHA is responsible for determining which of the standards proposed by NIOSH are feasible.

In developing standards, NIOSH proceeds with an in-depth analysis of existing literature, which can take six to twelve months. The agency can conduct research to fill in gaps revealed by the literature. The research can be done in-house or by contract to outside firms.

NIOSH also has the authority to conduct inspections and to question employers and employees. It has the authority—conferred by the courts—to issue warrants to gain access to establishments to gather information, to examine records, and to perform medical examinations and environmental sampling. If a hazard is discovered as a result of such inspection, NIOSH reports it to OSHA. This could trigger an inspection, but not an immediate citation.

The process of gathering all the information needed about a particular hazard can, of course, take years. A first draft of recommendations is reviewed by NIOSH staff as well as outside consultants, professional societies, and federal agencies. When approved by NIOSH and the general counsel of the Department of Health and Human Services, a recommendation goes to the Secretary of Labor.

To obtain information to initiate standards, OSHA has also relied on voluntary organizations that have produced a variety of national standards pertaining to workplace safety and health.

Other responsible federal agencies

Although OSHA is the dominant governmental agency responsible for occupational safety and health, a number of other agencies have related responsibilities. These agencies are primarily concerned with protecting the public, but they include provisions for protecting workers who come in contact with regulated products or substances in their manufacture, distribution, or use. It is not uncommon for the regulations of these agencies to overlap, resulting in duplication and even conflicting regulations. In other instances, the regulatory interface is blurred, which leads to confusion and uncertainty over which agency has jurisdiction in a particular situation. The other responsible agencies are:

- Food and Drug Administration (FDA) enforces the Food, Drug and Cosmetic Act (FDCA) and Radiation Control for Health and Safety Act (RCHSA).
- Environmental Protection Agency (EPA) enforces the Toxic Substances Control Act (TSCA); Federal Insecticide, Fungicide, and Rodenticide Act (FIFRA); and Noise Control Act (NCA).
- Department of Transportation (DOT) enforces the Hazardous Materials Transportation Act (HMTA).

The FDA has jurisdiction over foods, drugs, cosmetics, medical devices, animal feed, and biological and electronic products used for medical therapy or having a potential of harm for the human being. The thrust of the FDA's regulatory and enforcement activity is to make sure that such products are safe, wholesome, and effective and that they are manufactured, packaged, shipped, and stored under sanitary conditions. To carry out these functions, the FDA can shut down manufacturing or packaging of products and can seize such products *anywhere* in distribution. It can obtain court injunctions against the sale of such products. It can also initiate criminal prosecution of violators of the regulations.

The FDA is also concerned with the health and safety of workers in plants making and distributing products. For example, the FDA enforces permissible levels of pesticide residues in processed foods. But the worker on the processing and packaging line in a food plant needs protection against exposure to the pesticides in the food handled as well. EPA has primary responsibility to regulate the use of pesticides and to establish nonharmful pesticide levels in or on raw agricultural products. However, the FDA regulates the pesticides that remain in processed foods, based on EPA tolerances. This is an example of overlap of regulatory responsibility. Because the two agencies focus on two separate parts of the food supply, they complement each other.

The FDA brings about its general regulation through the vehicle of Good Manufacturing Practices (GMPs). GMPs detail requirements governing all aspects of sanitation and quality control of food, finished pharmaceuticals, blood and blood components, and medical devices. They address sanitation of processing and packaging lines, general cleanliness, control of communicable diseases, and appropriate education and training of staff. GMPs also regulate the buildings in which covered operations take place, dealing with ventilation to minimize fumes or vapors, adequate working space for health and safety, lighting, personal hygiene facilities, safe water supply, sewage and waste disposal, and storage of housekeeping chemicals and toxic substances.

The Radiation Control for Health and Safety Act (RCHSA), which the FDA's Bureau of Radiological Health enforces, regulates products that emit electronic radiation. It can require such products to bear warning of hazards and contain safe installation and use information. Such labeling protects workers in plants that manufacture the products as well as those in the distribution system. It is also important for workers who use the products.

The Toxic Substances Control Act (TSCA), which the EPA also administers, is the government's tool for regulating the output of the chemical industry. Its function is to control the introduction of new chemical substances into the marketplace and workplaces. Included in the coverage are existing chemicals that may have new uses, or whose hazards may only now be ascertained. The EPA has a range of regulatory actions it can take, from label warnings and restrictions on use, to a complete prohibition of manufacture. Under the authority of TSCA, the EPA works with other agencies in collecting information about the hazards of such substances and in disseminating this information. After notifying another agency of a particular substance hazard, the EPA must allow that agency to carry out the necessary regulation. At the same time, if the agency's action is not sufficient, EPA action is not preempted. That can result in duplication of regulatory action, possibly leading to conflict and confusion.

Any company that manufactures, processes, or distributes chemical substances must submit to the EPA data on health and safety effects of those substances. If such studies reveal a serious hazard, the EPA must be notified at once. The data must include studies of occupational exposure to such chemical substances.

The Federal Insecticide, Fungicide, and Rodenticide Act (FIFRA) covers the substances named in its title. The possibility of health hazards trigger EPA restrictions. Among the restrictions are limitations on the use of the product in certain localities and on the strength of substances themselves, as well as in combination with other substances. Protection is to be provided by employers for the public as well as for workers applying the substances. The agency also sets the degree of training and experience required of the applicators.

The Noise Control Act requires EPA to establish noise emission standards for mechanical products that are major sources and frequently found in the workplace. Included are motor vehicles, construction equipment, and other machinery using motors or engines and moving parts. Such noise control could affect members of the general public and operators of the equipment.

The Department of Transportation (DOT) enforces the Hazardous Materials Transportation Act (HMTA). It controls the packaging, handling, and shipment of substances that could become a hazard to workers in the distribution chain as well as to the general public. A major target of the regulatory process is to prevent accidental leakage or spillage of such substances during transportation and storage, particularly, if an accident occurs to the train, truck, or airplane carriers. The regulations cover warning labels as well as packaging and handling specifications and procedures.

Minimizing duplication

To try to minimize duplication and overlapping jurisdiction, OSHA has entered into written agreements called understandings with other federal agencies. They specify the spheres within which each agency is to operate. They should pinpoint for an employer whether OSHA or another agency is responsible for regulating and enforcing safety and health in his workplace.

Employers within the following industries may be exempt from OSHA's authority and responsible to other agencies, or may be responsible to both, depending on the understanding between the agencies.

Airlines. The Federal Aviation Agency (FAA) issues standards governing the transportation of radioactive material by air. It has jurisdiction over control of hazards to workers in airline facilities. For example, OSHA has also been barred from exercising authority over aircraft ground maintenance personnel changing aircraft lights because the FAA requires airlines to issue procedure manuals dealing with safety of ground maintenance employees. On the other hand, the FAA has issued an Airport-Facility Construction Safety Handbook, but OSHA still exercises jurisdiction over contractors.

Atomic energy. An employer having a contract with the U.S. Navy to build a nuclear-powered ship is exempt from OSHA, coming instead under the jurisdiction of the Nuclear Regulatory Commission.

Coast Guard. Coast Guard regulations cover seamen, but not longshoremen. The relationship of the employee to a vessel as opposed to a dock is the criterion. A work-related injury on a vessel is in the sphere of the Coast Guard rather than OSHA. As a result of amendments to the Outer Continental Shelf Act of 1978, the Coast Guard has authority over working conditions on the Outer Continental Shelf (primarily drilling rigs).

Customs service. Regulations of the Customs service do not take priority over OSHA because they are primarily for purposes of security, not protection of employee safety and health.

Explosives. The Bureau of Alcohol, Tobacco and Firearms controls the sale and storage of explosives, including safety and health of workers in arsenals and explosives factories. A distribution outlet for explosives does not have to meet OSHA requirements on storage facilities, but rather those of the bureau.

Meat packing. Meat packers are regulated under the Wholesome Meat Act, and therefore are under the jurisdiction of the U.S. Department of Agriculture, but must comply with OSHA standards since the Wholesome Meat Act is for the protection of consumers, not employees.

Mines and milling. Employees in mining or milling are regulated by the Metal and Non-Metallic Mine Safety Act and are exempt from OSHA regulations. The Mining Enforcement and Safety Administration (MESA) and OSHA have a Memorandum of Understanding which establishes their respective spheres. OSHA exercises authority over operations that are not directly involved in mining and milling. Therefore, employers who engage primarily in mining and milling work are not exempt from OSHA, when performing work unrelated to those two operations. For example, a company engaged in constructing underground caverns for oil storage and a company repairing a rig at the top of a mine shaft would each have to deal with OSHA. The Mine Safety Health Amendment Act of 1977, which created the Mine Safety and Health Administration within the Department of Labor, has reinforced the exemption from OSHA. One objective of the statute was to set up a vehicle for settling jurisdictional disputes between MESA and OSHA.

Pipe lines. If a company is regulated by the Department of Transportation's Office of Pipe Line Safety, then it is exempt from OSHA authority. However, a contracting firm repairing a natural gas tank is subject to OSHA because it is engaged in repair work, not transportation of gas or operation of a pipe line.

Railroads. The railroads do not have industrywide exemption from OSHA authority. The Federal Railroad Administration (FRA) exercises authority only in relation to working conditions. For example, the railroads have been exempt from OSHA recordkeeping because the FRA has comparable recordkeeping regulations. However, a 1982 decision of the OSHRC is likely to change this exemption unless overturned by a court appeal. The commission declared that a railroad must meet OSHA recordkeeping requirements because of the importance of such data in determining overall enforcement needs in the industry.

The importance of the interrelationship of federal agencies in connection with worker health and safety was highlighted in the winter of 1984 by the Bhopal, India, industrial disaster. A malfunction in a Union Carbide Corp. plant's processing system allowed the escape of deadly methyl isocyanate gas throughout the plant and into the surrounding countryside. Result: death and permanent health injury for thousands of people, making it the most disastrous industrial accident in history.

This event illustrated how a failure in control of the conditions which affect workplace health and safety can, under the right set of circumstances, lead to an environmental catastrophe of mammoth proportions.

THE RIGHT TO KNOW

Access to medical and exposure records

In 1980, OSHA published its controversial "Final Rule on Access to Exposure, Medical Records." The standard gives employees and their representatives access to their own medical records and to company records of exposure to hazardous conditions in the workplace. OSHA also has access to the same information, including medical records, without the consent of employees.

The promulgation of this rule culminated a battle that began formally on July 21, 1978, when OSHA first proposed its records-access standard. During the subsequent two years, the procedure for written comment on the proposed standard and for testimony from interested parties generated thousands of pages, much of it in opposition.

Opposition came from company management and from industry organizations and from medical directors and their professional organization, the American Occupational Medical Association. Labor unions were critical of the proposal because it did not go far enough in allowing union health and safety staffs access to company records on employees.

The battle is not over, having moved now from the hearing room to the courts. Industry associations in various parts of the country have brought suit, charging the standard is unworkable, that it goes beyond the statutory authority of OSHA, and that procedural irregularities in the development of the standard call for its invalidation. So far, the courts have upheld the standard.

Responding to the criticism subsequently, OSHA published proposals to modify the access rule. The revision made it less onerous to industry, while maintaining the basic objective: letting workers know what's happening to them in the work environment. The OSHA revision, in turn, has generated heated opposition from labor unions which see it as an attempt to sharply restrict access to records. Court challenges by unions have not been successful.

The standard has put new demands on the time of those involved in occupational health and safety management, although the oft-stated fears of a flood of requests for access to records and copies were unfounded. Employers must deal with requests for access to the employee records. And health professionals have to explain the content of such records. There are tests of discretion and judgment within the framework of the rule and certain limitations on access and disclosure. However, there are new conflicts now among employees, managers, unions, personal physicians, company health professionals, outside occupational health consultants and care providers, and family members.

Certain basic questions about the purpose and operation of the standard require consideration.

How does OSHA justify the access standard? "Based on experience, expertise, and the rulemaking record," the standard says, access to employee exposure and medical records is "critically important" to the detection, treatment, and prevention of occupational disease. Workers need direct access to this information as well as to any analysis of the records made by employers. Employee exposure records reveal the identity of an extent of exposure to toxic substances or harmful physical agents; employee medical records contain individual health status information which may indicate whether or not an employee's health is being, or has been, impaired by exposure to toxic chemicals or harmful physical agents. The overall objective of the standard, as set forth by OSHA, is to allow workers "to play a meaningful role in their own health management. . .as there are no assurances that anyone else will protect their health with equal vigor or determination."

Who is subject to the standard? Employers having employees currently or previously exposed to toxic substances or harmful physical agents, must comply. "Employer" means a current employer, a former employer, or a successor employer; "employee" is a current or former worker or employee being assigned or transferred to work where there will be exposure to toxic substances or harmful physical agents.

What records are covered? The records covered by the standard are those containing all relevant data on exposure, medical conditions and analysis of effect of exposure on the health of individual workers, whether or not they are related to specific OSHA occupational safety and health standards. Past, present, and future records are included. The record can be any type of information regardless of form or process used to gather it. Not considered part of the records covered are physical specimens, health insurance claims, and information concerning voluntary employee-assistance programs.

What is a toxic substance or harmful physical agent? Any chemical substance, biological agent, or physical stress (for example, noise, heat, cold, vibration, repetitive motion, ionizing and nonionizing radiation, hypobaric or hyperbaric pressure) for which there is evidence of harmful health effects is covered. The criteria for determining whether harmful health effects exist are: (*a*) the substance is regulated by federal law or rule; (*b*) it is in the NIOSH Registry of Toxic Effects of Chemical Substances; (*c*) the company has knowledge that the substance has shown a health hazard as a result of experience or research; (*d*) the substance is accompanied by a material safety data sheet when brought into the plant.

What does "access" mean? "Access" is the right and opportunity to examine and copy records. Access must be provided at a reasonable time and place and in a reasonable manner, but no later than 15 days after request. Employers are entitled to direct access to their own medical records, upon request. Or they can give specific written consent for representatives to have access. (This person is called a designated representative in the standard.) The standard explicitly encourages the physician maintaining the medical records to explain the content of the records to the employee. The physician can suggest—but cannot insist on—alternative forms of disclosure if that is viewed as desirable or necessary, for example, providing a summary report or transfering of the records to another physician.

Can direct access to records be denied for any reason? If the occupational physician believes in good faith that direct access to certain records would prove detrimental to the employee's health and well-being, access can be denied. Examples of such situations: When the records contain diagnoses of terminal illness or psychiatric conditions. The employee must be told that the records will only be provided to a designated representative. This limitation does not apply to direct access to other kinds of information, such as tentative differential diagnostic possibilities or biological monitoring data unrelated to a specific finding of terminal illness.

Under what circumstances can information be deleted from medical records? When information in the records was provided by confidential informants, the identities of the source can be deleted. The sources must be fellow employees, friends, or family members. Note that the employer or other physicians or medical staff employees are not included in this exclusion. Further, the information must be disclosed as long as doing so would not in itself reveal the identity of the source.

Can the medical records be restricted to a designated physician? The standard does not allow this discretion, stressing the position that any "responsible person," including family members or friends, could be designated to receive access to the records. OSHA based this position on its contention that not all employees have personal physicians. Furthermore, it could be costly and complicated for an employee to make arrangements for a physician to review the records.

Are there limitations on OSHA's access to and use of the records? Stating that it recognizes a conflict between the right to unconsented access to records and the privacy interest of the employee, OSHA has instituted stringent internal procedures to limit the circumstances under which identifiable medical records are examined or obtained by OSHA personnel and to control the use of such information once it is in the agency. These controls, however, are not described.

What limitations are there on third-party access to records? A representative must be designated in writing. No qualifications are put on who the designated representative can be. A recognized or certified collective bargaining agent (the union) is automatically considered a designated representative in respect to exposure records. The union must, however, have the employee's written consent for access to medical records. The consent must be specific; no blanket or perpetual authorization is allowed.

Can companies protect trade secrets which might be revealed by dis-

closure of exposure reports and medical records? The fact that disclosure of chemical and physical agent identity, level of exposure, and health status information at the same time gives the possibility of revealing trade secrets does not act as a bar to access. The employer may, under the standard, delete information disclosing manufacturing processes or the percentage of a chemical substance in a mixture, for example, as long as the records note this deletion. Employers can require written agreements from employees to protect the information. OSHA's position is that employee health and well-being outweigh the competing interest of protecting trade secrets.

OSHA claims broad power in issuing regulations dealing with occupational injury and disease. It has used a generic approach to rulemaking, rather than a toxic-substance-by-toxic-substance approach. Further, it rejects contentions that access can only be allowed in respect to specific toxic substances already regulated.

Privacy is a right of individuals guaranteed by states and the U.S. Constitution, as well as numerous specific statutes. However, no court has declared it an absolute right. When the need is strong, for example, a public interest in protection of worker health and safety, the courts have allowed—and would allow—an invasion of privacy as long as it is accompanied by administrative regulations to protect against abuse.

The standard gives employees the right to complain to OSHA about conditions they believe represent a hazard to their safety and health. They are entitled to prompt inspection of the worksite in question. Knowing this information in advance of an investigation supposedly will help OSHA respond immediately to severe problems, to prepare itself better for the inspections, and to have the right people and equipment in place to investigate the situation complained about.

The proposed revision of the access standard, published for comment in mid-1982, was promoted as an attempt to retain the objective of the standard, while removing requirements that have proved burdensome to industry and even counterproductive to worker health.

OSHA's former Administrator Thorne G. Auchter stated at the time of publication of the notice that "...the complexity and burden of the present requirements discourage employers from creating records which would then be subject to the standard." By simplifying the standard and making it less restrictive, more employers would be encouraged to monitor exposures and maintain records, thereby aiding in the detection, evaluation, and prevention of occupational disease.

The focus of the revised standard was on "those employees most likely to be at risk from occupational exposure to toxic substances."

The key features are:

- The scope of the standard was narrowed by limiting its coverage to records of employees who have *direct,* not incidental, exposure to hazardous substances.

• Exposure records consist only of environmental and biological monitoring results and material safety data sheets, not purchase orders or other documents that could identify the substances.

• The toxic substances to be covered by the standard are limited to the 3,492 substances in the NIOSH Registry of Toxic Effects of Chemical Substances. This coverage is in contrast to the some 39,000 substances covered currently. Further, any substances known by the employer to be hazardous or cited as toxic in Material Safety Data Sheets would be included.

• Medical records have to be retained by the employer for the period of employment plus five years, or for 30 years, whichever is longer. This contrasts to the original standard's employment-plus-30-years' requirement. Workers in the construction industry or those employed for less than a year would be given their records upon termination of employment.

• Trade secrets are given added protection: (*a*) records of biological monitoring results not defined as exposure records by specific occupational safety and health standards would not be covered by the access standard; (*b*) a liquidated-damages clause would strengthen confidentiality agreements between employers and workers or their representatives; and (*c*) detailed disclosure of constituents in a compound is restricted to highly toxic substances—carcinogens, mutagens, and teratogens, or those known to cause significant irreversible damage to human organs or body systems.

A distinction is drawn between "high chronic hazard" substances, whose identities and amounts must be disclosed, and other substances that may very well be hazardous. The latter's identity may be withheld if it constitutes part of a trade secret, provided: (*a*) the employer substantiates that a trade secret is involved, if challenged; (*b*) the substance is identified by a generic classification; (*c*) the employer provides nonsecret information on the properties and effects of the substance; and (*d*) the employer releases, under a confidentiality agreement, the withheld information to a treating physician who has stated in writing (except in emergency) that a patient's health problems may result from occupational exposure.

The modification to the standard appears to accomplish the policy objective of balancing the interests of worker health and safety with commercial advantage, skewed toward the human side.

Hazardous products in the workplace

As of late 1982, as mentioned earlier, 15 states and at least five cities had filled the void left by OSHA's on-again, off-again identification of hazardous substances in the workplace. (OSHA's Hazard Communication Standard only went into effect in 1985-1986.) These governmental bodies had enacted statutes and ordinances requiring employers within their jurisdictions to notify their employees of the hazardous substances with which they work or come in contact in the work environment. Other communities were known to be planning similar legislation.

The state of Connecticut's statute was typical of the legislation on behalf of workers' right to know the hazards they encounter while working. Its statute—which went into effect in July 1983—allows employers five working days to respond to a written request by an employee or an employee's representative for information about hazardous substances found on the job site. The employee may refuse to work with the substance until the information is furnished. Within one month after hiring new people, employers must give them training in the use of the hazardous substances with which they are or will be working. Hazardous substances under the Connecticut law are those listed in the OSHA airborne contaminants standard. Carcinogenic substances are not included, since a separate statute already regulates them.

The information, in easy-to-understand language, includes generic or chemical name, location in the plant, effects of adverse exposure, conditions for safe use, procedures for clean-up in event of accidents, and emergency treatment. The burden of providing the information to employers rests on the suppliers of the hazardous substances. They must also provide the necessary labeling. The state's Department of Labor will cooperate by providing employers with information about hazardous substances on request.

By January 1984, all Connecticut employers had sent lists of hazardous substances in their work areas to the state Department of Labor. Employers must post notices letting employees know their rights under this law. If any employer thinks the required information violates a trade secret, the Connecticut Department of Labor will use an identification number, rather than the name, to shield identity, except in emergency situations.

An employee, for purposes of the law, is a person "who may be exposed under normal operating conditions or foreseeable emergencies to toxic substances while engaged in service to an employer in a business of the employer." Emergencies are defined by examples: equipment failure, container breakage, and equipment malfunction leading to the uncontrolled release of hazardous substances. An employer is barred from asking for or accepting a waiver of the right to this information about hazardous substances in the workplace.

The rash of local legislation has convinced many manufacturers of hazardous substances that OSHA should have the sole authority to regulate identification of these products in the workplace. The reason is simple: the prospect of having to accommodate a variety of different sets of labeling and notification requirements in different markets is frightening. National marketers are staggered by the cost of preparing multiple labeling and multiple data sheets.

It was in response to the need for uniformity that OSHA published its Hazard Communication Standard in 1984 with effective dates of November 1985 for all suppliers and distributors of chemicals and May 1986 for all manufacturing employers. The standard requires the chemical suppliers to determine the hazard of their products, put warnings on all containers, and provide their distributors and customers with Material Safety Data Sheets containing detailed safety information about the product's handling and use. Manufacturing employers had more time to put into effect education programs aimed at helping employees learn about the hazardous products they work with and how to use them safely.

The existence, though, of an intricate patchwork of labeling and employee notification laws throughout the country (see above) and the federal Hazard Communication Standard has raised the issue of whether or not OSHA has the right to preempt local right-to-know laws.

OSHA has adopted the position that it alone has the right to require identification of hazardous substances found in the workplace. It bases this position on the concept of preemption, i.e., the right of a federal agency to exercise priority over a lower-level governmental unit. The problem is that the law creating OSHA does not contain a clear preemption clause with respect to local legislation. Lacking such authority, it is necessary to look at the legislative history. But clear-cut expressions by members of Congress reserving to the federal government the right to regulate occupational safety and health cannot be found.

There is ample precedent, however, in OSHA's sharing of regulatory and enforcement responsibility with the states to justify at least concurrent jurisdiction. Only if the state activity is less effective than the federal, or if it tends to defeat the OSHA objective of promoting workplace safety and health, can OSHA preempt. An example would be identification or hazard requirements so stringent or demanding that they are impractical.

The courts will undoubtedly have to decide the issue, in light of the fervor of the supporters of local regulation of workers' right to know.

FEDERAL PRODUCT LIABILITY ACT

The Federal Product Liability Act* preempts state laws dealing with product liability. Thus, a uniform standard would be created for deciding cases throughout the country. The defense of product liability lawsuits should be simpler and less expensive for these reasons:

- Economic loss due to a product failure cannot be a basis for a product liability lawsuit.
- Strict liability, under which the plaintiff does not have to prove the manufacturer responsible for a defect, is barred.
- Construction and design defects are defined in a way that narrows their use as the basis for lawsuits.
- Plaintiff's misuse and abuse of a product can lead to a reduction in an award.
- Allocation of damages among parties, including an employer, can take place.
- Nonmanufacturer parties in the chain of distribution are protected.
- Product liability awards mesh with workers' compensation.
- The ability to recover under both workers' compensation and the tort system is restricted.
- The plaintiff is limited in bringing a product liability lawsuit over old equipment.
- Punitive damages are restricted.
- Product improvements cannot be used as evidence of prior defects.

Product liability law is society's way of making its members whole again after injury due to negligence in the manufacture of products. It is a response to industrialization, which created mass assembly and distribution of products. Product liability law is a branch of tort law governing human contact with other humans. The tort is called *negligence,* in

---

*By early 1986, no federal product liability legislation had succeeded in getting to the floor of either house of Congress, despite extensive revisions and compromises in the wording of proposed bills. The fact that the proposed legislation did not die testifies to the strong support for it throughout industry and the widespread belief that changes in tort law are needed. All the evidence indicates that some form of federal product liability legislation will become law in the mid- or late-80s.

this context, negligence in some aspect of the design or construction of the product. The law recognizes that there is no way to make an injured person whole again, but the next best thing is money. Money damages also act as an incentive to the negligent party (officially called the tort-feasor) to avoid being negligent.

Product liability law is a creation of common law, arising in the mid-19th century in England and adopted in the United States, where it has been developing in its own way ever since. As in all tort situations, the person injured by a product must establish (*a*) that an injury occurred; (*b*) that a connection existed between the cause and the manufacturer of the product; and (*c*) that the manufacturer had a duty to prevent that injury.

At first, the law of product liability put many obstacles in the way of the injured party's recovering money damages. It was difficult to prove a connection between the injured person in Ohio, say, and the manufacturer in New York, since the product had gone through many hands. This was called the doctrine of privity, i.e., establishing a direct connection. Moreover, the user had to meet a heavy burden of proof that he or she had used the product properly and had not assumed a risk in its use. This situation began to change in the early 1900s with the growth of social and political liberalism and an early brand of consumerism. The balance shifted away from favoring the manufacturer to a position of equality. The courts operated on a principle of expectation. The manufacturer was expected to deliver what the buyer expected in all ways—value, performance, and safety. Both sides had obligations: the manufacturer to fulfill expectations; the buyer to use the product properly. No more *caveat emptor.*

The consumerism of the 1960s shifted the balance in product liability in new directions—toward the buyer and user. The epitome of the shift concerns the doctrine of strict liability. Strict liability means that a manufacturer of a product can be held responsible for an injury to a user, even if it is not possible to prove negligence on the part of the manufacturer. An early step in that direction was the doctrine of *res ipsa loquitur,* "the thing speaks for itself." "Res ips," as lawyers call the doctrine among themselves, is used in airplane crashes to pin liability on the manufacturer and the airline company. The fact of an accident means that somebody must have been negligent, even if guilt cannot be established.

A further refinement of strict liability is the erosion of the requirement to prove that a product causing injury was made by a specific defendant in a lawsuit. People who contracted cancer from exposure to chemicals long ago, have been awarded damages from a manufacturer or a group of manufacturers without having to prove a connection (i.e., DES, asbestos).

Strict liability made it easier to get a recovery. It encouraged people to sue. So did the contingency-fee arrangement on which personal injury lawyers operate. When prospects are good for a recovery, lawyers are willing to take a chance on winning. The hefty awards in product liability cases in the 1960's and 1970's also encouraged prospective plaintiffs and their lawyers. The aggressive consumerism then provided a rationale for juries who favored the little guy in a fight with corporate giants by encouraging large settlements.

Bringing everybody in the chain of distribution into the lawsuit became a useful device to locate the "deepest" pocket or pockets.

The workers' compensation system also began applying new pressures on product liability law. The pressures had been built up by the ingenuity of plaintiffs' lawyers and the courts in finding ways around the worker's compensation system. The system was designed to be an exclusive means of compensating individuals who were injured in the course of employment. Nevertheless, some believe that today, compensation payments do not produce an adequate award, and plaintiffs' lawyers have sought ways around the system.

Workers also began suing manufacturers for injuries sustained from the product in the workplace. Workers' compensation benefits previously obtained were not always taken into consideration or recovered by the employer or insurance carrier. Additionally, insurance carriers have made use of the legal procedure of subrogation, which allows them to sue a product manufacturer to recover money paid out under workers' compensation claims. (Also see Chapter 17, "Workers' Compensation.")

Workers' compensation laws have traditionally barred lawsuits against an employer for product liability on the basis of the employers' contributory negligence. Cases abound where a machine manufacturer has been successfully sued by an injured employee even though (*a*) the employer had modified the machine in a way that created a risk of injury; (*b*) either removed safety devices or they were removed with full knowledge of the employer; (*c*) inadequate instruction or training was provided to the injured employee; or (*d*) the equipment was not properly maintained.

The result was the product liability crisis of the mid-1970s. People sometimes were unable to sell their machinery businesses because buyers were afraid that they would also be taking on unwanted lawsuits. This possibility arose because of the growing number of cases holding buyers of business assets liable for injuries sustained on machines sold years before. The most dramatic evidence of the existence of a crisis was the rapid increase in premiums for product liability insurance, and the sudden cancellation of policies for companies having claims.

The crisis made good copy for business publications and the general media. Something had to be done. Trade associations set up committees, as did the Commerce Department, with the goal of reforming the tort system to prevent alleged abuses of the product liability law. Interestingly, word started filtering out of the committees about studies showing that the crisis was not completely the fault of the tort system. Insurance companies, as it turned out, were propagandizing to cover their vested interests. In short, the insurance companies were not satisfied with their return on the product liability premium business. Maybe they needed some competition, many decided.

One step in this direction was the Federal Product Liability Risk Retention Act of 1981, which permitted manufacturer groups to band together to offer such insurance. This introduced a new element of competition into product liability insurance rates.

Although the hue and cry has died down, proponents of reform continue to try to restructure product liability law. The first avenue has been to try to bring about changes in state law. Manufacturer groups supported this approach through lobbying. Changes in this direction have occurred in the laws of some 31 states. But not all. Uniformity, which major marketers needed, was still lacking. The tort reform laws passed at the state level are not consistent. In fact, state product liability laws have created new variances in tort laws, resulting in more confusion and uncertainty.

Another approach has been to convince the states to adopt a uniform product liability act. This failed because of the impossible task of getting 50 state legislatures to accept the same piece of legislation without any substantial changes in any reasonable amount of time. Proponents of product liability reform have given up on obtaining uniformity through state-by-state changes and have shifted their energies instead to the federal law approach.

The proposed Federal Product Liability Act seeks to reform the product liability law by addressing the following issues relevant to worker compensation:

How much can the liability be spread?

- The claimant's own negligence reduces any damages awarded proportionate to that negligence.

- Any award to a claimant from a manufacturer is reduced proportional to the liability of claimant's employer or coworkers and corresponding judgments are to be entered against them.

- If a judgment cannot be collected against one defendant, that defendant's share can be imposed proportionally on any other defendant.

Is the manufacturer responsible for misuse and modification of his product?

• Misuse of the product results in an allocation of the judgment against those whose misuse caused or contributed to an injury—such as against an employer or coworkers. Or, it can lead to a finding of reduced liability or no liability because of misuse.

• The same kind of allocation or reduction results from proof of alteration or modification that leads to injury.

• Such allocation or reduction will not occur, however, if the modifications or alterations were provided for in the specifications, if they had the express consent of the manufacturer or product seller, or such modifications or alterations could have been anticipated.

• Such modifications or alterations occur "when a person other than the manufacturer or the product seller changes the design, construction, or formula of the product or changes or removes warnings, instructions, or safety devices that accompanied or were displayed on the product; or when a product user fails to observe the routine care and maintenance required for a product and that failure was the cause of the claimant's harm."

Can product liability law mesh with workers' compensation?

• Workers' compensation benefits reduce any award from a product liability action.

• Employers' and workers' compensation insurance carriers have no right of subrogation against the manufacturer or product seller.

• The dual capacity doctrine, allowing an injured employee to collect workers' compensation and also sue the employer for negligence under certain fact situations, is restricted.

Will this new federal law—if and when it is passed—achieve its objectives: uniformity of product liability law, fewer lawsuits, more effective defenses, more dismissals for defendants, lower defense costs, and lighter verdicts?

Trial lawyers predictably rail against the proposed legislation, using such words as "drastic" and "devastating." They also call it anticonsumer and promanufacturer. Their evaluations may well be self-serving. Yet the lineup of supporters leads to a conclusion that it is, at least, promanufacturer. These trial lawyers point out that judges are adept at interpreting all laws to achieve an objective of fairness, fairness in their view. Furthermore, members of the Plaintiffs' Bar—as trial lawyers representing injured parties like to call themselves—are also adept at shaping persuasive arguments to sway judges and juries. A large measure of uniformity in product liability law throughout the country is certain to evolve. But as time goes on, judge-made or case law will develop. At least, it will start from a common statutory base.

STANDARDS FOR WORKPLACE HEALTH AND SAFETY

A current issue is whether employers have to or do not have to make their health records available to NIOSH. The main concern is the possible revelation of trade secret information. Protection of employee privacy is also emerging as an issue. Actually, the process of gathering all the information needed about a particular hazard can take years. A first draft of recommendations is reviewed by NIOSH staff, as well as by outside consultants, professional societies, and federal agencies. When approved by NIOSH and the general counsel of Department of Health and Human Services, a recommendation then goes to the Secretary of Labor.

To begin developing standards, OSHA has relied on voluntary organizations that have produced a variety of national standards pertaining to workplace health and safety. Three such organizations are now relied on for standards and background information:

- American National Standards Institute (ANSI),
- National Fire Protection Association (NFPA),
- American Society for Testing Materials (ASTM).

These private organizations assist OSHA in reviewing and rewriting existing standards and serve as a source for the development of new standards. They use the consensus method of developing and modifying standards, which means voluntary arrangement of the interested parties. Canvassing is one approach to the problem. In this case, a proposed standard drafted by an interested group is submitted to a vote of knowlegeable people in companies, trade associations, and professional societies. Another approach to consensus is by committee. Representatives of interested parties form a committee to draft the standard. The result of the consensus is a draft of a standard, which must then be approved by the standards organizations.

ANSI and OSHA have a working arrangement for developing standards. ANSI provides technical support through its committees of which OSHA representatives are members. The OSHA representatives provide safety and health information and research reports to the standard development and modification process.

Another vehicle to generate standards is the National Advisory

Committee on Occupational Safety and Health (NACOSH). Its 12 members represent management, labor, the public, and occupational safety and health professions. The objective of the group is to study all information available on relevant occupational safety and health topics, consider alternatives, and weigh the feasibility of proposed standards. Subcommittees of NACOSH explore significant topics relating to occupational health and safety standards. Some such subcommittees are permanent, like that for construction; others are set up to consider a specific problem, such as asbestos. Typically, the subcommittee consists of 15 members representing employers, labor unions, trade associations, state agencies, professional people (such as occupational physicians and safety managers), and finally someone appointed by the Secretary of Health and Human Services.

The resulting national consensus standards have been challenged legally. A principal objection has been that they impart the force of law to standards that have been modified by OSHA upon adoption. This partial adoption has occurred when a voluntary or advisory standard was made mandatory by changing "should" to "shall," for example.

The courts have generally viewed this as a substantial alteration and thus subject to rulemaking procedure. The first step in rulemaking is to publish a "Notice of Proposed Rulemaking" in the *Federal Register,* allowing 30 days after publication for interested parties to submit written comments. This provides the notice to interested parties that administrative law requires. The notice must include a statement or reasons for the action as required by the act. The courts have ruled that partial adoption of standards by omitting portions of them does not require the rulemaking procedure.

Standards issued must be technologically sound and economically feasible. The question of feasibility of standards—or lack thereof—has produced extensive litigation. The legal challenges focus on the meaning of the phrases "so far as possible" and "to the extent feasible" which the OSHAct uses to qualify its objective of creating safe and healthful workplaces.

The courts have required those challenging the technological feasibility of standards to show that they are "clearly impossible of attainment." The holdings have made clear that a standard can envision the possibility of improvement in technology or development of new technology within a reasonable period of time. Just because compliance with a standard is financially burdensome, however, does not mean that it is economically infeasible. Since all employers must meet the standard's requirements, none is at a competitive disadvantage.

When a standard will have an impact on the environment, OSHA must publish an environmental impact statement in the *Federal Register.*

If there is a question of whether or not the standard will have such an impact, then OSHA must publish its reasons for not preparing the statement.

Another type of analysis of the effect of a proposed standard is its inflationary or economic impact. Economic impact statements cover these factors: effect on markets, consumers, and business; a comparison of benefits with costs; and a review of the costs and benefits of all alternatives. The economic impact statement requirement has been criticized for placing too much emphasis on economic factors versus social and health and safety benefits, and delaying publication of the standard.

If employees are "exposed to grave danger from exposure to substances or agents determined to be toxic or physically harmful or from new hazards," OSHA can issue an emergency temporary standard. It remains in effect for only six months, when a permanent standard must replace it. The emergency temporary standard serves as the proposed rule in the necessary rulemaking procedure. The customary rulemaking procedure is bypassed for the emergency temporary standard. However, a statement of reasons for the extraordinary action must be published. The statement must indicate the facts being relied on for the action, in what way they show the harmfulness of the situation, and why the standard is necessary to protect employees. The facts must show "more than some possibility" of the existence of a grave danger. Grave danger has been defined by one court as "incurable, permanent, or fatal consequences to workers as opposed to easily curable and fleeting effects on their health. . . ."

Any party adversely affected by a standard may obtain a court review by filing a petition within 60 days of the issuance of the standard. The petition is directed to the U.S Court of Appeals in the circuit where the party resides or where his principal place of business is located. The court may issue a stay of the standard, but does not have to.

The court applies the "substantial evidence" test in reviewing challenges of OSHA standards. This is the typical test applied in administrative hearings.

In reviewing the standard, the court considers such questions as:

- Was the Notice of Proposed Rulemaking sufficient to adequately inform interested persons of the action to be taken?

- Did the promulgation notice adequately set forth the reasons for the standard or rule?

- Do the reasons given reflect factors that are relevant to the purpose of the statute?

- Have present alternatives been considered?

• Does the substantial evidence in the record as a whole support the agency's action?

Variances from the standards

There are four types of variances from the standards. One is a permanent variance when a company's safety practice does not strictly comply with a standard but still achieves the same result. The second is a temporary variance when the employer cannot meet the requirements of a standard within the time specified. This is basically an extension of the abatement period which is obtained as a result of an inspection. The third is a variance in the interest of national defense. The fourth is a variance in connection with research relating to safety and health standards.

An application for a variance must provide basic information about the reason for the request. In the case of a temporary variance, the application must also indicate the time period for which the variance is requested. The application must show that employees have been told about the request for variance, and the employer must establish that the variance will not increase or contribute to the hazard to the worker.

Requests for variance are published in the *Federal Register,* to allow interested parties an opportunity to comment. An employer can be cited for violations even while a request for a variance is pending. To establish the need for a temporary variance, an employer has to provide that the date for compliance cannot be met because of a lack of necessary staff, materials or equipment, or because of construction schedules. While the variance is pending, the employer must make every effort to protect the employees and to reach compliance as quickly as possible. A temporary variance is limited to one year, but can be renewed if the need is substantiated. A permanent variance requires the workplace be as safe and as healthful as it would have been through compliance with the standard. A plant inspection may be necessary in this case.

A clarification is another type of variance. For this variance, an employer requests OSHA to determine whether a standard applies to a particular situation. The formalities of a variance, such as publication in the *Federal Register,* do not exist, but the clarification has the same effect of staying any kind of action against the employer.

In September 1981, OSHA had received 1,450 variance requests, of which 1,401 has been processed to completion. It issued 97 permanent variances, 842 clarifications, and 188 denials. Seventy-six requests were withdrawn, 28 standards were modified, 38 revoked, and 132 were simply dismissed. Only one variance had been issued for purposes of research, and none for reasons of national defense.

Employers must comply with standards, and preinspection compliance with standards is the basic concept of OSHA. The civil penalties resulting from citations are supposed to act as an incentive to employers to prevent hazardous conditions in the workplace. This approach is the only practical one when there are millions of workers and hundreds of thousands of workplaces—and only a limited number of inspectors.

In light of the magnitude of the task of ensuring compliance, attempts have been made to schedule inspections on the basis of the number and severity of accidents occurring in a particular workplace. This approach has not been accepted because the purpose of the OSHAct is to prevent injury and illness by *removing* or *preventing* hazardous conditions that produce accidents.

The fact that an accident happens does not necessarily mean a violation has occurred. However, that fact could help to prove the government's case. For example, a fatal accident could be cited to prove that a violation that is subsequently found is a serious one.

More than 5,000 standards have been issued, of which almost half have been General Industry Standards. (The other three categories are Construction Standards, Maritime and Longshoring Standards, and Agricultural Standards.)

Standards are referred to informally as horizontal or vertical—horizontal standards are those that cover many employers in various industries—vertical standards are those that apply to a relatively small number of employers, usually in one industry. Standards can be divided into categories: specification or performance. Specification standards give detailed requirements for the physical environment of a workplace and usually call for specific equipment and materials; performance standards set forth the general procedures that workers are to follow in accomplishing their work. A vertical standard typically takes precedence over a horizontal standard, and a specification standard takes precedence over a performance standard.

Employers must comply with standards in every detail. Only by obtaining variances, can an employer bypass the need to comply. The fact that an employer is complying with non-OSHA (e.g., state) standards that differ from OSHA's is not a defense to a citation for violating a standard. Nor does a satisfactory state health and safety inspection serve as a defense against an OSHA citation.

Training programs

OSHA places strong emphasis on safety training. This government agency is supposed to provide training programs and consultation for employers, employees, and unions for preventing occupational injuries and illnesses. Employers have primary responsibility, however, for establishing such programs. A number of standards require the employer to provide safety training and failure to provide such training can be a violation and lead to a citation, if discovered during an inspection. Simply making safety information available to workers is not viewed as training. The employer has to take precautionary steps by instructing and training employees to protect themselves against hazards.

Part of the training is to make sure that the employee knows how to do his or her job safely. The adequacy of safety training is evaluated on the basis of the employee's experience and background, the particular job function, and the extent and nature of the hazards. In one company, therefore, meetings for general safety discussions may be adequate, but in another company very specific instructions and printed work rules may be required to meet the adequacy requirement.

Safety program

A safety program can help to determine whether a violation was foreseeable and whether or not it was preventable. It can also protect an employer against a situation where an employee is guilty of an isolated case of misconduct. It can also influence any penalty imposed for violation. See Chapter 8, "Safety Programs," for a detailed discussion.

Adequate supervision of the worker on the job is part of the duty of the employer to provide a safe and healthy workplace. It is a dynamic requirement: less supervision is necessary if employees are well trained, competent, and experienced, and if the work is not overly hazardous.

Compliance with a number of OSHA standards requires setting up safety procedures, modifying equipment to make it safer, and improving safety devices. One of the duties of an employer is to provide employees with tools and equipment that can be used in a safe way. Employers are not liable, however, for defects in machinery or tools that were manufactured or repaired by others, if the defects are not known to the employer or are latent. The test of an employer's liability is reasonable diligence in providing safe tools and equipment and encouraging systematic examination and testing.

Employers also must provide safety equipment and devices and insist on their use. A common category is personal protective equipment, such as safety shoes, safety belts, goggles, and hard hats. Compliance with this responsibility is tested by the "prudent man rule"—would a reasonably prudent person familiar with the circumstances of a particular industry or a particular job recognize the need for protection? The standard specifies that the requirement for using personal protective equipment must be specific, not vague. Employers are responsible for insisting that their employees use the safety and health protections provided. But OSHA still must prove that such equipment is needed. One issue that has arisen is whether the employer or employee pays for such personal gear as safety shoes. Typically, the cost is borne by the employer, but collective bargaining agreements can provide for an allocation of cost between employer and employee.

But personal protective equipment may not be adequate, because citations have been given for failing to set up administrative controls or engineering controls to reduce a hazard, such as noise. Administrative controls include the rotation of work schedules (also see Chapter 13, "Work Scheduling"), shortening of the time an employee is exposed to a hazard, and scheduling of testing operations when the fewest people would be exposed. Engineering controls include mechanical or other methods for preventing the existence of a hazard or removing it, or minimizing its effect on those exposed to the hazard.

The use of personal protective equipment is often justified by its lower cost compared with the use of administrative or engineering controls. But some standards require the use of these controls despite their higher cost. For example, with respect to dust from various processing or manufacturing operations, the courts have upheld a standard barring the use of respirators because of a finding that respirators are ineffective—they do not fit or they adversely affect some workers breathing and ability to work.

The highest priority of the employer should be to protect employees from health hazards. The evidence is clear that such substances as asbestos, polyvinyl chloride, and lead represent grave dangers. The employer has the responsibility to conduct periodic tests to determine the amount, if any, of hazardous substances in the working environment. The extent and frequency of the monitoring may be established in the standard, which must meet a test of reasonability.

Detailed warnings are another aspect of the employer's responsibility. Failing to inform employees, purchasers, and users of the dangers of a product can be a violation of duty under the OSHAct and could subject the employer to a lawsuit from someone outside the company who is injured. Company officers must also comply with safety requirements and can be cited for failure to comply. Even shareholders, partners, and employees who are not working for money, but for payment in kind are covered by the OSHAct.

Supervisors and foremen are also obliged to comply with safety and health requirements. One court has said that a supervisor has a "heightened duty" to set a good example to other employees. When a supervisor's conduct leads to a violation, is the employer responsible? If he is aware of a hazardous condition and does not report it, is the employer responsible? The general rule of agency applies—that an employee's actions become the responsibility of an employer, if the employee is acting within the scope of his employment. Another view is that if the supervisor's conduct was not foreseeable, then the employer is not liable. Will this rule be used to allow employers to avoid compliance? Employers will avoid liability only if they have a safety program, including training that includes supervisory staff, and do not tolerate any violation of safety practices.

Special circumstances

Employees subject to special circumstances must receive consideration. The non-English-speaking worker is an example. Safety signs must be in languages the workers understand if they cannot read English. Illiterate employees must receive special training to compensate for their deficiency. Employees with physical and mental handicaps may need special training or facilities to enable them to work safely. They may not be able to use personal protective equipment. Women of child-bearing age and the elderly present their own challenges because of the potential of harm to a fetus and the general impairments of the elderly. Such an employee may create a hazard to safety but termination may not be permitted by a court, if challenged by the affected worker. OSHAct compliance can lead to conflicts with nondiscriminatory statutes. For example, refusal to hire diabetics for welding jobs because of potential hazard to other employees drew a citation as a violation of a state nondiscrimination statute.

The employer is only responsible for its own employees, unless part of the business is a common undertaking with another employer. "Common undertaking" is interpreted to mean a common worksite, not a joint venture. Passersby and visitors are not considered employees for purposes of protection under OSHA.

RESPONSIBILITY WHERE NO STANDARDS EXIST

The OSHAct contains a catch-all provision to cover hazardous conditions for which no standard has been issued. This is Section 5(a)(1), the General Duty Clause:

> Each employer shall furnish to each of his employees employment and a place of employment which are free from recognized hazards that are causing or are likely to cause death or serious physical harm to his employees.

General Duty Clause

As new standards are issued, occasions to apply the General Duty Clause (see below) become less frequent. They often arise out of an unusual situation or conditions leading to a freak accident. But since this provision represents a way to deal with such unusual situations and thereby serves as an incentive to management to prevent their occurrence, it is important to understand the workings of this clause.

The General Duty Clause states that the elements of any crime (a violation of a law) are the conditions that must be met to establish that a person has violated a law. Unless all the elements are present, a person cannot be convicted of a crime. In the case of a violation of the General Duty Clause, the facts must show the presence of all elements contained in the clause.

The first element is place of employment. It is the place where at least one employee is working. The location need not be owned by the employer. It can be another employer's workplace, but the cited employer's employees are on assignment there. The employer is responsible under the General Duty clause only for his own employees, not others who happen to be working there. If his employees are working on another party's property, or even public property, the employer is still responsible for their health and safety.

The second element concerns "recognized hazards." A hazard is considered recognized in one of three cases: (*a*) if it is generally accepted as a hazard in the industry; (*b*) it can be detected either through the human senses or through instrumentation; or (*c*) if the employer has or had knowledge of the hazardous condition. A hazard is generally accepted as such in the industry if safety experts familiar with the industry view a condition or situation as hazardous.

This general recognition of hazards throughout the industry can make the employer liable for constructive knowledge of it, even if he or she can establish lack of actual knowledge. Constructive knowledge comes about when the facts show that a person should have known something or could have known something, if behaving in a reasonable way. Thus, an employer could in all honesty not know about a hazardous condition recognized as such by others in the industry, but if a reasonably competent and knowledgeable person in that position would have known, the employer has constructive knowledge.

Actual knowledge can come from an employee reporting a condition or reports from other employers in similar circumstances. An OSHA citation, for example, can be a source of actual knowledge. The employer may be able to prove lack of knowledge of the existence of the hazard prior to the citation, but the citation then becomes the source of recognition of the hazard. To support the citation, OSHA must show that the employer has knowledge of the hazard or that it is generally recognized in the industry. The fact that no accidents have previously occurred does not support a claim of lack of recognition of the hazard.

The third element is called "free of recognized hazards." The only duty of an employer under this clause is to eliminate preventable or foreseeable hazards. To prove that the place of employment is free of recognized hazards, employers must show that they provided good supervision, that training took place to teach workers how to work safely, and that use of safety equipment was required. An employer would not be responsible for hazards created by the recklessness of an employee or an employee's criminal actions, as long as the employer had made good faith efforts to provide adequate supervision, training, and equipment.

The fourth element is called "causing or likely to cause." OSHA cannot support a citation merely by showing that an accident *could* happen and serious injury or death could occur as a result. It must prove that such an accident either has occurred or is likely to occur. An accident must be probable, not merely possible. The distinction is subtle. A court's decision may well be based on the quality of OSHA compliance officers' expertise and experience. A fatality is *prima facie* (on its face) evidence of the likelihood of a condition leading to worker injury, however.

As part of the proof that a condition is causing or likely to cause death or serious physical harm, OSHA must show the steps an employer should have taken to avoid the hazardous conditions. It must also show that such steps were feasible, both technologically and economically, and that they would definitely have worked.

An employer who is cited under the General Duty Clause has several affirmative defenses to challenge the citation. An affirmative defense is a substantive reason why the charge made is not true or completely true, in contrast to a defense based on a technicality of the law or defect in procedure. One defense is the existence of a specific standard covering the hazardous condition. OSHA is obliged to bring an action under a standard, which may have more specific elements that have to be met than the General Duty Clause, which has broad elements. Another defense is that the hazard is not recognized as such by the employer or by the industry. A third is that an accident was unlikely to occur. A fourth is that avoiding the alleged hazardous condition was not feasible by training or by application of technology.

The typical condition covered by the General Duty Clause is the unusual occurrence, so it is not surprising that a standard does not exist. However, as more standards are issued, the applicability of the General Duty Clause is likely to decline, as said earlier.

The defense of isolated employee misconduct is a common way of dealing with a General Duty Clause citation. The accident is said to have been caused by a drunken employee, one who went berserk, or one who acted in direct disobedience to orders or to a clear company policy. This defense must meet three tests: (*a*) the recognized hazard was created as a result of an isolated, brief occurrence caused by an employee; (*b*) the employer had no knowledge of it, either actual or constructive; and (*c*) the employee's action was against instructions and general policy of the employer. The employer has the burden of proving these contentions. Heavy emphasis is placed on whether or not the event was foreseeable. An employer aware of an employee's violent temper, irrational behavior, or drunkenness, but doing nothing about it could have trouble escaping liability. This defense has been most successful where a freak accident has occurred. The existence of active safety training programs and good supervision enhance the effectiveness of this defense. Conversely, lack of safety training and questionable supervision can destroy it.

IMPORTANCE OF RECORDKEEPING

Recordkeeping is an important part of complying with OSHA requirements. Most employers, with the exception of those employers with ten or fewer employees, must maintain certain records pertaining to health and safety of employees and make certain reports. An employee is considered to be any person who works either full- or part-time.

In addition to small employers, there are certain exempt categories. Farmers do not have to keep records on family members or workers exchanged with other farmers from time to time to help one another. Employers of domestic servants are also exempt when the workers are in the employers' residence or religious organizations.

Employers regulated by the Federal Coal Mine Health and Safety Act or the Federal Metal and Non-Metallic Mine Safety Act are also excused from OSHA record-keeping. So are railroads. However, railroads, as well as all other employers not covered by safety legislation, have to report fatalities.

OSHA records cover "work-related deaths, injuries and illnesses other than minor injuries requiring only first aid treatment and which do not involve medical treatment, loss of consciousness, restriction of work or motion or transfer to another job." (Section 8(c)(2)). The record is only applicable to occupational illness or work-related injury. To fall into that category, the injury or illness must be directly related to the workplace and occur in the work environment. Any work-related or occupational illness or injury resulting in a fatality or in lost work days must be recorded regardless of recordkeeping exemptions.

The OSHA recordkeeping system consists of three forms: a log, a supplementary record, and a summary. The completed forms are kept at the work location and must be available for inspection and copying by

OSHA, and must be kept at each "establishment," that is, a single physical location where business is conducted or where services or industrial operations are performed.

The log for occupational injuries and illnesses classifies each case and indicates the nature of the injury or illness and the outcome. Supplementary records contain the name and address of the injured or ill employee, his or her occupation, and certain related pieces of information, where and how the accident occurred, the extent of the injury or illness, whether it involved a fatality, and the name and address of the physician who provided treatment or of the hospital where treatment took place.

At the end of each calendar year, the employer must summarize all of the injuries and illnesses recorded in the log and post this summary in the work establishment for one month. The summary must include the number of fatalities, the number of lost workday cases, transfer or termination cases, number of lost workdays, nonfatal cases without lost workdays, number of occupational illnesses, and number of workplace injuries.

An employer must report a job-related fatality or an accident requiring the hospitalization of five or more employees within 48 hours of the occurrence. The report must be sent to the nearest OSHA office.

Failing to keep these records can lead to citations and penalties, beginning at $50. Good-faith errors, mistakes, and first-time offenses are generally viewed as mitigating circumstances. Falsification of OSHA records is a criminal offense that could lead to imprisonment of the responsible person for up to six months and a fine up to $10,000 or both.

EMPLOYEE RIGHTS

Employees have a number of rights under OSHA. They can file a complaint with OSHA, they can force OSHA to conduct an inspection where imminent dangers exist, they can participate in the inspection tour, they can request an informal closing conference, they can oblige the employer to post all citations received, they can contest the abatement period required by OSHA, they can oppose settlements, they can seek judicial review, they are entitled to notification of applications by employers for variances, they can participate in the standards-setting process, and they can insist upon the posting of annual summaries of injuries and illnesses. If there is the possibility of exposure to health hazards, they have additional rights including participation in monitoring of exposure to toxic materials, access to the information obtained as a result, and information about their exposure. Employees also can ask directly or through authorized representatives (typically unions) for a determination of whether or not toxic substances are present in the workplace.

The employer is obliged to pay for periodic medical examinations. However, an employee may refuse the examination on religious grounds unless such an examination is necessary to protect other workers from harm.

One of the most important rights that employees have is being protected against employer retaliation for exercising these very rights. An employee who files a complaint, testifies, or exercises any of the other rights cannot be discharged or otherwise discriminated against as a result. In the event of any of these, he or she can file a complaint with the OSHA area director.

OSHA ENFORCEMENT

Most employers have no contact with OSHA until the time of an inspection. (Also see Chapter 14, "Plant Inspection and Worker Protection.") The inspection is the start of enforcement activity and can result in citation and penalties. Since only a small fraction of all citations are contested, the determinations of the inspecting compliance officers are usually not reviewed at a higher level and are binding. The compliance officer has to comply with safety regulations in the workplace he or she is inspecting. Compliance officers are also obliged to respect confidential information they have access to as a result of the inspection. Obviously, employees must cooperate with the compliance officer during the inspection.

Employers are not entitled to advance notice of inspections. Criminal charges, including a fine and imprisonment, can be imposed on any individual who forewarns an employer of a prospective visit. The only time advance notice is allowed is (*a*) when there is a situation of imminent danger to allow immediate abatement; (*b*) when inspections are to be conducted after normal business hours; (*c*) where the nature of the workplace requires special arrangements to conduct the inspection, and (*d*) where advance notice is necessary to make sure that employer and employee representatives are available.

Whether or not an OSHA officer can inspect a workplace without a warrant, if the employer does not consent to it, has been a subject of controversy since the law went into effect. The Supreme Court has ruled decisively on this issue. It has upheld the right of OSHA to obtain a warrant under its legal authority. These warrants, which are available through U.S. magistrates and are relatively simple to procure, can be issued *ex parte* (unilaterally, without notice to the other party) and can be used for unannounced inspections. It is not necessary for OSHA to show probable cause. An inspection is simply part of a general administration plan for the enforcement of the OSHAct.

If an inspection reveals a violation, a citation follows. The citation does not have to state in detail the nature of the violation, only a description of it and a reference to the standard that has been violated. The citation also does not have to state how the employer can abate the condition cited. The employer is entitled to fair notice, and is given sufficient information to know what the alleged violation was and where it took place.

The citation, together with a notice of proposed penalties, is sent by certified mail, or can be delivered personally to the employer. Either an individual in the case of a sole proprietorship, a partner in the case of a partnership, or a corporate officer in the case of a corporation will suffice.

An employer has 15 days after receiving the notice of a citation and proposed penalty to file a notice of contest. The objective of this short time period is to promote quick response. Failure to meet the 15-day time period leads to an uncontested citation becoming a final order and not subject to review.

It is not uncommon for an employer to decide not to contest a citation. The penalty may be so small that it is not worth the time and possible legal fees to contest. To protect against the uncontested citation being used as an admission of negligence if a civil action should arise, the employer should file a notice of no contest with OSHA. This places an employer on record for not admitting the existence of a violation, but simply paying the penalty for practical reasons.

The citation carries an abatement period, which is the time an employer has to remedy a violation. The typical abatement period is 30 days or less. The statute requires the abatement period to be "a reasonable time." It is supposed to be based on considerations of the seriousness of the violation, the number of employees exposed to the hazard, what it takes to remedy the violation, and any other circumstances that affect the ability of an employer to deal with the problem, including the employer's financial condition. When a notice of contest is filed, the abatement period is stopped until a final order is issued. Because of this, OSHA deals harshly with employers who file notices of contest not in good faith, but solely to delay. In such situations, additional penalties may be imposed.

Where more than 30 days is needed to accomplish the abatement, OSHA accepts progress reports every 30 days. This is sometimes called multi-step abatement. OSHA can require short-term remedies, such as administrative controls (work scheduling) and personal protective equipment, if a violation can only be abated with long-range action, such as installing equipment to reduce noise levels. If the abatement cannot be completed within the required time, the employer can petition for a modification of abatement to obtain an extension. But the employer still must act to safeguard employees during the extension period.

Violations are classed according to their degree of seriousness. The classes are: *de minimis,* nonserious, repeated, willful, and imminent danger.

• The *de minimis* violation is one that has no direct connection with employee health or safety. It is typically a technical violation of a standard.

• The nonserious violation is one where a direct and immediate relationship exists between the condition that produced the citation and the safety and health of employees, but not to the extent that it would produce injury or illness leading to death or serious harm.

• A serious violation is one that creates a "substantial probability that death or serious physical harm could result." It is not necessary to establish that the condition could lead to an accident, but rather that the result of an accident would be very serious. Two or more nonserious violations in the same workplace can be grouped to create a serious violation if the combination of the two conditions would prove serious. A serious violation requires that the employer knew or was in a position to know about the conditions leading to the violation.

• A repeated violation arises out of a situation where an employer has received at least one previous citation for a violation of the same standard and the original condition was abated. This type of violation raised questions as to whether the employer was the same one, whether the citation had previously been given, whether the citation had been confirmed as a final order, whether the similarity in the citations was sufficiently close, and whether the original violation had been abated. It is usually issued only to those employers who are indifferent and unresponsive to OSHA requirements.

• A willful violation is one in which the employer had knowledge of the standard being violated, but despite this knowledge allowed a hazardous condition to exist with intentional disregard of the standard or demonstrated indifference to it.

• An imminent danger violation is issued for conditions that could "be reasonably expected to cause death or serious physical harm immediately or before the imminence of such danger can be eliminated through the enforcement procedures" available to OSHA. This type of violation requires immediate action to protect employees and quick abatement. In this case, OSHA has the power to obtain an injunction if the employer refuses to take voluntary action to abate.

Employers who fail to abate a cited violation are subject to severe penalties of up to $1,000 a day. To establish a case of failure to abate, however, the original citation must have become a final order; the condition must not have changed since the original inspection; and the condition must still be a hazard. Criminal sanctions are available under the OSHAct against an employer who willfully violates the act in such a way that death occurs. Only one such criminal proceeding has taken place. It involved the death of a construction worker in a cave-in.

The Occupational Safety and Health Review Commission (OSHRC) has the power to modify a violation to a different level of seriousness. Like a court of law, however, the commission may reduce the level of seriousness, but rarely increases it. OSHRC is designed to separate enforcement from adjudicatory functions pertaining to the OSHAct. It consists of three members appointed by the president. The commissioners review cases decided by administrative law judges who hear contests throughout the country. The judge's decision becomes final 30 days after the hearing unless a commissioner orders a review of the hearing record. The commission can then reconsider the matter and issue a new decision. Once the final order has been entered by the commission, the next step is judicial review through the U.S. Court of Appeals.

Controversy has arisen about the role of the commission. Should it be neutral or act as another vehicle for protecting the public? It has attempted to stay on middle ground by being flexible in interpreting citations and allowing amendments; by minimizing technicalities; by allowing employers contesting citations on their own, without lawyers, leeway in presenting evidence without complying with formal procedures, and by reviewing cases and challenging settlements to protect the interest of the public.

All cases that have been subject to a final order by the commission can receive a judicial review, and it is important to have that final review. If a citation is heard by an administrative law judge and decided in favor of OSHA, and the commission does not decide on its own to review the decision, the employer must file for a review. Otherwise, the employer has lost the opportunity to take the case to court for review. This is called exhaustion of administrative remedies.

Judicial review covers a wide range of issues regarding the handling of the case at the administrative level and the decision. The court looks at these questions: Was the decision arbitrary or an abuse of discretion on the part of the administrative law judge or the commission? Was the decision against a constitutional right of the appealing party? Did the administrative law judge or the commission go beyond its statutory jurisdiction or authority in making the decision? Was proper procedure followed? Was the evidence insufficient to support the decision?

OSHA'S EFFECT ON LABOR-MANAGEMENT RELATIONS

Labor and management have both felt the impact of OSHA. It has led to the inclusion into collective bargaining agreements of provisions for safety and health; safety and health disputes have gone to arbitration under such agreements; and unions have made safety and health issues part of contract negotiations. Employees are protected against retaliation for filing complaints about safety and health conditions and against termination when they walk away from a job because of a dangerous condition.

An employee cannot bring a legal action against an employer because of injury resulting from a safety and health hazard. Such legal action is barred by the exclusive-remedy feature of the workers' compensation law (but, see below). A line of cases has established the right of employees to seek injunctions against their employer because of the existence of hazardous conditions. This right has been upheld by the courts on the basis of a common law right to a safe working environment.

The existence of a violation of an OSHA standard has proven important in actions of injured employees against manufacturers of the equipment or materials causing the injury. Conversely, some courts have decided in favor of the equipment or material manufacturer because an OSHA standard did not require certain safety features.

Labor unions have been the target of lawsuits brought by members on the basis of negligence. The theory has been that the union assumed responsibility either for securing a safe working environment as a result of collective bargaining or maintaining that environment as a result of its right to make inspections or to call workers off the job if hazardous conditions arose.

Civil actions have used OSHA standards to establish negligence on the part of a defendant employer. But the standard of conduct of a reasonable person must also be proved. Some courts have held that proof of the existence of an OSHA violation shows negligence without qualification. Other courts have held that it is only evidence of negligence, not unqualified proof. OSHA standards may be introduced by defendants in such law suits. The fact that an employer complies with an OSHA standard, or that no OSHA standard exists, does not mean that negligence did or did not exist, but it could be relevant in determining whether the defendant in such a lawsuit acted in a reasonable way.

THE FUTURE: MANAGER VULNERABILITY

By early 1986 the framework for regulation of workplace health and safety was firmly in place. This included mechanisms for providing workers with access to company records pertaining to workplace hazards and the effect of those hazards in the day-to-day working environment. The focus of proponents of increased protection for workers then shifted from broadening and intensifying coverage to stricter enforcement of the existing law and supporting regulations.

The perception in some quarters that enforcement was lagging in the early years of the decade of the 1980s brought a response at the community level that points to a significant and far-reaching impact on industry and individual managers. That response is criminal charges brought against companies *and* their top and middle managers by local prosecutors.

Following instances of worker death and injury, two employers in the Chicago area and their top and middle managers were indicted for the crimes of murder and aggravated battery (reckless disregard for the welfare of others resulting in serious injury). Their criminal trials were pending in early 1985.*

Another development reflecting the dissatisfaction with governmental enforcement of health and safety law was an increase in personal-injury lawsuits filed by injured workers against their employers and individual managers. Such lawsuits are only possible as a result of an exception to the exclusive-remedy provision of state worker-compensation laws that ordinarily bar legal action against employers and co-workers. The exception is available in instances of criminal action or intentional torts (i.e., reckless or grossly negligent conduct) by employees of a company that lead to injury of a fellow employee.

This development is particularly significant to the managers being named personally as defendants in the criminal actions and personal-injury lawsuits. Their companies' liability insurance won't cover any award of damages and only part, if any, of the cost of defending the lawsuit. Middle managers, in particular, are vulnerable because they may be charged with liability for working conditions that are the result of exercise of reasonable (in their view) judgment in allocation of resources for health and safety protection. Or they may be carrying out policies established by higher management and in use for years.

This personal impact on the manager puts a new dimension on the role of government in assuring health and safety in the workplace.

---

*The company and its managers were convicted of both crimes, and substantial fines were imposed. Most importantly, the managers were each sentenced to 40 years in prison. The convictions were on appeal in early 1986. Because the convictions were unprecendented, the appeals are being closely followed.

SUGGESTED READINGS

Federal Register. All regulations and notices of proposed regulations appear in this daily publication printed in Washington, D.C. While ponderous, it is the authoritative source for agency regulations. The notices of proposed regulations are valuable because they usually provide extensive background for the rationale for the regulation. They also give summaries of the principal points of view of parties interested in the subjects under review. The notices indicate the time available for comment and dates of public hearing. They also give the names of staff people at government agencies to contact for information. Available from Superintendent of Documents, U.S. Government Printing Office, Washington, D.C. 20402.

"United States Code Service." Title 29 contains Sections 651-678 of that portion of the United States Code (federal laws) dealing with occupational safety and health. A section-by-section annotation of each statute contains references to explanatory sources including law review articles, judicial decisions, and applicable supporting and related regulations. Available from Lawyers Cooperative Publishing Co., Rochester, N.Y. 14694.

Code of Federal Regulations. This publication is a compilation without annotation of all regulations published in the *Federal Register.* Title 29—Labor, Chapter XVII, Parts 1900 et seq. deal with OSHAct. Standards are covered in Part 1910. Available from the Superintendent of Documents, U.S. Government Printing Office, Washington, D.C. 20402.

"Occupational Safety & Health Reporter." This weekly newsletter highlights the news about occupational safety and health from government agencies, the courts, and industry. It reports actions of OSHA and includes excerpts of statements and talks by government officials and leaders from management and labor. From time to time, interviews with newsmakers appear. It also provides complete texts of standards and regulations or summaries, and capsule summaries of citations and cases coming before the Occupational Safety and Health Review Commission, and lists of meetings. A supplement provides texts of commission decisions. Available from Bureau of National Affairs, Washington, D.C. 20037.

OSHA Compliance Guide This one-volume loose-leaf publication informs employers and employees of their rights and duties under the OSHAct by explaining requirements under the law and interpreting OSHA directives, Occupational Safety and Health Review Commission rules, and court decisions. It contains the full OSHAct and selected regulations. Subscribers automatically receive OSHA standards and record-keeping forms. Supplements are mailed to subscribers monthly. Available from Commerce Clearing House, 4025 W. Peterson Ave., Chicago, Ill. 60646.

Employment Safety and Health Guide This three-volume, loose-leaf publication covers the OSHAct and the Federal Mine Safety and Health Act. It treats in detail the legal and management aspects of complying with the safety and health laws at federal and state levels. The publication reproduces all current regulations and standards. It includes a report on state OSHA activity. Weekly reports keep the information current. Administrative law and judicial decisions are provided in a separate publication. Available from Commerce Clearing House, 4025 W. Peterson Ave., Chicago, Ill. 60646.

Occupational Safety and Health Law, Second Edition, edited by Mark A. Rothstein, published in 1984 by West Publishing Co., St. Paul, Minn. Structured like a textbook or manual, this work is a comprehensive review of judicial interpretation of the OSHAct. It is organized according to the major sections of the statute with succinct discussions of the requirements for compliance with the law. While intended primarily for the attorney and law student, it is written in clear language, although it may tell some readers more than they want to know about the law's workings.

Occupational Health Law, edited by Joseph LaDou, M.D., published in 1981 by Marcel Dekker Inc., New York, N.Y. This survey of the field of occupational health laws looks at the subject from the perspectives of the regulatory agencies, workers' compensation, and legal issues having an impact on occupational health such as medical malpractice and employment discrimination. It is designed particularly for the manager in industry and the health professional needing an overview of the main current and developing issues in the field.

For those interested in delving into case law, knowing how to find printed decisions is indispensible. Decisions of the U.S. District Court, the U.S. Appeals Court, and the U.S. Supreme Court are published in hard-cover annually, with soft-cover "pocket" supplements bringing the reports up to date during the course of the year. These publications are available at county courthouse, law school, and bar association libraries. The three groups of publications are: *Federal Supplement* (abbreviated as *F. Supp.*)—covering U.S. District Court cases. (Where cases are brought on appeal from the Occupational Safety and Health Review Commission.) *Federal Reporter* (abbreviated as *F.*)—covering decisions of the U.S. Court of Appeals, and *U.S. Reports*—covering decisions of the U.S. Supreme Court.

Whenever a case is referred to in literature, it carries a citation, i.e., 501 F. Supp. 73(1980), 462 F. 2d, 1292(1981), 439 U.S. Reports 350(1978). The first number refers to the volume, and the second refers to the page number within that volume. The compilation may be in a series, in which case it is numbered: 2d. Giving the volume number and the abbreviation, with or without the date, to a librarian will enable him or her to direct you to the particular book, where you can find the page.

(For additional references, see Chapter 15.)

Chapter 17

Workers' Compensation

by James Cinney

Workers' compensation as a form of social insurance came into being around the turn of the century in the United States. Prior to that time, principles of English common law prevailed. Then, the employer was responsible for job-related injuries or death, resulting from his sole negligence. Employees had to sue for damages and prove that the employer was, in fact, negligent. It is easy to imagine what that did for morale, and job security. For the injured employee, this approach was slow, uncertain, expensive, and generally inadequate. Major reform was needed.

During the early 1900s, many states passed employer liability laws, which tended to modify common law defenses. Since these laws still required that employer negligence had to be proven, many loopholes remained. The courts awarded sizable damages, therefore, the system became a costly one for employers.

The first workers' compensation laws were passed in 1911. Some states allowed employees the choice of suing under employer's liability, with potentially unlimited awards, or accepting workers' compensation benefits that were fixed by law as an "exclusive remedy."

Under workers' compensation, the employer assumes the cost of occupational disabilities without regard to who is at fault. An injured employee can recover hospital and medical expenses and also receive weekly indemnity (partial wage replacement) payments while unable to work.

In addition to compensation benefits, most states allow for uninsurable punitive damages against the employer for "serious and willful misconduct." For example, if an employer knowingly violates a safety statute and an employee is injured as a result, a separate penalty may be assessed against the employer. This penalty is usually a percentage of compensation paid by the carrier. Some states have unlimited penalties while others have a ceiling or maximum.

Workers' compensation is required by state law as are its benefits. A common misconception among employees is that workers' compensation is a state-provided benefit and is unlimited in terms of damages awarded. Not true. The overriding principle of the system is to help the injured employee cover expenses, including a portion or all of the lost wages. No "profit motive" was intended because that would, possibly, encourage malingering and increase expense.

All 50 states and U.S. territories now have workers' compensation laws. The laws and benefits vary by state, but many are similar. In recent years, state legislatures have enacted new laws to help improve the system by increasing benefits, coverage, medical and vocational rehabilitation, also reducing waiting periods. In 1981, nearly 250 new laws were passed affecting almost every aspect of workers' compensation. From 1979-80, more than 1,100 bills were introduced in the legislatures of the various states and in the U.S. Congress.

No two jurisdictions have laws that provide identical administrative procedures, coverage, or benefits. Each jurisdiction, however, agrees on the basic principles of workers' compensation. They are:

- Victims of work-related accidents receive sure, prompt, and reasonable income and medical benefits. In case of death, their dependents receive income and burial benefits.

- Negligence and fault are largely immaterial and do not bar an employee's right of recovery. The employee, however, gives up the right to sue for damages.

- The workers' compensation appeals process settles disputed claims.

- Fees to lawyers and witnesses, as well as the cost of time-consuming trials and appeals, are minimized. Court delays, costs, and workloads are reduced.

- The amount of benefits is basically predetermined by law except for employer's liability and punitive damage actions.

- The system relieves the financial drain that uncompensated accidents would cause.

- The system encourages employer interest in safety and rehabilitation and provides a monetary incentive to improve safety records.

The passage of the Occupational Safety and Health Act in 1970 included a requirement for establishing a national commission on state workers' compensation. Commission members subsequently made 84 recommendations, 19 of which were considered essential for improvement of the system. They also concluded that state law should be preempted by federal law, if states did not meet the essential recommendations within a specific time period. The majority of the recommendations were for increased and uniform benefit levels. Since then, a number of bills have been introduced in Congress to involve the federal government in the state workers' compensation system by mandating minimum standards.

THE COST OF THE SYSTEM

Estimates are that workers' compensation premiums and self-insurance costs in the United States reached more than $20 billion in 1982. The entire cost is absorbed by the employer community and passed on to the consumers as a cost of doing business; employees do not contribute to the payment of this insurance.

In 1980, the average cost per $100 of payroll was $2.72, up from $2.64 in 1979. Recent increases in insurance rates have been caused by benefit increases, also by inflation. Rates are usually adjusted annually on the basis of past industry experience and projected benefit levels. In recent years, there has been a trend toward rate reductions. These reductions frequently have been coupled with benefit increases, causing net increases in insurance rates. All of which means that injured employees are receiving greater benefits and employers are paying more for coverage.

Rates charged by insurance carriers generally anticipate that about 65 percent pays for claims. The balance is used for operating expenses including claims handling, profits, and dividends. Most states allow an up-front premium discount based on the size of the account and the historical profit picture of the carrier.

Rates are promulgated by rating organizations; some states have their own. The National Council on Compensation Insurance, an association of workers' compensation insurance companies, is the licensed advisory organization. The main functions of this and similar organizations are to collect statistics and calculate proposed rates, establish policy, develop experience and retrospective rating plans, and serve as the filing organization with the various insurance commissioners for its member companies.

Some states allow carriers to set and file their own rates, which is referred to as open rating. The major problem with open rating is that individual carriers do not have adequate and complete historical data on which to base future projections. Other states allow carriers to deviate from published rates, and still others publish minimum rates that all carriers must charge.

All rates and rating programs are subject to state insurance department scrutiny regardless of the pricing mechanism. One of the major functions of such insurance departments is to determine that carriers remain solvent. Otherwise, the injured employee might become a victim of an inadequate social system and a burden to the taxpayer because someone will have to compensate the injured.

The net result of increased benefits and liberalization of statutes has been increased product cost to consumers. Many employers have chosen to insure themselves in an attempt to reduce costs and improve their cash flow. Inflation has had some impact, but most of the increased costs are associated with increased benefit levels. Many states now automatically increase maximum indemnity benefits annually, based on average weekly wages. Most set the maximum benefit at two-thirds of the state average wage; however, some are at 100 percent—and that benefit is tax-free income. Not all workers receive the maximum benefit, and for some, the benefit is less than two-thirds of their preinjury wages.

FEDERAL WORKERS' COMPENSATION ACTS

The federal government has enacted several laws to cover its own employees, as well as railroad, maritime and admiralty employees, longshore and harbor workers.

Federal employees

The Federal Employer's Liability Act covers employees of interstate rail carriers only. Workers are compensated for injury or death on a comparative negligence basis. Juries have generally granted substantial awards because injuries are usually serious.

The Federal Employees Compensation Act (FECA) governs the workers' compensation system for civilian employees of the U.S. government. This act does not apply to military personnel.

District of Columbia Workers' Compensation Act

Enacted by the District of Columbia City Council in July 1982, this act replaced the Longshoremen's Act (see below) that was in effect prior to that date in the District of Columbia. The act excludes military and federal government employees.

U.S. Longshoremen's and Harbor Workers' Compensation Act (L&HWCA)

The L&HWCA was passed by Congress in March 1927 in order to close the coverage gap for employees engaged in maritime employment, including longshoring, stevedoring, ship repairing and building—those workers who were outside the scope of state compensation laws. The act, however, excludes members of the crew of any vessel. There have been many legal disputes, mostly jurisdictional in nature, between the scope of the federal act and the state acts. Many precedents have been set that have added to the confusion. Further, the U.S. Supreme Court has recently granted jurisdiction to either state or federal law depending on which of the two was sought for remedy. Furthermore, although the L&HWCA was thought to be the "exclusive remedy," the Court did not always view the act in that light.

In 1972, Congress amended the L&HWCA as follows:

- Hold harmless agreements between stevedore companies and vessel owners were held contrary to public policy and unenforceable.
- The use of the "unseaworthiness" doctrine was disallowed as a basis of recovery against third parties by those who came under the act.
- L&HWCA was the exclusive remedy to injured employees coming within its scope.
- Benefits were increased substantially. The minimum and maximum weekly indemnity payments in January 1984 were $144.92 and $579.66 respectively.
- The jurisdictional scope of the act was expanded onto land and to employees engaged in maritime employment.

Since the law extended coverage onto land it became widely subject to interpretation. Moreover, the amendment did not define the term "maritime employment," which has caused further confusion. The net result is that more workers are covered under the act than originally anticipated. The confusion and the expansion of coverage with benefit levels beyond state acts are causing hardships on employers and making it difficult to find carriers who will accept this coverage. It can be a very expensive coverage because ultimate costs are unknown by the carriers. Significant legislative revision is needed and hearings are under way at this time.

STATE FUNDS (MONOPOLISTIC AND COMPETITIVE)

Simply put, state funds are insurance companies owned and operated by the state. There are two types of state funds—monopolistic and competitive. In a monopolistic fund, the state is the only carrier authorized to write workers' compensation insurance. Some monopolistic states, however, do allow self-insurance. Competitive state funds, in contrast, compete for business along with private carriers and those who self insure.

Initially, many of these state funds were created to fill an important need. Private carriers sometimes refused to cover certain classes of risks to which workers were subject. Since coverage was mandated by federal statue and had to be provided, certain states went into the business themselves. Originally, many funds did not compete for business but wrote only those risks that could not find coverage, at a reasonable price, in the private marketplace.

SELF-INSURANCE

During the past ten years, employer self-insurance has become an attractive alternative to conventional forms of insurance with private carriers or state funds. This trend has now leveled off and may be reversing itself. Currently, about 10 percent of the nation's workers' compensation premiums has been diverted to self-insurance.

The switch to self-insurance is a logical reaction for businesses to reduce costs and maximize cash flow. In recent years, benefits, claim reserves, litigation, and associated administrative costs have risen dramatically. So, in general, have interest rates and rates of return on invested income. Risk managers for the large sophisticated accounts perceived that they could do as well as the insurance industry by handling their own claims, while enjoying the advantage of the investment income and cash flow themselves.

Workers' compensation claims have a relatively long pay-out period. The more serious claims tend to be paid over a long period, sometimes five years or more. Insurance carriers tend to reserve for the ultimate potential of each case. Self-insureds, on the other hand, tend to be less conservative in reserving and increase reserves as the need becomes apparent; some even operate on a pay-as-you-go basis. The difference is the cash that the self-insured account can apply directly to operations. The cash flow advantage is usually lost after about five years, however. At that point, claims payments for the current and prior years, plus reserve increases, excess insurance, and administrative costs, are about the same as an insured program.

Few companies are truly self-insured. Most purchase excess insurance above a self-insured retention limit, and these policies usually are written on a per-injury or per-occurrence basis. Although some states allow aggregate excess policies for the policy term, others do not. The ultimate exposure could be nearly unlimited. For example, if a company were to purchase an excess policy for $500,000 to $10 million and have ten $400,000 claims, the excess policy would not come into play and the cost to the self-insured account would be $4 million. When administrative services, claims, loss control, and the like are purchased, the lowest price is not always the best bargain.

Experience modifications can usually be maintained if an administrator or self-insured continues to make unit statistical filings with the jurisdictional rating bureau. This can make a significant difference in cost.

Buying back into a standard insurance program can be expensive. The state agency having jurisdiction will require that securities or a bond be posted to cover liabilities for outstanding claims. Consequently, those claims will have to be paid, claims-handling expense will be incurred, and new premiums will have to be paid all at the same time.

COVERAGE—STATUTORY SCOPE OF THE LAWS

Two basic coverages are provided by workers' compensation policies: workers' compensation and employer's liability.

- Workers' compensation provides benefits to injured workers and their dependents for injuries and illnesses related directly to their employment. Coverage and benefits are required by law, and there are no limits on this coverage.

- Employer's liability protects employers from suits in tort. But employees or their dependents may sue the employer if their injuries are such that the Workers' Compensation Act is not applicable, or if there are special exceptions. This provision also provides employers with defense coverage if the allegation is within the scope of coverage. Some examples might be:

1. Suits by third parties, such as dependents of the injured .
2. In action-over cases, when the injured brings suit against a third party and that third party in turn brings an action against the employer.
3. Dual capacity issues which, basically, the employee having collected workers' compensation benefits, also brings an action against the employer as the manufacturer of a product or provider of a service. This situation may be covered under employer's liability in the future, although it is debatable, since such coverage was never intended by the workers' compensation system. Coverage provided for employer's liability is limited by the policy. Policy limits generally vary from $100,000 to approximately $1 million. They may be increased at the carrier's option for an additional premium.

4. Recently, there have been a rash of wrongful termination actions presented for coverage under Employer's Liability. It is too soon to predict the outcome.

Who is an employee?

The test for employee status may vary by state, indeed, in some cases by the circumstances of an individual claim. As a general rule, an employee is defined as "... every person who performs a service for an employer expressed or implied, oral or written, whether lawfully or unlawfully employed including aliens and minors." Generally, officers and general partners are considered employees. In some states and under certain circumstances, however, they may elect to be excluded from coverage.

The employee must receive some form of remuneration and the employer some benefit from the services performed. The employer must control and direct the employee's activities, not only as to result but also as to method (what they want done and how to do it).

Bona fide independent contractors are not employees. There are many tests for independent contractor status, but the primary test is control. The surest method to protect against this exposure is to require certificates of insurance from all independent contractors. Usually, individual independent contractors cannot purchase coverage for themselves.

When the carrier cannot determine independent contractor status, it will probably charge a premium for this exposure, usually a percentage of the contract cost.

AOE/COE

Workers' compensation statutes apply only to accidents that result in injury or illnesses that arise out of employment (AOE) and, in some states, occur in the course of employment (COE). The event must be an accident to the extent that it was not planned for and unexpected, was fortuitous, and occurred without any intent to injure by the employee or employer.

Other forms of insurance provide coverage for off-the-job injuries and illnesses, for example, accident and health, state disability, income protection plans, private short- and long-term disability plans, unemployment insurance, and Social Security.

OCCUPATIONAL DISEASE

Occupational disease is rapidly becoming a major coverage and benefit concern to employees and employers. Most states allow for this coverage, provided the disease was caused or aggravated by the employment. Some states have specific lists of diseases that are covered, but most include diseases that are related to employment.

Occupational disease can be caused by toxic materials, radiation, heat, vibration, noise, air contaminants, and skin irritants, all of which can result in adverse effects on the human body. Because of the significant latency period for some occupational diseases, they may not be detected for 20 to 30 years after exposure.

Most workers' compensation carriers provide occupational health services to policyholders that test, monitor, and advise on control of hazards and exposures.

POLICY EXCLUSIONS

Most policies have standard exclusions, which generally pertain to job type, multiple policies, contractual liability, punitive damages, statute of limitations, and workers' compensation claims.

Employment. Some jobs are excluded, and some are covered on a voluntary basis. These exclusions vary by state but frequently include domestic workers, children employed by parents, voluntary service, ski patrols, and, in a few states, farm employees.

Other insurance. To avoid duplicate coverage, carriers exclude coverage when there are multiple locations and separate policies have been purchased for the other locations or when other locations are self-insured.

Contractual liability. Contractual liability exclusions apply solely to employers' liability. The policy states that liability assumed by the employer under any contract is not covered. This coverage may be specifically provided under the contractual liability part of a general liability policy.

Punitive damages. Any penalty or punitive damage assessed against an employer for violations of law is not insured under the workers' compensation policy. Examples would be (*a*) fines and penalties for illegal employment of a minor; (*b*) failure to insure; (*c*) failure to report an injury; (*d*) or serious and willful misconduct.

Statute of limitations. No coverage is provided for claims presented after the statute of limitations has expired—usually a period of five years after an injury was discovered and three years for employer's liability.

Workers' compensation exclusion. Under employer's liability, coverage is excluded for awards obtained under workers' compensation or occupational disease laws, unemployment insurance, disability benefits laws, or any other similar law.

BENEFITS

The level of benefits for workers' compensation varies significantly among states. Each state legislature determines the appropriate benefits and benefit levels. One state may be generous whereas another may be closefisted. The quality and makeup of the labor force, wage levels, and availability of medical facilities all contribute to these differences.

Workers' compensation laws place a definite but limited liability on the employer for employee disabilities caused in the workplace. Benefits payable to the disabled worker, under the law, attempt to compensate for part of the economic loss of earnings and extra expenses associated with the injury. The types of benefits provided are income or cash benefits, medical benefits, rehabilitation benefits, and death benefits.

Income or cash benefits replace an injured employee's loss of income or supplement his or her reduced earning capacity. These benefits are provided for both physical impairments and disability. Disability benefits are available whenever there is an impairment and a wage loss.

Medical benefits are provided for injured workers without dollar or time limitation. Most states use a medical fee schedule which is somewhat less than the customary charge for compensating industrial injuries and illnesses. Physicians and pharmacies are obligated to accept the fee schedule as payment in full. Insurance carriers have been known to make exceptions in the case of specialists and pay them their usual and customary fees. Medical benefits include all medical, surgical, and hospital treatment, including medicine, prescription drugs, medical and surgical supplies, crutches, and other apparatus including orthopedic replacements.

In some states, the injured has a free choice of physician; in others the employer can designate the physician. Some states allow a selection from a panel, and some allow the employee or the employer to request or honor a change of physician.

Vocational rehabilitation benefits are provided for those cases involving severe disabilities and include both medical rehabilitation and vocational rehabilitation. This distinction was not made until vocational-rehabilitation came on the scene. As a matter of practice, they are provided in all states even when not specified by statute (currently three states). The provisions of the benefits vary significantly between states. As a general rule, benefits are available to a worker whose industrial injury or illness permanently precludes him or her from engaging in the former occupation. Benefits may include temporary disability while undergoing rehabilitation, medical treatment, living expense, vocational and psychological counseling, or retraining and job placement assistance.

The goal of a rehabilitation plan is to return the injured to suitable, gainful employment.

Death benefits are provided to dependents or, in some cases, to the state if no dependents exist.

DISABILITY CLASSIFICATIONS

Permanent total disability (PD)

When the injured is totally disabled (not able to do any kind of work) and will not improve regardless of treatment, he or she is classified as "totally disabled." For this category, the benefit in most states is two-thirds of average wages, subject to a minimum and a maximum. In most states, the benefit is for the duration of the injured person's life. Some states provide additional amounts for dependents. A few provide permanent total disability benefits for a maximum number of weeks. For example, Indiana provides a benefit for 500 weeks with an additional 150-week benefit from a second injury fund. An example of a permenant total disability would be the loss of sight in both eyes.

Temporary total disability (TD)

Most injuries fall into the temporary disability category. In this case, the injured is expected to recover but is temporarily unable to work. The benefit is payable, subject to a minimum and a maximum and based on the average wage, until the individual is able to return to work. Benefits begin after a waiting period, and the waiting period is retroactively covered after a certain number of days off work have elapsed. An example of temporary total disability might be a broken leg.

Permanent partial disability

A permanent partial disability causes only a partial disability. The injured may even be able to perform the customary job, but earning capacity and ability to compete in the open labor market have been impaired. These benefits are designed to compensate for the effect of the injury on the future earning capacity and are calculated on a scheduled or nonscheduled basis.

- *Scheduled injuries* are compensated for, according to a specific schedule. The schedule is fixed by statute and the benefit is paid whether the injured party is working or not. Most schedules provide payment over a set number of weeks (based on the body part affected), multiplied by the weekly benefit amount (based on by a percentage of earnings at the time of injury), and subject to a minimum and a maximum.

- *Nonscheduled injuries* are compensated for on the basis of a replacement of a percentage of the wage loss. The percentage used is the difference between wages earned before and after the injury.

Temporary partial disability

Temporary partial disability means the worker is disabled, but still capable of working. However, the employee can perform only part of the usual duties of the job until recovery is complete. This type of injury is not common. The injured would be compensated a percentage of the difference between the previous and current wages.

Death benefits or survivor's benefits

Benefits are paid to the dependent spouse or children if an employee is killed on the job. If there are no immediate total dependents, benefits may be paid to partial dependents such as mother, father, grandchildren, brothers, sisters, or others who were receiving support from the deceased employee. Currently, there are significant differences in the amounts and methods of payment among states.

Social Security disability benefits

The Social Security Act provides monthly benefits that may supplement workers' compensation, to those under age 65, who are disabled beyond the required waiting period, which is five months after an application has been filed. Benefits are calculated as though the injured had reached the age of 62. Under the act, the term "disability" means inability to engage in any form of gainful employment, and the disability is expected to last for a period of not less than 12 months or to result in premature death.

Combined Social Security disability and workers' compensation benefits may not exceed 80 percent of the "average current earnings" prior to the disability. Social Security benefits are reduced from what they would otherwise be so that the combination of the workers' compensation benefit and the reduced Social Security benefit equal 80 percent. In some states, workers' compensation benefits are reduced by all or part of Social Security benefits. This is known as a reverse offset.

Subsequent (second) injury funds

A pre-existing injury may combine with a second injury to produce a disability greater than the second injury alone. For example, a worker loses a hand in an industrial accident and later suffers loss of the other hand, thereby becoming totally disabled.

The amount of workers' compensation benefits is considerably increased when the injured receives benefits for the combined disabilities. Because of this potential increase in the cost of compensation benefits, an employer might refuse to employ handicapped workers.

The subsequent (second) injury funds were established to solve this problem. The employer or insurance carrier pays compensation for the second injury alone, even though the employee receives benefits relating to the combined disability. The difference is made up from the subsequent (second) injury fund of the state. Most states, but not all, currently have subsequent (second) injury funds. In the states that do not have such funds, employers would be liable for the combined benefits.

INSURANCE SUBSIDIARIES OR CAPTIVES

The concept of an insurance subsidiary (this term is preferred by some, but the word "captive" has become a part of our insurance language) goes back 70 or more years. In Europe and England, the shipping industry had captives for marine insurance around the turn of the century, some of which have developed into substantial insurers covering marine risks and eventually other transportation areas. In the 1950s and 1960s, the large multinational companies formed wholly owned insurance subsidiaries or captives for many valid reasons, few of which were related to workers' compensation. Some of the reasons were to:

- Combine an otherwise unwieldy insurance program into a single placement source.

- Obtain broader/standard coverage than the multinational company enjoyed in its own country.

- Make premium and loss allocations on a more equitable basis to its various profit centers.

- Manage cash flow better.

- Apply a standard deductible, through reinsurance, in countries where standard forms would not allow deductibles.

• Take advantage of tax havens for overseas business so that money did not have to be returned home and subsequently taxed.

• Provide for huge capacity (property) where the conventional market could not.

• Purchase various forms of insurance that were simply unavailable through conventional markets.

The captive concept eventually spilled over into workers' compensation. These insurers (captives) were originally associated with workers' compensation self-insurance programs. The self-insured employers accumulated funds to pay losses out of pretax dollars. According to the Internal Revenue Service, self-insurers are allowed to deduct losses only as the losses are paid, whereas premiums paid to an insurance company are deductible in the current period.

By forming a subsidiary (captive) insurance company, the parent corporation could take a deduction for premiums paid to the subsidiary. The subsidiary, in turn, would follow insurance accounting and deduct loss reserves from income. Thus, on a consolidated basis, the loss reserves were effectively deducted from net taxable income. In a series of cases, the most famous being the Carnation case, the Internal Revenue Service has successfully defeated this scheme, to take full premium deduction. Currently, a whole series of tests has been devised and must be met to allow premium payments to be deducted.

Major corporations with captives have two alternatives. They can either disband the captive as an unnecessary complication to their self-insurance program or expand the business of the captive to make it a true risk-bearing insurance company with substantial business from sources other than the parent company.

Generally, the captives that write workers' compensation today appear to be performing two functions: they are acting as small specialty companies and they are functioning as a means to invest in the insurance business. The latter aspect arises since, under several schemes, the policyholder (or insurance agent) becomes an investor in the captive. In evaluating captives, it is important to clearly separate the two roles—insurance and investment.

In recent years, a tremendous number of offshore captives have been created. Most were chartered in Bermuda or in the Cayman Islands. A U.S. company would work just as well except that the state insurance departments impose higher standards for capitalization and surplus than do the foreign countries. The state insurance departments also impose other regulations designed to protect the solvency of domestic carriers. Thus, the foreign captives have greater freedom (investors would also normally translate this to mean higher risk). Since the offshore captive is not licensed in any state, it cannot directly provide insurance coverage in the United States. It must contract with a licensed company to provide such coverage. The licensed company then cedes, via reinsurance, the premiums and losses back to the captive. The licensed company is referred to as a "fronting" company. Viewed as a specialty insurance company, the captive would have reason to exist if it could provide broader coverage, lower costs, or better service than is available through normal insurance market channels.

Each captive company and captive program would need to be evaluated separately to determine whether any of the above reasons apply. However, coverage for the primary insured can only be provided on forms and at rates authorized for use by the fronting company. Most captives would appear to be at a distinct disadvantage in providing service. They must either contract for this service with the fronting company or with an outside vendor.

The investment aspect of captives would appear to be a major justification for using them at the present time. The policyholder or insurance agent becomes a stockholder in the captive and many derive a profit through the ownership—investments may be referred to as stock, preferred stock, redemption certificates, and so forth. (The issue of soliciting and selling unregistered securities, however, should be looked into closely because it may not conform to security statutes.) To the extent that an investment is made, the policyholder accepts the risk of investing, including becoming a risk bearer. If no investment or only a nominal investment is made and "stock" is offered as a consideration to insure, it might be considered an illegal rebate.

It is frequently stated that captives operate at a lower expense ratio than primary carriers. But captives actually incur additional expenses that are not required of normal direct carriers, such as:

- *Fronting company fees.* The captive must pay a fee sufficient to cover services and provide the fronting company with a margin for profit and contingencies.

- *Service vendor fees.* Whether provided by the fronting company or other service vendor, the captive must pay for these services including a markup and profit to the vendor.

- *Reinsurance costs.* Reinsurance costs would normally be higher or coverage lower or placed with less sound reinsurers than could be obtained by an established direct company.

- *Cost of letter of credit.* The captive must provide the fronting company with a letter of credit equal to the loss reserves and unearned premium reserves.

• *Transfer tax (offshore).* One percent of money transferred in and out of the United States is subject to a transfer tax.

CASH-FLOW PLANS

Essentially, cash-flow plans are a means of deferring payment of a substantial portion of the workers' compensation premium beyond the policy period. While the expense element of the full premium plus interest is required, a substantially longer period of time for payment is provided. Since the cost of borrowing from an insurer is less than through bank or other sources, the plan generally provides a positive cash management feature.

There are a number of different types of plans, but the most popular is the *deferred premium plan.* With this plan, the insureds finance some percentage of their policy premium over some period of time that is longer than their normal policy term and do so at an interest rate generally below the prime. The mechanics are that the employer pays some stated percentage of the normal deposit and subsequent premium billings during the life of the policy while the balance is financed. Interest expense may be added to the premium for dividend calculations, potentially reducing the net effective interest cost.

The deferred premium is usually guaranteed by a promissory note or some form of security and frequently backed by a letter of credit or bond. The promissory note usually bears a fixed rate of interest for the life of the note. In some cases, the interest may be a variable rate tied to prime rate.

Another popular cash flow plan for very large accounts is the *paid loss retro.* Under this plan, the insured pays the administrative expense portion of the premium, plus the claims costs, as they are paid by the carrier. Since the carrier foregoes investment income on loss reserves, an interest charge is added to the expense charged. The policyholder may also be required to set up a replenishing fund from which claim payments can be made.

COMPULSORY OR ELECTIVE COVERAGE

When workers' compensation laws first came into being, the employer had the choice of coming under the act or not. By purchasing coverage, the employer would elect to come within the law. The advantage was that workers' compensation then became the exclusive remedy for compensation of the injured with benefits set by statute. But the employer gave up many of the common law defenses that were used in tort law at that time.

Similarly, in some states, employees could reject the act by filing a rejection at the time of hire or prior to an accident. Since workers' compensation is a no-fault system, however, the employee was certain to collect benefits if injured and therefore it was generally to his or her advantage to accept the provisions of the act. If the employee rejected the act, the employer retained the common law defense. The basic reason for elective coverage was to avoid the constitutional issue that the employer might be deprived of the right of due process. New Jersey, South Carolina, and Texas currently have elective laws.

Under compulsory laws, the employer is required to provide coverage and will be penalized for failing to do so. Moreover, the employee cannot reject benefits and sue in tort.

Employment of minors

Because of possible penalties and criminal sanctions, every employer should know the child labor laws in his or her state. Moreover, all employers should verify the age of any minors hired. The hiring of minors is affected both by labor laws and workers' compensation law. Labor laws spell out the type of work that minors may perform. None of the states allow minors to work on machinery with dangerous moving parts or in hazardous occupations. Employers who do not abide by these laws are subject to fine and possible imprisonment. If a minor is injured on the job, compensation laws provide for uninsurable penalties against the employer. Some states provide unlimited double indemnity, which means the employer must match what is paid by the carrier. Others, such as New Jersey, give the injured minor the right to sue for civil damage in addition to collecting workers' compensation benefits.

Injury outside the jurisdiction

As a general rule, when an employee is hired in one state and generally works in that state, but is injured in another, he or she has a choice of where to file a claim. Employees will usually select the state that provides the greatest benefits. However, if no work is performed in the state where the contract of hire was made, that state will not normally accept jurisdiction.

When the contract of hire is in one jurisdiction and the employee is injured in an area of federal jurisdiction, the employer may have a liability under either the state or federal jurisdiction. An injured employee cannot make a double recovery by filing more than one jurisdiction. However, the employer, not the carrier, may be liable for benefits in excess of those provided in the state of hire when coverage was not purchased in the state of injury.

Failure to insure

In those states where workers' compensation is compulsory, failure to insure, or to qualify as a self-insured, will result in an uninsurable penalty (fine) and possible imprisonment. Moreover, the employer's operations may be stopped until coverage is secured. Some states provide an Uninsured Employer's Fund from which the injured may collect benefits and still bring an action against the employer for penalties and attorney fees. Additionally, the Uninsured Employer's Fund has a right to recover benefits paid the injured from the uninsured employer.

ADMINISTRATION

Industrial/Medical facilities

It is the primary responsibility of the employer to arrange for and provide initial medical treatment for any work-related injury. This duty includes all medical, surgical, nursing, and hospital care reasonably required to cure or relieve the effects of the injury. This responsibility may also include therapy and various other healing techniques. The employee is not required to ask for medical treatment, but to notify the employer of the injury. It is the employer's responsibility to provide or offer treatment. Given this responsibility, the employer would be well advised to make arrangements with a quality medical facility, in advance, to accept injured employees. The facility should be within a reasonable distance from the worksite and should be able to accommodate emergencies.

The employer should meet with the administrator of the facility to discuss the types of work done by employees (skills, physical requirements), the types of injuries expected, and the opportunities for light-duty work following an injury. Job descriptions and duties also should be provided to the facility. The physician who will treat injured employees should tour the work location to become familiar with the exact job requirements.

Reporting of injuries

The employer must report injuries to the carrier and usually to the state agency having jurisdiction within certain time limits. The time limits for reporting vary from immediately to 30 days. Failure to report may result in fines and, in some cases, imprisonment. The carrier will usually file on behalf of the employer with the required state agency.

In some states, it is not necessary to report first aid cases treated in-house or by a physician, provided that the injury did not result in lost time. Not all state compensation laws require recordkeeping on injuries.

OSHA, however, does require recordkeeping on *all* work injuries. The same form used to report the injury to the carrier will suffice for OSHA.

The employer's first report to the carrier is confidential and in most states is not admissable in the appeals process. The treating physician is also required to make an initial report; however, that report *is* admissible.

Discrimination charges

Most states have a provision to protect the injured employee from discharge or harassment because of filing a claim or testifying in support of another employee filing a claim. Discharge of an employee for these causes will almost always result in a litigated claim that will probably cost the employer more than an unlitigated claim. In some cases, uninsurable criminal penalties and fines may be charged against the employer, and the employee may also bring a wrongful discharge suit. If a terminated employee's case is upheld, the employee will most likely be reinstated with back wages.

THE CLAIMS PROCESS

Once an injury report has been filed with the carrier, it will either accept the claim, reject the claim, or investigate to make a determination. If the claim is accepted, medical payments begin and are usually paid 100 percent by the carrier with no limits as to total payments paid. In some cases, fee schedules are set by state legislatures. Disability benefits are paid by a scheduled figure based on earnings at the time of injury after the required waiting period.

Permanent disability is determined by medical evaluation without a "pain and suffering" concept, but rather with a medical determination of the known and/or subjective disability that affects the injured's work performance. The commission, board, or agency having jurisdiction makes a final decision, and payment is made either as a weekly benefit for a prescribed period of time or in a lump sum settlement, usually indicating a final conclusion to the claim and eliminating the possibility of further recourse on the part of the injured worker. If a claim is rejected, usually over an issue of AOE/COE (previously described), the injury may be covered by some other form of insurance. If the injured believes that he or she has a right to workers' compensation benefits, he may appeal to the board, commission, or agency having jurisdiction. When a claim is investigated by a carrier, it is usually to obtain enough data to make a determination or to evaluate the possibility of a third-party subrogation action.

As a general rule, the rank-and-file worker does not understand the workers' compensation system. When injured, employees frequently think that they must sue their employer to recover benefits. They do not understand that workers' compensation is a no-fault system and that they are due benefits, but they generally have no idea what benefits are due them. Usually, they think the employer is upset with them because they had an injury, and they feel guilty, unsure, and worried about financial matters. A visit by an employer or his representative to the hospital, to the home, or even a phone call to explain that the employer is not upset and wants the worker back as soon as possible and to explain benefits that the injured can expect will frequently accelerate the recovery, reduce the cost of the claim, and minimize litigation.

In its Report to the Industry, the California's Workers' Compensation Institute concluded that most people seek legal assistance because they do not understand the system and the benefits due them. It would seem that by simple communication, employee needs would be better served and the cost of workers' compensation would be reduced.

Unit statistical reports

Each year following the expiration of a policy, carriers are required to file with the rating organization, a report of payroll and claims costs by classification. This information is used to promulgate industry rates and the experience modification (credit or debit charged against the premium) for the individual account. Prior to filing that report, the employer may wish to review all open claims with the carrier as to status and reserves because, once those figures are filed, they are frequently irreversible.

FINANCIAL INCENTIVES

One of the basic objectives of the workers' compensation system is to encourage safety in the workplace. This objective can be achieved by a number of financial incentives (ability to reduce the ultimate cost), primarily by preventing and controlling occupational injuries and illnesses. For self-insureds, the incentive is reduced claim payments and costs and, for insureds, it is reduced net premiums. To the extent that the cost of prevention is perceived as cost effective, management will usually respond positively. The larger the potential savings, the greater the potential preventive effort.

Cost incentives of self-insureds

There is a direct one-to-one relationship between the reduction of costs and preventive effort. For every injury prevented, the self-insured employer saves the costs of such benefits as medical, indemnity, claims handling and legal fees, permanent disability, and rehabilitation, plus the loss of use of reserves and the tax credit on those reserves.

Cost incentives of insureds

Experience rating. Experience rating is one of the most influential cost incentives for insureds. It can be an up-front discount or surcharge for the individual employer based on past loss performance. Losses for an entire category of insureds can be predicted with a reasonable degree of accuracy. However, it is more difficult to forecast or predict losses for the individual risk.

Experience rating attempts to predict the losses of the individual risk by comparing the individual insured to a homogeneous group on a statewide basis. If the experience of the individual risk is better than average for the group, insureds will receive a credit rating, or if worse, a debit rating. The rating, expressed as a percentage, is applied to the "manual premium." Some modifications can be as low as 30 percent, some as high as 300 percent.

Not all insureds qualify for experience rating. Loss experience for smaller firms is not statistically reliable for forecasting purposes. There is usually a set premium level to qualify and not a sufficient premium to cause a financial incentive for smaller insureds. These insureds use manual rates without a modification applied.

Because of the formulas used to calculate the modification, the very large account may become self-rating. Their losses are not discounted in the calculation, and the effect is similar to that of self-insurance. In effect, they pay 100 percent of their losses in their future projections. For those insureds, loss experience has a significant impact on premiums and that becomes a strong financial incentive. As a general rule, for any experience-rated insured, lower losses will ultimately mean lower premiums. Three years of past payrolls and losses are used in the calculation, excluding the most recent expired policy year. The net result is that good or bad year(s) will have an impact on the calculations for three future years.

The financial incentive of experience rating often exceeds the financial incentive of self-insurance, and self-insureds should study each carefully.

Retrospective rating. There are a number of different retrospective rating plans approved by the various insurance departments. Basically, a retrospective rating plan is a contractual agreement whereby the insured's premium includes carrier expense and the insured's own losses and loss-adjustment expenses and is subject to a minimum and a maximum premium. With excellent experience, the minimum will be less than the standard premium. However, with poor experience, the maximum is charged and the premium will be higher than the standard premium. Standard premium is calculated as the manual premium multiplied by the experience modification factor.

Again, the financial incentive is greater for the larger insured, and to the extent that this insured can reduce injuries, it can save money. With a retrospective rating plan, premiums usually must be adjusted several times after the policy has expired before the true cost can be determined.

Dividend plan. A dividend plan is basically a statement of intent by the insurance carrier to allow the insured to participate in a refund of excess profits, if earned, from the premium they have paid the carrier based on that insured's own losses. Dividends are paid from accumulated earned surplus (profits), must be declared by the board of directors, and cannot be guaranteed. To a large extent, dividends are declared at the discretion of the carrier and based to some extent on the carrier's overall profitability as well. It's hard to justify dividends when a carrier is losing money.

Dividends can be significant financial incentive, and again are materially affected by the insured's own loss results.

Cash-flow plans. There are a number of cash-flow plans in existence today. Some are directly related to paid losses, and others to deferred premium payments. All have built-in financial incentives to reduce injuries and illnesses (previously discussed).

Cost relationships

Management has historically made its business decisions in terms of "the bottom line," in other words, the effect on its profits. There are some truly significant relationships between workers' compensation costs and profits. Once some of these relationships are realized, management will tend to be more responsive to its safety priorities. Table 17-A shows two identical examples in terms of manual premium, gross sales, and planned net profits to illustrate some of these relationships.

For this example, the same net profit percentage has been used for both companies. Actually, the net profit for Company B would have been reduced by the standard premium differential. It would also have increased the allocation to sales, the cost per man-hour, and required sales to support that cost. The required sales are calculated by dividing the cost of insurance by the net profits (Cost/Profit = Required Sales). From required sales, the number of units, given the cost per unit, and/or

TABLE 17-A
THE IMPACT OF EXPERIENCE MODIFICATION ON CALCULATION OF INSURANCE PREMIUMS

Manual premium	$ 100,000	
Annual sales	$5,000,000	
Net profit (planned), percent	5	
Manhours worked	200,000	
	Company A	*Company B*
Experience modification	0.50	1.50
Standard premium	$ 50,000	$ 150,000
Variance	$ (50,000)	$ +50,000
Percent of sales	1.0	3.0
Required sales	$1,000,000	$3,000,000

the months of production required to support the cost of the insurance can be projected.

SUMMARY

Workers' compensation represents a significant business expense for employers, in some cases, 20 to 30 percent of payroll. The system, however, is critical to the economic well-being of the injured worker. There is a clear trend of increased benefits to the injured and increased cost to the employer which are passed on to consumers, in terms of higher priced products. Only one thing is certain: workers' compensation will cost more, not less, in the future.

To a large extent, employers can influence their net costs for this coverage, positively or negatively, through their own efforts. There are sufficient cost incentives in the system to motivate management positively in that direction. The enormity of the cost and the means of controlling these costs must be understood, accepted, and applied. Management is the key. When health and safety becomes a high enough priority for top managers to pay attention to it, they, their employees, and the public will realize the direct benefits.

ADDITIONAL READINGS

"Analysis of Workers' Compensation Laws, 1982." Washington, D.C.: U.S. Chamber of Commerce, 1982.

Grimaldi, J. V., and R. H. Simmons. *Safety Management,* 3rd ed. Homewood, Ill.: Richard D. Irwin, 1975.

Hanna, W. *The Workers' Compensation Laws of California.* San Francisco, Calif: Matthew Bender, 1981.

Herlick, S. D. *The California Workers' Compensation Handbook,* 6th ed. San Bernadino, Calif.: Herlick Publications, 1979.

Malecki, D., J. Donaldson, and R. Horn. *Commercial Liability Risk Management and Insurance,* Vol. 1. Malvern, Pa.: American Institute for Property and Liability Underwriters, 1978.

The Producer's Guide to Workers' Compensation. Encino, Calif.: Zenith Insurance Co., 1980.

Tarrants, W. E. *The Measurement of Safety Performance.* New York; N.Y.: Garland STPM Press, 1980.

Zenz, Carl. *Occupational Medicine: Principles and Practical Applications,* 2nd ed. Chicago, Ill.: Yearbook Medical Publishers, 1986.

Chapter 18

Malpractice

by Gerald R. Williams

Considerations of malpractice in occupational medicine are somewhat complicated by the relationships of the involved parties. The consumer of services is basically a third party to the interaction. The employer pays a health-care provider* to meet with his or her employees and provide for their well-being, which is only indirectly beneficial to the employer. Health-care providers are charged with the task of providing benefits directly to employees and only secondarily to the employer. Employees receive medical attention from a person not necessarily of their own choosing, but with whom they are expected to have a trusting relationship. In each of these relationships, the beneficiary of the contract is not specifically the individual paying for it. These are difficult contractual relationships and can confuse the issues of liability in situations where malfeasance is alleged or established.

CORPORATE LIABILITY

The benefits to a corporation for providing its employees with care for injuries or accidents experienced while they are working are many. One justification for providing such care is that union contract or labor negotiations require employers to pay the direct cost of services needed by an employee in the day-to-day conduct of his or her business. Such contracts may include not only illness and injury which arise directly out of employment, but health matters in general for which the employer provides health insurance.

---

*In this chapter, health-care provider is inclusive of the physician, nurse, or other individuals licensed in health care delivery.

Another reason for employers to provide health care to their employees is to control costs. Prenegotiated contracts or employing in-plant health care providers to provide the care will reduce or eliminate paying an outside profit for such care. A significant justification for providing in-plant health care is the potential decrease in the amount of lost work-time involved when employees are forced to go outside the working location to receive medical care for problems that are not severe.

When employers provide health care, the responsibility will generally fall upon the employer for any damages that result if such care is negligent. Responsibility to prove liability in another party will then generally fall to the corporation.

The justifications for this liability of the corporation are numerous, but some that have been suggested are: (*a*) the control that the corporations or owners have over their employees in accomplishing their tasks; (*b*) the corporation as the owner of the enterprise is responsible for the consequences of the enterprise; (*c*) the corporation had the ability and the responsibility to select its employees so that if a selection is inappropriate or inadequate, the corporation would suffer the consequences; and (*d*) the losses caused by the enterprise are charged against the enterprise itself as an additional operating cost. Therefore, the costs are absorbed by the consumer who needs or wants the products that the enterprise produces. Ultimately, according to W. T. Prosser, the situation is reduced to the doctrine of the "deep pocket," that is, the person or persons most able to pay for the damage are called on to make recompense.[1] As a result, the injured or aggrieved party has the right to look to the owner of the enterprise to pay ultimately for any damages that result from the activities producing the products.

LIABILITY OF THE HEALTH-CARE PROVIDER

The liability of a health-care provider practicing occupational medicine depends on his or her legal relationship to the employee-patient. One must consider whether the provider of health care is an employee of the corporate employer and duty-bound to the employer, or an independent contractor who is primarily duty-bound to the employee-patient. Generally, employers are liable for the negligence of employees, whereas independent contractors are sued separately.

The nature of the duty determines who has to pay for failure of the parties to meet that duty and, generally, whether liability is based on no-fault workers' compensation or common law negligence. Common law recoveries for damages can be much larger than those under workers' compensation acts; however, a greater burden of proof is required of the injured party.

Physician as employee or independent contractor

Courts have been variable in deciding whether or not a physician can be held liable as an employee or as an independent contractor. Issues that may be considered are: is the physician on a salary, or does he or she charge a fee for service? How much control of the physician's activities does the employer have in terms of interacting with the employee-patient? Does the physician practice in the employer's facility or in his own office outside the employer's facility? Does the physician work for one employer only or have several corporate clients or others who benefit from the interaction of employer or employee?

The Indiana court held that a physician does not accept from the employer direction or instruction for diagnosis or treatment. He cannot therefore be considered an employee regardless of the extent of his relationship to a corporation. It was further felt that "fellow employee" immunity provisions of Workers' Compensation Act laws could not protect a company physician from a separate suit for medical malpractice because it was not likely that the legislative intent was to interfere with the traditional doctor-patient relationship.[2]

The courts of California,[3] Michigan,[4] New York,[5] and Ohio[6] have held that the traditional doctor-patient relationship does not occur between occupational physicians and employee patients. The physician's duty, therefore, is more to the benefit of the employer, the implication being that the physician is an employee. An Ohio court decision stated that because the efficiency of the corporation's operations are directly served by providing readily accessible medical care to corporate employees, its physicians were performing work that furthered the employer's business venture and hence were employees.[7]

It is generally held that no physician-patient relationship occurs at a preemployment physical exam.[8,9] Thus, the physician would then be more likely to be considered a fellow employee to other employees. The judicial trend would seem to be that if a physician *only examines* the employee for an employer, the physician is considered an employee for duty and liability purposes. However, if the physician extends the relationship by *treating* or *advising* the employee-patient, the duty shifts to a more traditional relationship for malpractice liability.

Occupational nurse as employee or subcontractor

The occupational nurse, almost without exception, is considered an employee of the corporation or physician for whom he or she works. The procedures and means used to accomplish nursing tasks are essentially controlled by another person, whether it be the corporation or a physician. Sometimes a nurse can be an independent contractor, for example, a nurse who is retained to do a glaucoma survey and report the results may be functioning in a discretionary fashion as an independent contractor. A private-duty nurse who cares for a patient in a hospital and is paid by the patient may be more at liberty to use her own personal professional judgment and could also be considered an independent contractor.

If a nurse is considered an employee, as opposed to an independent contractor, his or her liability for negligence will fall on the employer (under the doctrine or *respondant superior*) where such acts were committed within the scope of the nurse's employment. The employer is not responsible for acts committed by an employee nurse outside the scope of the nurse's employment. If the nurse is deemed to be a fellow employee by the courts, the injured party may be precluded from suing for common law negligence. If the nurse is found to be a nonemployee or an independent contractor he or she can be sued for damages not restricted by workers' compensation. The damages awarded for negligence are generally much larger than they would be under workers' compensation, but the burden of proof to a jury is generally much more difficult.

Medical care consent

Consent to medical care by the employee-patient must be given to the health-care provider before medical treatment is begun. If there is no evidence of such consent, the patient has the basis for an action in battery against the provider. Consent must include the type of care and its extent of involvement.[10] Except in an emergency, if an anticipated procedure requires substantial additional unanticipated treatment, the consent may require modification before the health-care provider proceeds. There is an additional requirement of adequate disclosure prior to any procedure involving risks. The provider must give the patient those facts concerning the nature, extent, risks of care, and alternatives to care he or she needs to make a reasonable decision. Inadequate disclosure is negligence.

In the modern climate with a plethora of required screening examinations, it is interesting that the California Supreme Court has ruled that a physician must also inform a patient of the risks he or she takes by refusing a "risk-free" examination.[11]

Consent to medical care by the employee-patient may be manifest by word or deed. The provider can reasonably rely on employee-patient conduct that would indicate a consent to a reasonable person. Silence or inaction may imply consent when a reasonable person would otherwise have objected. The consent of the employee to care by the occupational provider of care is to be implied from at least four factors: (*a*) the acceptance of employment by the employee-patient implies consent to at least immediate care from the employer's health-care providers; (*b*) usage and custom dictate that ordinarily in an emergency medical situation, medical persons would provide care to prevent further damage; (*c*) in an emergency situation, a patient's consent to care is ordinarily not necessary if immediate action is necessary to prevent irreparable damage to life or health; and (*d*) statutes often provide a period of time when an employer may control the provisions of medical care. For example, California allows 30 days of control by the employer for the provision of the medical care to an injured employee, unless the employee prior to the injury has named in writing a preference for a health provider for emergency care.

A nonlucid person cannot give express consent to actions he or she does not understand. In an occupational setting, this would include a patient in the workplace sufficiently intoxicated by drugs or by chemicals.

Negligence can be based on failure to treat a nonlucid patient by implied consent in an emergency, provided there was a legal duty to give care, such as where the injured is a fellow employee or where a contractual relationship exists.

NEGLIGENCE

Common law presumes that each person has a right to assume that no other person or thing will harm him. If he or she is harmed, he or she is to be compensated by the party causing the harm for any damages experienced, as long as the harmful conduct was in violation of the common law.

Negligence is based essentially on three elements. First, there must be a legal duty owed by the actor.* Usually this is just a duty to act as a reasonable person would act under the same set of circumstances. Second, it must be shown that the actor in some way breached that duty. Third, it must be shown that the breach of the duty caused the damage that resulted to the plaintiff.

Every licensed health professional is charged with the duty of acting with the same degree of care in discharging his or her responsibility to patients as an average individual would in the same field of practice. This duty or standard of care does not require that the individual display above-average performance. With some exceptions, the standard of care is generally not limited by geographic location, and for physicians and nurses, it is essentially the same for any part of the country. If a practi-

---

*The word "actor," in legal terms, defines a person or thing that causes a set of circumstances to occur either by action or inaction. Actor here refers to the health-care provider.

tioner specializes, he or she is held to a standard of care that would be expected of the "average specialist" in that area of practice. When a practitioner does not meet the standard or duty of care in a given situation, it is considered to be negligence.

Proof that a practitioner has not met the duty- or standard-of-care requirement is not grounds in itself for a suit. The plaintiff patient must show that actual damage resulted because of the act. The committed act or acts must have substantially contributed to the harm the plaintiff experienced.

Health-care providers are generally not held liable for errors of judgment provided they have used the current methods of diagnosis and treatment that are available under the circumstances. A provider may be held liable, however, for failure to refer when the situation warrants referral to another provider with more training or experience in the relevant area of practice, for example, an occupational nurse who did not refer a wound that failed to heal to a physician,[12] or an occupational physician who failed to refer a complicated fracture to an orthopedic surgeon.

Negligence per se is a legal doctrine that applies to a practitioner who is acting in violation of a statute or regulation and is therefore presumed not to be satisfying the required duty of care. Being able to advance the negligence per se doctrine places the plaintiff at advantage in a malpractice suit, if the violation of a statute in fact caused the injury. For example, failure to report an infectious disease as required by department of health regulations, causing another employee to contract that disease, can lead to liability of the health care provider under negligence per se.

The following additional concepts should also be considered. It is not an adequate defense for a nurse to claim that he or she was following a physician's orders when such orders would appear negligent to a reasonable nurse. A physician or supervisor nurse who orders a subordinate to do a task that is known, or should be known, to exceed the subordinate's ability can be liable for resulting injuries. An unlicensed person may not practice a licensed medical skill under protection of a physician's license. For example, an unlicensed physician may not practice medicine under the supervision of a licensed physician.[14]

Defenses to an allegation of negligence

Good Samaritan immunity generally applies by statute to a health-care provider in an emergency, outside the usual scope of activities. The California Business and Professions Code, as it applies to registered nurses, is a good example. Section 2727.5 of that act provides: "A person licensed under this chapter (a registered nurse) who in good faith renders emergency care at the scene of an emergency which occurs outside the place and course of her employment shall not be liable for any civil damages as a result of acts or omissions by such person in rendering the emergency care." The purpose of the act is to encourage nurses to give aid in emergencies.

A nurse who renders care on the employer's premises, even though it is not directly related to her duties, would not be likely to have Good Samaritan immunity because it is her place of employment.

Some states, such as California, provide statutory immunity for anyone engaged in cardiopulmonary resuscitation.[13] Under California statutes, there is also immunity from civil liability for those acting as part of a hospital rescue team trained in cardiopulmonary resuscitation and who are actively resuscitating persons who are in immediate danger of loss of life.

Another defense to a suit is a passage of time greater than the statute of limitations would allow for bringing such suit. The prescribed period of time defined by law to bring an action to litigation may begin at the time of the negligent act or the time of discovery of injury and its negligent cause by the injured party.

MEDICAL RECORDS

There are two broad areas of concern regarding medical records of employees. (Also see Chapter 16, "Government Impact on Occupational Health and Safety.") The immediate concern is the responsibility to notify the employee of the results of his or her examination and to determine the information from the examination that must be given to the employee. The second concern is publication, which involves stored records of examinations for use for various purposes by the employee, other people, or agencies.

Informing the employee

The Code of Ethical Conduct for Physicians Providing Occupational Medical Services adopted by the American Occupational Medical Association Board of Directors (Part nine) provides that physicians should: ". . . communicate understandably to those they serve any significant observations about their health recommending further study, counsel, or treatment when indicated."[15] (See pages 477-481.)

The courts have held that there is no duty on the part of the medical examiner to inform the employee of any specific deviation from health that he or she manifests, but has held the employer who sends the employee for examination liable for failure to inform. The issue is based on whether or not a physician-patient relationship exists. Courts in California,[2] Michigan,[3] New York,[4] and Ohio[5] have held that the occupational physician and employee-patient do not enter into the kind of consensual, voluntary relationship that gives rise to typical contractual obligations. "A different nonconsensual relationship comes into being if a physician examines a job applicant or employee for the benefit of a prospective or actual employer. . . ." And ". . . [C]ourts have uniformly imposed a lower duty of care than that owed to regular patients."[16]

A California court clearly indicated that a physician might be held liable for damage resulting from failure to inform a man who on repeated examinations was found to have evidence of disease.[17] In this case, several X-rays had indicated the presence of asbestosis in the employee-patient and the examining physician had not informed him. A federal court held that "When a doctor conducts a physical examination, the examinee generally assumes that 'no news is good news' and relies on the assumption that any serious condition will be revealed."[18]

The best reasoning would probably be that medical personnel should inform the employee-patient of any abnormalities discovered during the actual examination and that the employer is responsible for informing the employee-patient of abnormalities in test results subsequent to the examination.

Medical records are in the custody of the company and its failure to inform the employee allows that employee "to rely on tacit assurance of safety despite its knowledge of danger." The Sixth Circuit Court stated that although the employer was not obligated to provide examinations to its employees, when it did so, the employee has the right to assume that he or she would be informed of any dangerous condition actually found upon the examination.[19] If a physician does give an employee-patient a misdiagnosis or misinformation relative to his examination, he or she may be held liable for damages either in malpractice,[20] or perhaps in fraud.

Publication of medical records

Employee medical records can be misused in any number of ways, depending on who has access to the files. Except as these records are needed for job placement, periodic evaluation relative to the workplace, or treatment of the employee-patient, information should not be obtained without the consent of the subject.

Various forms of consent, actual or implied, are used to gain information from employee records. Actual consent leads to no particular problems although its form and extent may be questioned. Implied consent from custom, contract, relationships of the parties, public safety, or legal requirements may cause problems.

If a physician-patient relationship clearly exists without involvement of a third party, such as an employer, the duties are relatively clear as to the publication of medical records. The privacy of the communication between physician and patient, whether verbal or nonverbal, is protected by both the Hippocratic oath and the law. In almost one-half of the states, a physician's license can be revoked for releasing information from a patient's records without the patient's permission or a legally justified need.

Before records are released to the employer relative to an employee-patient, his or her consent whether implied or actual, the type of information, the purposes for which it will be used, who will gain access to the released information and the breadth of dissemination must be considered. Courts have found unauthorized publication of patient record information to be a breach of implied contract by the physician.[21,22]

In the occupational setting, where third party relationships exist and the physician-patient relationship is either not very clear or nonexistent, questions about the publication of record information are common. The codes of ethics of the various disciplines responsible for health and safety of workers address directly or indirectly the need to respect confidentiality. Although codes of ethical conduct are not law, they may contribute to the development of standards of care. (See the examples in the Appendix at the end of this chapter.)

The seventh clause of the Code of Ethical Conduct for Physicians Providing Occupational Medical Services states that a physician should: "... treat as confidential whatever is learned about an individual, releasing information only when required by law or by overriding public health consideration or to other physicians at the request of the individual according to traditional medical ethical practice and to recognize that employers are entitled to counsel about the medical fitness of individuals in relationship to work but not entitled to diagnosis of a specific nature."[15]

The code of ethics of the American Association of Occupational Health Nurses states, "Occupational Health Nurses should ... 2. Safeguard the employee's right to privacy by protecting information of a confidential nature; releasing information only as required by law or upon written consent of the employee."[23] The American Academy of Industrial Hygienist's code of ethics also provides under "Professional Responsibility" part 5, "Protect confidential information."[24]

The code of ethics of the American Society of Safety Engineers does not directly address the matter of confidentiality, but Section II, part A, provides that: "The member will be guided by the highest standards of integrity when acting in professional safety matters for his employer, client, or customer."[25] The customer or client is presumed to include the employee. State-of-the-art considerations for a professional relationship as practiced by the other disciplines (hygienists, nurses, and physicians) would suggest that the relationship should include protection of the employees' and employers' privacy.

When considering the duty or standard of care owed to the employee, the courts have the relationship between employee or job applicant and the health-care provider to be such that a lesser duty or standard of care is owed by the health-care provider than would be given to private patients.[16] Legal decisions imply that if the physician or other health care provider goes beyond examination to treatment of the employee-patient, his or her duty is more that of the traditional doctor-patient relationship for standard or duty of care and confidentiality.[26,27]

The physician-patient relationship is not formed in the usual sense during preemployment physical examinations.[8,9] Thus, the information obtained during such examinations can be given to the prospective employer, but should be limited to those factors necessary to make a placement decision. Essentially it should be limited to answering the following questions:

1. Does the prospective employee have any discernible limitations of activities?
2. Could the prospective employee correct the limitations that are found?
3. If the limitation can be corrected, approximately how long will it take?

As to information to be given to an employer by a physician, the Alabama court noted: "Nor can it be said that an employer is necessarily a person who has legitimate right in knowing each and every detail of an employee's health. Certainly, there are many ailments about which a patient might consult his private physician which have no bearing or effect on one's employment.[28] This is the logical judicial trend and the best rule for those who practice occupational medicine. There are elements of a medical history that are not needed to answer employer questions about employees' ability to work and consequently should not be reported beyond a confidential medical record. A history of treatment for syphilis is an excellent example.

Information concerning a current employee that should be divulged from an examination includes:

- Whether the employee is ill or has been injured.
- Whether the illness or injury is related to employment activities.
- Whether the illness or injury is serious.
- When it is estimated that the employee can return to his job.
- Whether activities should be limited upon all employee's return to

work and for what length of time.

The right of the employee to medical records is relatively moot in that the OSHA Regulation of August 1980 requires all employers who deal with toxic materials or harmful agents to make records available to the employee within 15 days.[29] Records that must be given to the employee include medical tests, environmental data about the workplace, medical opinion relative to the employee, any medical complaints, and any prescriptions that are given. Since it would be extremely difficult for an employer to rule out the possibility of employees' being exposed to any toxic substance or harmful agent (for example, computer noise), the rule would seem to apply to all employees. Because the records must be kept for 30 years, it would also apply to past employees. Under the rule, the custodian of the records may take exception to divulging information that may be felt to be detrimental to the employee's health. If this happens, the information must be given to a representative of the employee.

Employees are also given the right to examine the toxic exposure records of another employee that may also apply to them or affect their work place. Union or employee representatives may, with the employee's consent, examine the employee medical records. State and federal governments also may compel publication of medical records by statute requiring the reporting of occupational injury or illness, by judicial process, or by quasi-judicial administrative process. Before turning over health information under compulsion of process, the employer's representative should determine the appropriateness of the process itself and whether the health-care provider's response is correct. The federal courts have suggested that NIOSH can only obtain individual medical records without the employee's consent when the employee cannot be identified from the material.[30]

Case law suggests that employers who permit defamatory or private information about employees to circulate indiscriminately throughout a corporation may be liable to the aggrieved employee.[31] If false health information is distributed that damages an employee's career or reputation there could be a basis for an action in defamation against the employer and the individual responsible for distributing it. Liability could also arise if persons distributing the information knew that it was inaccurate or if the information was given to parties who had no legitimate interest in the material. Disclosures that can be demonstated to be gratuitous or irrelevant can also be subject to legal action.[32] Moreover, disclosure of medical information made widely and indiscriminately, violates the employee's right of privacy, which is recognized in most states as a basis for civil action.

Case law and judicial trends indicate that revealing employee health records outside the corporation, unless specifically for the care of the employee or with a carefully worded release by the employee, can be the basis of an action against whoever releases the records. In California, it is considered a misdemeanor. The trend of the law at present is to suggest that employee health records are not to be disclosed to individuals outside the corporation without a properly signed authorization from the employee. On the other hand, in the event that the information obtained by a physician is reasonably required to protect the public safety, the physicians (including occupational physicians) could be held liable for failure to report it.

STANDARDIZED PROCEDURES

Nurses in occupational settings are likely to be considered more autonomous than their hospital-based counterparts. Often they act as primary health-care providers when specific physician care is not needed, and they are regularly called on to make that decision. Nurses are allowed to practice occupational nursing under statutes of various states that authorize some extension of traditional nursing duties into the areas that border between the practices of nursing and medicine. Standardized procedures are written authorizations, limitations, and guidelines for such activities. The California law in this regard (found in the California Business and Professions Code) is a good model because it is representative of the general legislation of the various states.

A state's business and professional codes grant the privilege to practice a profession or trade, and, at the same time, limit the nature or extent of the activities a licensed person may use to accomplish his or her respective tasks.

Business and Professions Code Section 2725(d) defines standard nursing procedures allowed in California.

> "Standardized procedures," as used in this section means policies and protocols developed through collaboration among administrators and health professionals, including physicians and nurses, by an organized health care system which is not a health facility licensed pursuant to Chapter 2. (The specific licenses needed to collaborate in the development of procedures are physician and nurse.)

Section (c) of Section 2725 defines the practice of nursing under standardization procedure.

> The performance, according to standardized procedures of basic health care, testing, and prevention procedures, including

but not limited to skin tests, immunization techniques, and the withdrawal of human blood from veins and arteries.

The key wording is "including but not limited to." Under standardized procedures, nurses are called on to do screening and basic health-care testing of patients and decide whether or not the patient needs to see a physician. Such activities done without standardized procedures could be held to constitute the unlicensed practice of medicine.

The California Business and Professions Code 2725 subsection (d) includes the practice of nursing:

> Observations of signs and symptoms of illness, reaction to treatment, general behavior, or general physical condition, and (1) determination of whether such signs, symptoms, reactions, behavior or general appearance exhibit abnormal characteristics; and (2) implementation, based on observed abnormalities, or appropriate reporting or referral or standardized procedures, or changes in treatment regimen in accordance with standardized procedures or the initiation of emergency procedures.

Thus, the nurse is authorized to evaluate signs and symptoms, to diagnose whether such signs, symptoms, reaction, behavior, or general appearance exhibit abnormal characteristics, and react to patients' needs for reporting, referral, standardized procedures, or initiation of emergency procedures. The nurse is able, under standardized procedures, therefore, to diagnose and give some types of care without a specific order from the physician.

Written protocols

Written protocols for nursing standardized procedures should:

- Be in writing and show date or dates of approval.
- Specify the standardized procedure functions that registered nurses are authorized to perform and under what circumstances.
- State any specific requirements registered nurses are to follow in performing all or part of the functions covered by the particular standardized procedure.
- Specify and experience, training, or special education requirements for performance of the functions.
- Establish a method for initial and continuing evaluation of the competence of those registered nurses authorized to perform the functions.
- Provide for a method of maintaining a written record of those persons authorized to perform the functions.

• Specify the nature and scope of review or supervision or both required for the performance of the standardized procedure functions. For example, if the function is to be performed only under the immediate supervision of a physician, that limitation must be clearly stated. Conversely, if physician supervision is not required, that fact should be clearly stated.

• Set forth any specialized circumstances under which the registered nurse is to communicate immediately with a patient's physician concerning the patient's condition.

• Specify any special requirements for procedures relating to patient recordkeeping.

• Provide for periodic review of the standardized procedure.

Nurses' activities under standardized procedures exclude activities that exceed their skills, require a more detailed medial work-up than they can provide, or require prescribing or dispensing of medication not in the standardized procedures. Under California Law, the nurse may, under standardized procedure, order such examinations and treatments as: X-ray, occupational therapy, physical therapy, respiratory therapy, or therapeutic diets.

CONCLUSION

As a part of compensation for working, the employer usually provides payment for a major portion, if not all, of the employee's personal health and accident insurance, as well as occupationally related health and accident insurance. There is, therefore, little reason to distinguish between care given to the employee for occupational or personal reasons based on the identity of the party paying. As a practical matter, the only difference to the employee is that medical costs for personal matters are paid as a percentage of the total cost, while occupationally related problems are fully paid. Payment is compensation to the employee for work done in either case. There is no logical reason for the legal system to treat the occupational medical practice in a different manner from private care except for identifying the party responsible for paying the costs.

The relationship between the health-care provider and the employee-patient should not be unique for considerations of standards of care or confidentiality when compared to private care situations. The right of a third party to bring an action to enforce rights stipulated in a contract is relatively new in common law. The individual's ability to enforce these contract rights is growing steadily and, in judicial terms, quite rapidly. The employees' rights as third party to enforce a contract for his or her health and environmental protection will continue to expand. In the future, it is likely that employees will have the same rights in relationships with health-care providers who are retained by employers as they would if they were paying directly for these services. The federal privacy laws and the states' increasing interest in guaranteeing the individual's medical privacy will probably lead to the concept that occupational medical records are indistinguishable from records held by private practitioners.

REFERENCES

1. Prosser, W.T. *Law of Torts,* Chapter 13, "Imputed Negligence." St. Paul, Minn: West Publishing, 1964.
2. *Ross v. Schubert,* 338 N.E. 2d 623 (1979).
3. *Hoesl v. United States,* 451 F.Supp. 1170 (N.O.Cal. 1978).
4. *Rogers v. Hovarth,* 237 N.W. 2d 595 (Mich. 1976).
5. *Mrachek v. Sunshine Biscuit, Inc.,* 123 N.E. 2d 801 (N.Y. 1954).
6. *New York Central R R Co. v. Wiler,* 177 N.E. 205 (Ohio 1937).
7. *Proctor v. Ford Motor Co.,* 302 N.E. 2d 580 (Ohio Sup. Court 1973)
8. *Wilcox v. Salt Lake City Corp.,* 484 P.2d 1200 (Utah 1971).
9. *Lotspeich v. Chance Voight Aircraft,* 369 S.W. 23d 705 (Texas Civil App. 1963).
10. *Vladex v. Percy,* 35 C.A. 2d 485 (1939).
11. *Truman v. Thomas,* 27 C3d 285 (1980).
12. *Cooper v. National Motor Bearing Co.,* 288 P. 2d 581 (1955).
13. *California Health and Safety Code,* Sec. 1317
14. *Magit v. Board of Medical Examiners,* 57 Cal.2d 74 (1961).
15. "Code of Ethical Conduct for Physicians Providing Occupational Medical Services," adopted by the Board of Directors of the American Occupational Medical Association, July 1976.
16. *Anhert v. Wildman,* 376 N.E. 2d 1182 (Ind. 1978).
17. *Vela v. Wise,* (California Superior Court, Contra Costa County Docket No. 116083 Nov. 1, 1973).
18. *Betesch v. United States,* 400 F.Supp. 238 (D.C. 1974).
19. *Union Carbide and Carbon Corp. v. Stapleton,* 237 F 2d 229 (6th Circuit 1956).
20. *Hoover v. Williamson,* 203 A. 2d 861 (Md. 1961).
21. *Anken v. Brodritz,* 413 N.Y.S. 2d 582 (Sup. Court 1979).
22. *Logan v. District of Columbia,* 447 F.Supp. 1328 (D.C. 1978).
23. "Code of Ethics." American Association of Occupational Health Nurses. Adopted by the Executive Committee January 14, 1977.
24. "Code of Ethics for the Professional Practice of Industrial Hygiene." American Academy of Industrial Hygiene.

25. "Code of Ethics for the Safety Profession," adopted by American Society of Safety Engineers Assembly, June 16, 1974.
26. Annas, J.G. *The Rights of Hospital Patients.* New York: Avon, 1975.
27. *Duprey v. Shane,* 39 C.2d 781 (1952).
28. *Horne v. Patton,* 287 So.2d 824 (Ala. 1973).
29. See also Privacy Act 1974 Implementation: 40 F.R. 47405 (Oct. 8, 1975).
30. *General Motors Corporation v. Finklea,* 442 F.Supp. 821 (S.C. W.Va. 1977).
31. *Tumbarella v. Krueger Co.,* 271 N.W. 2d 284 (Mich. 1978).
32. *Rogers v. Horvath,* 237 N.W. 2d 595 (Mich. 1975).

Appendix

Codes of Ethics

The codes of ethics for several organizations are presented here. Although codes of ethical conduct are not law, they serve as guidelines for professionals in their relationships with clients and employees of clients, with colleagues, and with the public, as well as advancing the quality of professional work.

AMERICAN ACADEMY OF INDUSTRIAL HYGIENISTS

Code of ethics for the professional practice of industrial hygiene was published in 1979 by The American Academy of Industrial Hygienists.

Purpose

This code provides standards of ethical conduct to be followed by Industrial Hygienists as they strive for the goals of protecting employees' health, improving the work environment, and advancing the quality of the profession. Industrial Hygienists have the responsibility to practice their profession in an objective manner following recognized principles of industrial hygiene, realizing that the lives, health, and welfare of individuals may be dependent upon their professional judgment.

Professional responsibility

1. Maintain the highest level of integrity and professional competence.
2. Be objective in the application of recognized scientific methods and in the interpretation of findings.
3. Promote industrial hygiene as a professional discipline.
4. Disseminate scientific knowledge for the benefit of employees, society, and the profession.
5. Protect confidential information.
6. Avoid circumstances where compromise of professional judgment or conflict of interest may arise.

Responsibility to employees

1. Recognize that the primary responsibility of the Industrial Hygienist is to protect the health of employees.
2. Maintain an objective attitude toward the recognition, evaluation, and control of health hazards regardless of external influences, realizing

that the health and welfare of workers and others may depend upon the Industrial Hygienist's professional judgment.

3. Counsel employees regarding health hazards and the necessary precautions to avoid adverse health effects.

Responsibility to employers and clients

1. Act responsibly in the application of industrial hygiene principles toward the attainment of healthful working environments.

2. Respect confidences, advise honestly, and report findings and recommendations accurately.

3. Manage and administer professional services to ensure maintenance of accurate records to provide documentation and accountability in support of findings and conclusions.

4. Hold responsibilities to the employer or client subservient to the ultimate responsibility to protect the health of employees.

Responsibility to the public

1. Report factually on industrial hygiene matters of public concern.

2. State professional opinions founded on adequate knowledge and clearly identified as such.

AMERICAN ASSOCIATION OF OCCUPATIONAL HEALTH NURSES

This code of ethics was adopted by the AAOHN Executive Committee in 1977.

This code of ethics has been developed to guide the Occupational Health Nurse to maintain ethical conduct in providing Occupational Health Nursing Services.

Occupational Health Nurses should:

1. Provide nursing service in the work environment with respect for the dignity of man, unrestricted by considerations of nationality, race, creed, color or status.

2. Safeguard the employee's right to privacy by protecting information of a confidential nature; releasing information only as required by law or upon written consent of the employee.

3. Maintain individual competence in Occupational Health Nursing practice, recognizing and accepting responsibility for individual actions and judgment.

4. Act to safeguard the health and safety of the employee in the work environment.

5. Use individual competence as a criterion in accepting and assigning responsibilities.

6. Participate in research and case finding activities related to Occupational Health when assured the employee's rights are protected.

7. Contribute to the efforts of the profession to define and upgrade standards of Occupational Health Nursing practice and education.

8. Participate actively in the establishment of Standards for Competent Occupational Health Nursing Services through the specialty professional nursing organization.

9. Cooperate with community health agencies and maintain sound

ethical relationships with members of allied health professions.

10. Refrain from allowing their names to be used in connection with the advertisement of products or with any other forms of self advertisement.

AMERICAN OCCUPATIONAL MEDICAL ASSOCIATION

This code of ethical conduct for physicians providing occupational medical services was adopted by the Board of Directors of the American Occupational Medical Association in 1976.

These principles are intended to aid physicians in maintaining ethical conduct in providing occupational medical service. They are standards to guide physicians in their relationships with the individuals they serve, with employers and workers' representatives, with colleagues in the health professions, and with the public.

Physicians should:

1. Accord highest priority to the health and safety of the individual in the workplace;

2. Practice on a scientific basis with objectivity and integrity;

3. Make or endorse only statements which reflect their observations or honest opinion;

4. Actively oppose and strive to correct unethical conduct in relation to occupational health service;

5. Avoid allowing their medical judgment to be influenced by any conflict of interest;

6. Strive conscientiously to become familiar with the medical fitness requirements, the environment and the hazards of the work done by those they serve, and with the health and safety aspects of the products and operations involved;

7. Treat as confidential whatever is learned about individuals served, releasing information only when required by law or by over-riding public health considerations, or to other physicians at the request of the individual according to traditional medical ethical practice; and should recognize that employers are entitled to counsel about the medical fitness of individuals in relation to work, but are not entitled to diagnoses or details of a specific nature;

8. Strive continually to improve medical knowledge, and should communicate information about health hazards in timely and effective fashion to individuals or groups potentially affected, and make appropriate reports to the scientific community;

9. Communicate understandably to those they serve any significant observations about their health, recommending further study, counsel or treatment when indicated;

10. Seek consultation concerning the individual or the workplace whenever indicated;

11. Cooperate with governmental health personnel and agencies, and foster and maintain sound ethical relationships with other members of the health professions; and

12. Avoid solicitation of the use of their services by making claims, offering testimonials, or implying results which may not be achieved, but they may appropriately advise colleagues and others of services available.

AMERICAN SOCIETY OF SAFETY ENGINEERS

This code of ethics for the safety profession was adopted by the ASSE Assembly in 1974.

A professional must by worthy of the trust and confidence of the public and his peers.

The Profession is defined in the Society By-Laws as follows:

"The Safety Profession is composed of individuals engaged in one or more of the specialties within the various sciences for the primary purpose of developing and/or implementing methods, procedures, systems, devices, and/or standards toward the reduction, control, or elimination of hazardous exposures to people or property."

"It is intended that the term 'Safety' be interpreted within the framework of one or more of the four areas A, B, C, and D, defined in the 'Scope and Functions of the Professional Safety Position,' as adopted by the Society Board of Directors in October, 1966, and within this context, engineering should be viewed as only one of the specialties practiced in the Safety Profession."

This Code of Ethics defines the ethical relationships of the members of the Profession, so that their conduct will provide sound judgement in pursuance of their professional duties and will uphold the honor and dignity of the Profession.

The following standards are considered obligatory for members of the Society and advisory for nonmembers.*

*Masculine gender words shall also include the feminine.

I. Relations with the public

A. The member when in public service as an advisor to or as an employee of a governmental body or department shall not participate in consideration of, or actions on, matters concerning purchased services to be provided by him alone or by any organization in which he is owner or have a principal or vested interest, and he shall inform the governmental body or department of the reason for his non-participation. Members shall endeavor to avoid any form or appearance of a possible conflict of interest.

B. The member will endeavor to extend public knowledge and appreciation of accident, injury, and illness control work of safety professionals.

C. He will limit his involvement in such endeavors, however, to only those areas wherein he has specific knowledge and competence.

II. Relations with employers or clients

A. The member will be guided by the highest standards of integrity when acting in professional safety matters for his employer, client, or customer.

B. The member shall not abuse his professional affiliations or any public office to secure personal advantage or to promote his own business interests.

C. The member will endeavor to avoid conflict of interests with his employer or clients. If such conflict is unavoidable, he shall at the earliest possible moment disclose fully the circumstances to all affected parties before proceeding.

III. Relations in seeking employment

The member shall not compete unfairly with another member by attempting to obtain employment, advancement, or independent professional employment by utilizing improper or questionable methods.

IV. Promotional practices

A. The member shall represent his work and/or capabilities in a manner which will not discredit or reflect unfavorably upon the profession.

B. The member shall not allow nor participate in employment listings which contain exaggerated or misleading statements of his qualifications.

C. Use of the Society name and/or emblem by the member shall be only in accordance with Article XIX, Section 2, of the By-Laws and shall not in any way imply Society endorsement of him, his firm, or activities.

D. Professional papers or articles prepared for publication shall be factual, dignified, and free from self-laudatory implications. Such articles shall convey fully the author's degree of participation in the work or program described, and shall give appropriate credit to others who shared in the activity.

V. Obligations as an employer

A. He shall provide prospective safety professional employees with complete information on the position and their proposed status.

B. The member shall provide equal opportunities for professional development of safety professionals working under his supervision.

VI. Relationships with other members

A. The member shall not attempt to injure, maliciously or falsely, directly or indirectly, the professional reputation, prospect, or practice of another member.

B. The member shall not take credit for the safety work of others, including designs, programs, suggestions, accomplishments, or proprietory interests.

C. Each member will cooperate in extending the effectiveness of the profession by interchanging non-confidential or non-proprietory information and experience with other Safety Professionals and encouraging fellow members to improve and/or update their education.

Part VI

Outlook for Occupational Health and Safety

Chapter 19

The Future of Occupational Health and Safety

by Steven D. Herron

The 1980s began with turmoil and upheaval as the nation's economy dropped into prolonged recession. One result was stringent budget-cutting mandated by the incumbent administration. These cuts have severely affected enforcement of occupational health and safety laws, by governments on all levels. Relieved of the threat of enforcement of these regulations, many companies reduced their occupational health and safety budgets and staffs. Such occupational health and safety professionals as industrial hygienists, occupational health nurses, occupational physicians, and safety engineers, who had been in short supply during the 1970s, were no longer in demand by industry. Many who were technically underqualified or merely "riding the occupational health and safety wave" were caught unprepared and, in some cases, terminated.

Occupational health and safety professionals were not the only ones affected. Labor unions were forced into a holding position, of either maintaining or retreating on demands for salary and employee benefits in contract negotiations. This union retrenching, however, will probably benefit the unions over the long term. Union management has been forced to become more cooperative and realistic in bargaining with management. Many perceptive union officials are now including provisions for worker health and safety in current negotiations. Such new terms as "labor-management committees," "quality circles," and "epidemiological/environmental study participation" will become more and more evident in union contracts over the next ten years.

Unions are not the only groups having to learn new methods of doing business. Occupational health and safety professionals are also having to develop strategies for survival and operation. Shrinking professional staffs and increasing volumes of health and safety monitoring reports emphasize the need for more effective time and task management.

Reduced budgets will require increased justification for health and safety expenditures and implementation. Employers will need to place greater emphasis on educating employees to recognize and foresee potential hazards as a means of reducing dependence on occupational health and safety professionals. Many occupational health-care departments and clinics will be working harder to maintain medical confidentiality and restoring the traditional physician/patient roles. Consequently, the occupational health and safety professional can be expected to be in turmoil throughout the decade, with no definitive end in sight.

This chapter explores expected ramifications in the occupational health and safety community over the coming decade. Of course, the future of this group is as unknown as the future of any other profession. A point to remember about any set of predictions is that "Truth is an elusive prediction, but the best predictions are always true."

GOVERNMENT IMPACT ON OCCUPATIONAL HEALTH AND SAFETY

Reduction in federal spending was emphasized in Washington during the early 1980s. The Congress placed very low ceilings on federal spending and enacted a large number of substantive law changes and non-budgetary actions, but rarely abolished federal programs. It is important, furthermore, to realize that these actions were taken by a bipartisan Congress. Concurrently, there was a strenuous pressure by the administration to eliminate or reduce federal regulation to alleviate excessive cost and compliance burdens and remove unnecessary barriers to economic expansion. Reduced federal spending, authority, operations, and regulations were consequently expected to be accomplished by:

- Returning federal authority to state and local governments.
- Assigning various problems to the private sector.
- Keeping tight reigns on the nation's money supply and reduced interest rates through the substantial power of the Federal Reserve System.

Each of these goals will have a dramatic impact on the safety and occupational health-related industries. The industries must evaluate and restructure their existing programs and activities to be responsive to changing federal priorities.

Returning federal authority to the state and local governments

The budget-cutting law passed by the Congress, the 1981 Omnibus Budget Reconciliation Act (P.L. 97-35), defused federal authority and cut funds to the states. The results were:

• Abolition or scheduled phase-out of some federal-state programs.

• Move block grants, or a merging of a whole series of loosely related federal grant programs into a single grant, that gives spending discretion to the states.

• Loosening, sometimes cessation, of some forms of federal regulation.

The return of program control to the states by the block grant route was accompanied by an overall reduction of 25 percent in federal funds. That means that the state and local governments must either raise, by state or local taxation, the money required, or reduce or eliminate certain programs. The block grant device also subjects state and local governments to a bitter competition for scarce funds. Thus, major governmental responsibilities and programs, and their consequent funding and budget controls, are being shifted from Washington to the 50 state capitols.

Will the states take on these new financial obligations? Some suggestion of what may occur stems from the experience with state plans under OSHA. Former Assistant Secretary of Labor Thorne G. Auchter stated that although, "OSHA was designed to be a partnership between the federal government and the states, yet only twenty-four states and territories have their own plans."

If safety, occupational health, and consumer safety programs become involved in the intense struggle for fiscal and program survival on a state-by-state basis, will they have the support of the private sector, organized labor, and the voluntary safety movement? Or will these forces look the other way because of the new state and local taxes that would be levied to support them?

The private sector is expected to assume some of the responsibilities that the federal government is rescinding. Efforts toward this end have been only partial, for example, the OSHA proposal that labor and industry develop self-inspection plans for specific companies. These plans would then be used as the basis for inspection and enforcement rather than OSHA's general standard and regulations. The complications from vested interests in the operation of this plan could be numerous and severe. Furthermore, the welfare of both workers and consumers may not always be the uppermost concern in such inspections.

The economy and the safety and health industries

The general state of the economy has a significant effect on safety and health programs of companies, as well as in "safety and health industries." These are usually individual consultants or private firms and associations in the safety and health business who obtain much of their revenue from providing services or selling instructional materials and other safety-related products. Other revenues come from membership dues, grants from private and governmental sources, and fees for arranging professional seminars and conventions. Most of these products and services have, as an underlying goal, to help safety and occupational health professionals guide their companies in providing safe and healthful working environments for employees.

On careful analysis of these revenue sources, it becomes clear that the safety and health industry is relatively dependent on the implementation of voluntary safety programs and their ancillary promotional and educational materials within companies. Employers are free to decide whether or not they will institute such safety and health programs.

In times of true economic growth, sales of most goods and services increase, thereby reducing or depleting inventories. Since businesses are motivated by profit, they will need more labor and raw materials to produce more products to meet increased demand. This increases the workers' income and decreases unemployment, so workers have more money to save and spend on other goods and services. The portion of that increased income that is spent on goods and services keeps production high and the cycle is complete. In this situation, the economy is healthy and spiraling upward toward economic equilibrium.

Each company, therefore, is in competition with others for labor resources. Since safety and general health are primary human needs, employers who offer more healthful and safer jobs than other employers, all other factors being equal, will attract better employees. As an added benefit, an employer who has fewer accidents than the average for that industry pays significantly less for workers' compensation insurance than do his competitors. Because of these two factors, a more effective safety and health program can actually reduce the cost of goods sold. The concerned company ultimately is more competitive and can still have an increased profit margin.

With such strong incentive, why don't all companies have outstanding safety and health programs? The reasons are complex, but primarily, industrial safety and health programs are costly and their benefits are long term and therefore somewhat discounted. In an expanding economy, some managers consider the cost of such programs to be necessary employee benefit overhead expense.

In a recessionary rather than a growth economy, expectations are poor and people save more of their income, which decreases the demand for goods and services. Company inventories rise, so managers reduce their resources. Their income decreases and unemployment increases, thereby further reducing demand for goods and services. The economy slows down. Companies want to keep their good people but must tighten their belts and hold costs to a minimum. In an attempt to reduce their overhead, some concerns pare their safety and health programs to the bone during recessions.

Following the above reasoning, the safety and health industry should have wild fluctuations depending on the current economic trends, but this is not altogether true. In reality, the industry is sensitive, but wild fluctuations are also rare. As with all issues that are influenced by people, emotions are a tempering factor. Providing a safe and healthful place for work should be an American ideal.

Many forward-thinking companies, on the one hand, realize the economic consequences of reducing or stopping in-house safety and health programs. On the other hand, if a company eliminates or minimizes such programs during recessionary times, it will realize a short-term gain by a significant reduction of overhead. In most companies, current year profits are the key measurement criteria of a department or divisional manager's performance. Company managers, then, might want to cut back on safety and health expenditures. As an unfortunate result, injuries and illnesses might then occur more frequently, causing employee unrest and higher insurance rates.

As the economy turns around and starts upward again, the company that took a short-term gain at the expense of safety and health may well be in serious trouble. Suddenly, it may find that competitors who gained the reputation of caring for their employees during the recession can easily lure their best employees away. The short-sighted company then is forced to raise wages well above prevailing rates to slow its personnel turnover rate and attract new employees to simply maintain, much less increase, production. This increased labor cost, combined with potentially skyrocketing workers' compensation insurance costs, result in a significant competitive disadvantage.

Of course, with reduced health and safety surveillance, there is always the potential for devastating occurrences like the 3-Mile Island accident or Bhopal, India, disaster. A comprehensive safety and health program will attempt to detect and minimize the possibility and occurance of such situations that could threaten the very existence of an organization. But, just as important, such a program will help to minimize the multitude of minor injuries and illnesses that cause employee pain and suffering.

Deregulation opportunities

Because of present economic conditions, the safety and health industry as well may be in a boom-or-bust situation. It would be foolish for occupational health and safety organizations to ignore the lagging GNP and falling corporate profit figures.

The dream of American business—"to get the federal government off its back"—is materializing. What does this mean to the safety and health industry? Perhaps, part of the answer lies in the remarks of Eugene Hartwig, associate general counsel of General Motors Corporation, when addressing the American Bar Association. Hartwig thinks that the last 20 years have been instructive with regard to society's expectations of a safe and healthful working environment. Therefore, he warned employers to heed this message, ". . . or invite even more oppressive regulation when the political wheel of fortune takes another turn." U.S. Senator Patrick Moynihan (Dem., N.Y.) has said that the experience of federal motor vehicle safety regulation proved that "if industry would not regulate itself, it would eventually end up being regulated." Such has also been the history of federal safety and occupational health programs.

The success and viability of shifting responsibility for health from the public to the private sector, and from the federal to the state governments, depend on the responsible behavior of state governments and the private sector. The present situation could well represent the last opportunity, for many years, for the industrial segment of the private sector to achieve its goal to "get the federal government off its back." Failure of the private sector and state governments successfully to undertake the safety and health responsibilities devolved upon them might result in a return to more restrictive approaches.

WHAT DOES SAFETY AND HEALTH COST?

Occupational safety and health funding by American industry indicates that mature companies invest between 2.5 to 3 percent of corporate revenues in safety and health-related items and issues. This may serve as a goal for industries that do not currently meet this commitment. Younger industries with a heavy burden of research-and-development costs (microelectronics and bioengineering companies are good examples) may typically invest less than 1 percent of corporate revenues in safety and health-related issues.

Regardless of the near-term appearance of safety and health, less than a full commitment will ultimately be judged inadequate and the long-term costs may be punitive. The activity of government and the cycles of economic gain and retrenchment should not alter this baseline measure of economic commitment to safety and health in the workplace.

THE ERA OF THE COMPUTER

The use of computers has grown at an astounding rate during the past few years. In 1975, the number of computer terminals in the United States was about one million. In 1985, the number reached more than ten million, and experts predict that almost all workplaces will have at least one computer terminal by 1990. This technological change has a scope at least as great as the Industrial Revolution of the 1800s, and the nature of local, national, and international business procedures are being radically changed. Many "blue chip" capital-intensive manufacturing companies are being bested by smaller, more flexible companies that utilize high-technology business aids. The companies that do not use the more effective and efficient computerized office tools that are available will soon perish.

This new technology is bringing automation to offices and workplaces throughout the world and is changing the workplace, the work environment, and the way people do their jobs. The seeds of this change were sown when the silicon microprocessor chip was invented and put to its first mass-market use by Hewlett Packard Corporation and Texas Instruments Company in the form of small, relatively simple, hand-held calculators. Almost overnight the engineer's slide rule became obsolete. Soon, the huge temperamental mainframe computers that have resided only in computer science buildings on college campuses and in huge insurance companies shrank to desktop size, while increasing their computing power. Then, the microcomputer or "personal computer" was born.

The development of computerized occupational health and safety information systems followed a similar developmental route. The passage of the Occupational Safety and Health Act in 1970 was a major impetus for these early developments. As the regulatory scene changed with the passage of the Toxic Substances Control Act and the Resource Conservation and Recovery Act, as well as other environmental laws, automated systems increased tremendously. (See Chapter 11, "Occupational Health and Safety Information Management Systems.")

Originally, these efforts were oriented toward the large mainframe computers because of the excessive quantity of data storage necessary. As microcomputers became more and more efficient, however, the pioneering mainframe programs that were developed by Standard Oil Company (Indiana) and Diamond Shamrock Corporation, as well as others, were modified and upgraded by other companies into mini- and microcomputer-based program software. To try to recover some of the tremendous developmental programming costs expended by the big corporations to develop the mainframe computer programs, some companies decided to market these programs to other companies through direct sales and computer timeshare rental agreements. Such marketing efforts have met with only limited success to date because of the tremendous quantity of historical and current data entry time involving very significant up-front computer time connection charges as well as many input person-hours. This marketing has been successful only when one company sells previously developed systems to companies in similar industries at a fraction of the cost of the total development. Thus, a fledgling market for occupational safety and health software systems was created.

This new market attracted the attention of the occupational health community and companies specializing in software development for occupational health services. The result has been the entry into the field of many new commercially available systems, especially during the past few years. Obviously, these developers learned from the successes and failures of their predecessors and adapted their systems to new hardware and software technologies. As a result, these new systems offer occupational health and safety professionals many more options and new capabilities.

Recordkeeping—The shrinking paper syndrome

At the same time that computer systems were becoming more compact, more efficient, and less expensive, U.S. regulatory agencies were evolving more and more requirements for recordkeeping. The two regulatory agencies that have been most responsible for the increase in health, safety, and medical records are the Environmental Protection Agency (EPA) and the Occupational Safety and Health Administration (OSHA).

Computerized occupational safety and health recordkeeping systems can help considerably in shrinking the "paper tiger" down to manageable size—for example, in the following areas:

- OSHA logs—Accident records
- Workers' compensation—Injury/illness records
- Industrial hygiene monitoring records
- Medical records
- Environmental agents—EPA effluent monitoring and Resource Conservation and Recovery Act (RCRA) records
- Maintenance of confidentiality
- Data base management techniques.

Of course, OSHA requires that accident/illness data be maintained in the form of OSHA logs. State workers' compensation laws usually require similar injury/illness records, including direct medical and salary costs of accidents and illnesses related to the individual company's workers' compensation insurance forms. When combined by computer, the incidence and costs values can be summarized to provide statistical cost-effectiveness program justifications.

Data relating to the physical description of the work environment can also be handled effectively by computer. Industrial hygiene, health physics, environmental, or other sampling and inspection data describing conditions in the workplace should be maintained as part of the data base. The data should not be limited to industrial hygiene data, but should include other environmental data such as EPA effluent monitoring data and Resource Conservation and Recovery Act (RCRA) hazardous materials handling records. If desired, the company can also maintain in the computer data base information on inventory, toxicological material safety, and toxic agents within the workplace, including chemical, radiological, and physical hazards. Records on these agents are required to identify the toxic substances to which employees are actually or potentially exposed and to maintain reference information on their properties, use, storage, and disposal.

Another excellent use is to maintain information on regulatory requirements such as OSHA standard Permissible Exposure Limits or action levels or internally adopted standards, such as ACGIH Threshold Limit Values. The system can also be expanded to include the tracking of ongoing problems or events that require follow-up, such as hazard abatement plans and schedules or accident reports.

Another major function of the occupational health and safety computer system is to train employees to recognize and use protective measures to be taken in hazardous areas. Such measures might include engineering controls such as hoods for ventilation, administrative controls for health hazards, and personal protective equipment like respirators or safety glasses to be worn in all designated hazardous areas within a facility. Safety and health training course documentation can also be computerized for use as a potential legal protection as well as data for casual comparison with injury/illness reports.

Perhaps the most important use of such systems is by occupational health specialists who manage records related to employee health surveillance. These records should include all physical examination data, clinic visits, injuries and illnesses, including workers' compensation information and results of biological monitoring. In short, records should include all information that relates to the health of individual employees.

The great advantage of maintaining such records via the computer is the ease with which medical confidentiality can be maintained. Reports can be generated for different purposes within the organization without including individual identifying factors such as employee names or identification numbers. These records might show only the pertinent data plus an internally generated record number that means nothing to anyone except the computer in numbering of its hierarchial files.

Another specialized value of computerized occupational safety and health systems lies in the ability to manipulate and correlate data across functions to support comprehensive occupational safety and health surveillance. Such systems make possible fairly extensive epidemiological research into injury and illness specific to each company's workplace. However, more basic but equally important administrative functions such as workers' compensation claims management, litigation defense, regulatory compliance, and records management can be automated and compared by the managers who approve the costs associated with the implementation and operation of the system.

The future of employee health and safety training

One of the major problems that has plagued safety and health professionals is the common conception by rank-and-file employees that the subject of safety and health is boring. This concept unfortunately has been continuously reinforced by the traditional form of training—safety meetings.

Almost everyone who has sat through such weekly or monthly meetings knows how thoroughly boring they can be. Even when the material is germane to the task at hand, it is sometimes technical and difficult to understand or else is too script-like. As any good teacher knows, student involvement in such subjects is very difficult to accomplish, requiring considerable preparation and training skill. Most safety professionals, industrial hygienists, occupational health nurses, and occupational physicians have never been formally taught such skills.

The recent entertainment explosion of video games provides an extremely interesting and novel medium that is achieving universal acceptance. For many people, but particularly for younger men, these arcade video games are an irresistible lure and an integral part of their entertainment activities. If such a medium could be used by health and safety professionals to teach health and safety concepts to employees, the "boring material" problem might be significantly reduced if not eliminated.

Utilizing a computerized safety message trainer is potentially a revolutionary new method of teaching and reinforcing safety and health concepts among employees. The safety message trainer might consist of a single microcomputer that could be placed by the company at a company party, picnic, safety meeting, lunch room, or wherever employees spend some of their free time. A software package should be provided that would allow an employee to play a "space invaders" or similar arcade game, but only after that employee has correctly answered a short series of safety questions. Such questions may be changed at will to suit the company's current health and safety emphasis.

Another more conventional computerized health and safety training idea would be to provide formal on-the-job training with computers in a classroom setting. Much software is currently available that can assist trainers in the writing of custom training programs. These programs are currently being used as "programmed instruction" throughout more modern formal education from grade school through college. In essence, these programmed instruction modules are programs that write programs. All the instructor must do is to type in the lesson, then type in relevant questions and multiple choice answers to be answered after each lesson. In most cases, the instructor need not know any computer programming language, but merely follow the instructional prompts that the programmed instruction module provides. Each employee using such a system can be uniformly trained and work at his or her own comfortable learning pace. A thorough understanding of the material presented is mandatory to complete the educational segment. This will become more and more important in corporate control of the quality of employee training. Moreover, in litigation situations where employees claim they were not taught the correct procedures for a given operation, the computer program and attendance records can prove otherwise.

The technology exists to substitute short video segments into computerized training to illustrate motion-intensive training. This "interactive video training" will also become more common as technology miniaturizes the currently bulky hardware required.

In essence, the new microcomputers will provide a unique method to make safety and health training more palatable for employees. Because of these very real advantages, within five to ten years computerized training will be the norm—not the exception.

THE PIVOTAL ROLE OF THE LABOR UNION

Preventing injury and illness on the job has long been one of organized labor's most important tasks. During the early part of the 1980s, this task has taken on even greater significance. Economic pressures have forced the labor unions into de-emphasizing the traditional salary/wage issues and emphasizing peripheral benefit programs that companies are more likely to concede. Since organized labor has found this to be such a fertile ground for negotiations, it is likely to continue. Labor will probably be engaged in convincing rank-and-file members that safe and healthful workplaces are more valuable than benefit programs covering medical expenses.

Because the total responsibility for a union's health and safety efforts may be too large a job for a single individual, many unions have begun to organize committees to correct job hazards. These committees may work in conjunction with, or be totally separate from, management. They have become more and more favored recently and will continue to receive much attention and effort throughout the decade of the 1980s. In fact, many unions have discovered that forming two safety and health committees, one for coordination with management and one limited to union members, works even better for their purposes. Unions are discovering that a safety and health committee can provide the means, not only to deal with workplace hazards, but also to involve more members with the union.

Workers concerned about hazards in the shop or plant often do not know where to turn for help or information. The union steward or president may not know much about correcting these problems; even the contract may make no provisions for handling them. By setting up a safety and health committee, unions establish a regular procedure for dealing with such problems. Members know where to go first to have matters resolved. Many problems can be resolved immediately without having to rely on outside governmental agencies and the delays that are inevitable when going through a bureaucracy. As these committees become commonplace in the workplace of tomorrow, they could represent a continuing commitment to and responsibility for improving the workplace on the part of both management and labor.

Labor-management committees (also called joint union-management committees) are by no means a substitute for traditional collective bargaining, but contract language has been established by collective bargaining that can serve as an important tool for unions in providing safety and health in their workplace. Besides providing for specific safety and health goals—for example, company-paid safety equipment and clothing, physical examinations, and paid safety training for new workers—contract language can also provide for broader procedural rights, establishment of joint safety and health committees, and guidelines for refusal of hazardous work. Currently, about 90 percent of all union contracts covering manufacturing workers contain some sort of stipulation for safety and health conditions. By the end of the decade, virtually all union contracts may address such health and safety issues.

Future issues will also include quality circles, independent medical surveillance and staffing, and epidemiological/environmental study. (Each will be discussed in successive paragraphs.)

Quality circles are functionally oriented production teams in which production and other decisions are made by consensus instead of by management decree. Of course, the quality-circle concept is an idea imported from Japan that coincides well with Japanese culture. Obviously, social and cultural differences between the United States and Japan may well preclude quality circles from becoming too popular in the U.S.

The concept of using independent medical surveillance and staffing is fairly new and appears more often in collectively bargained contracts. The use of independent services helps to provide objectivity and fairness in the administration of medical examinations and gives some measure of employee medical confidentiality. Because independent medical arrangements are so rare, there is no typical surveillance and staffing agreement that can be discussed here. However, in some contracts, provisions are included for the unions to assume an active role in the selection and the supervision of in-plant medical departments and/or outside occupational medicine clinics. Usually, such activities are coordinated through an existing labor-management committee. Because it diverges from established norms, this concept will probably not be readily accepted by the medical profession or management.

One of the more recent occupational health activities lies in epidemiological/environmental research. The Proportional Mortality Ratio (PMR) study is a typical one and was undertaken by union members. The study examines the causes of death among a selected group of workers in percentages, and compares those figures with the national average mortality rate for a particular cause of death. Of course, such a rough study is not able to pinpoint a specific cause of the disease, but the PMR can be a usable tool. PMR studies could well be the impetus for more detailed epidemiological studies. Progressive unions, such as the AFL/CIO, educate their members in the performance of these PMRs. Because of their relative simplicity, PMR studies should become more prevalent throughout the decade.

PRODUCTIVE MANAGEMENT STRATEGIES FOR SURVIVAL

Cost-benefit analyses of overall programs and similar analyses for project justification will also become more important during sensitive economic times. Methods of calculating health and safety costs vary substantially from one industry to another and even from one company to another. Perhaps the easiest to identify is the direct workers' compensation insurance cost which varies directly with the individual company's experience modification rate (EMR). The EMR is the multiplier applied to the basic insurance premium based on a formula integrating the individual company's accident and illness incidence and severity rates.

The EMR can make a tremendous difference in the actual workers' compensation premiums that a company pays. For example, the Department of Civil Engineering at Stanford University recently attempted to quantify the difference a good safety record can make in relation to the overall job cost for a construction company. They found that an EMR multiplier for national contractors ranges from 0.5 to 2.05. If you assume an EMR of 0.5 and direct labor costs totaling 30 percent of the job cost, a $100 million job would result in workers' compensation premiums of $1.1 million. With an assumed EMR of 2.05, the same job would require premiums of $4.3 million.

Workers' compensation insurance premiums are but one direct cost of injuries and illnesses. Other direct costs include such items as lost work time, direct damage to plant facilities, and product loss. However, most health and safety experts agree that these direct costs are but a small fraction of the total costs associated with workplace injuries and illnesses. According to these experts, such indirect costs usually vary from three to ten times the direct costs. Indirect costs might include downtime associated with the accident or illness; unit production disruption; recruitment, selection, and training of replacement personnel; and increased employee turnover rates in undesirable or hazardous jobs. Using the $3.2 million difference cited above, we can multiply that figure by four to arrive at an estimate of differential total cost equaling $13 million.

The computerization decision above is a good example of the kind of cost-benefit justification that will become increasingly important over the next decade. In most industries, the days of running the business by "the seat of your pants" have vanished forever. Hard statistical and cost data are becoming more and more necessary to convince today's business managers who are more knowledgeable in the financial aspects of risk management.

The occupational health and safety professionals of today are using such cost-benefit analyses to justify current and future program expansion. This trend toward the hard-business-dollar approach should continue to increase throughout the decade.

During the 1960s and 1970s, "me generation" thinking has done much to erode the old-fashioned work ethic. The nation is beginning to feel the pinch of such attitudes in a disturbing decrease in American industrial productivity. Subsequent increases in "welfare state thinking," where chronically unemployed workers feel that the government literally owes them a living is, unfortunately, increasing. But these trends can be turned around and a new work ethic established within this country. However, such an ideological work revolution will not happen with a drastic reconstructing of American business theory. Over the next few decades, this country must observe the lesson that the Japanese have taught us over the past 40 or 50 years.

There are already some signs of American business adopting certain styles of Japanese management to fit American societal values.

CAREER OPPORTUNITIES

As in any changing profession, career opportunities will arise. The "computer revolution" will be the major impetus behind changing the way in which industry works. The computer revolution and subsequent automation will open many doors to new opportunities in these fields.

Increasingly, the function of American industry will turn from manufacturing to information services. Obviously, such a tremendous conversion will not take place overnight nor over the next several years, but over several decades. The application of computer technology to existing knowledge bases will be the challenge of the 1980s. It will also be where additional employees are needed. As we convert more and more of our technical and scientific knowledge bases into computerized data bases, we will need people who are dually trained. Currently, these dually trained individuals are few and far between, usually have a formal educational background in a specific scientific specialty, or have learned computer application methods only peripherally as a hobby or interest.

Holistic approach

The trend of federal and state governments in cutting back on social services will probably continue. Thus, the federal government has already begun to shy away from its protective role. It is the wise businessman who will recognize that this basic human need for economic and health security is going unfulfilled within his employees. Such companies may well change their ways to guarantee their employees more employment security, while providing a safe and healthful place in which to work. The concept of company-sponsored holistic safety, health, and medical services may not be far away.

The holistic approach to safety-and-health-related professions will gain popularity in the future. Industrial safety professionals will become more concerned with reducing off-the-job injury and illness through training and counseling programs. Industrial hygienists will quite likely become much more concerned about ergonomics and applying human factors engineering to make workplaces more comfortable, wholesome places in which to work. Occupational health nurses will quite likely increase their effort to improve medical confidentiality and to introduce holistic nutrition education. Occupational physicians will quite likely become more concerned with preventive medicine, computerized illness trend analyses, and employee stress-management training.

Training programs in occupational safety and health are now commonly retitled "Environmental and Occupational Safety and Health." This better reflects the ultimate job responsibilities that trainees will accept. The safety and health community finds itself involved in issues of community awareness and concern for the environment, the safety of the products they buy, and the continued good health of the neighborhoods surrounding industry and its chemical and fuel storage areas. The safety and health professional of today inherits a difficult task in translating scientific information for lay understanding. In light of the frequent lack of reliable scientific fact, this job can be among the most arduous for the new breed of safety and health professionals.

Nicholas A. Ashford of the Center for Policy Alternatives at Massachusetts Institute of Technology points out that "The need for quantatative risk assessment has been thrust upon the scientific community. It will be a great challenge for scientists to respond to the increasing call for simple numbers. The scientific community must appropriately convey (*a*) the information available or deducible about particular risks; (*b*) the uncertainties in the quantatative information; and (*c*) unquantifiable but policy-relevant facts about the circumstances surrounding specific risks."[1] American industry is not able to deal with the complexities of worker and community safety and health without the participation of trained and experienced safety and health professionals. The era of "on-the-job-training" is over. The future demands professionalism.

Nowhere has the impact of the industrial environment on the community been better stated than by Phillip L. Polakoff of the Western Institute for Occupational and Environmental Sciences.

> Protecting the health of the worker is a first step toward protecting the health of the general public. The workplace is the early warning system for the rest of the environment. The hazards created by our industrial process show up first in those people who work with them on a daily basis. Many illnesses extend beyond the factory gate to the family. Occupational hazards present us with a serious task. We are faced with a monumental problem that is likely to increase. Government, labor, health professionals, industry, insurers, and environmental groups must cooperate to rid us of these problems. Making the occupational environment safe is not important just to workers, it is a critical issue for all of us.[2]

Labor

As previously mentioned, organized labor will play a major role in redefining American industry throughout the next decade. Naturally, this too, will provide opportunities within organized labor for safety and health personnel. It is undeniable that union staffing in these professional fields is currently limited but will almost certainly expand greatly within the next few years.

Government

Opportunities for the safety and health professions within government will probably continue to be limited. The present economy and the general trend away from the government-as-protector concept should steadily reduce the necessity for additional regulatory activities.

Our past employment experience with occupational safety and health professionals has been one of frequent compromise in the demand for technical training and background.

The vast majority of company safety officers are selected from the ranks of middle managers without specific training in the field, and all too often without any significant experience. In the minds of many corporate executives, safety and security are similar activities and the responsibility for the two can be combined. Many companies fail to appreciate the value of experienced occupational medicine specialists, industrial hygienists, and occupational health nurses.

The complexities of occupational safety and health program development, and the potential costs to the company for failing to institute a balanced program, may ultimately hurt those companies that continue to allocate only small sums for safety and health and employ the least expensive personnel to provide the service.

Trends

The future of occupational safety and health employment opportunities should reflect the growing demand for adequately trained and experienced people in each of the disciplines of safety and health. The time has come to replace the corporate physician who merely does annual physical examinations with a fully trained occupational health specialist who provides clinical judgment in areas ranging from epidemiological research to corporate product safety. The occupational health nurse must now be recognized as a vital link between worker health and worker morale and productivity and no longer be judged on the basis of cost effectiveness in a role limited to keeping people at work. The industrial hygienist is becoming a vital part of corporate safety and health strategy as toxicity monitoring of the workplace reverses the outdated notion that medical surveillance of workers is enough.

As Eula Bingham, former director of OSHA, so aptly stated, "...while it is useful in some workplaces to monitor workers, any attempts to substitute biological monitoring for environmental monitoring automatically concedes that prevention is not the hallmark of this field. Environmental monitoring, either area and/or personal, will signal any breakdown in control before an event so that preventive measures can be instituted and biological surveillance can become unnecessary."[3]

The vast majority of workers are employed by companies who appoint an individual to serve as both safety and health director. A willingness to assume responsibility will no longer be satisfied by attendance of an occasional course and the reading of lay journals. A true professionalism is sweeping the safety community and generating better educational opportunities, textbooks and journals, and meetings and conferences.

Occupational safety and health, which until recent years has been largely a field for nonprofessionals, has become a legitimate career opportunity for those willing to undergo extensive training. Such individuals will ultimately benefit from experience and background, as will their corporate employers, the workers, and the community.

There is a delicate balance of training and responsibility maintained by members of safety and health teams. Together, they can provide valuable services that no single member can provide alone. Petty practices of territoriality exercised by various team members or management can easily destroy the efforts of the entire group. It is the most important objective of corporate safety and health programs that professionals be given the opportunity to continue overseeing of safety and health in American industry.

In summary, the professional world of the safety engineers, industrial hygienists, occupational health nurses, and occupational physicians will undergo many changes during the next decade. The exciting changes toward computerization and holistic employee services will leave few positions unchanged. The changes may be massive and a bit frightening to some, but remember that with change comes opportunity.

ADDITIONAL READINGS

Hollinshead, A. C. "Point of View. Occupational Safety and Health: The Years Ahead." *Am. J. Ind. Med.* 2:273-291, 1981.

Robbins, A. "New Frontiers in Occupational Medicine," in *Environmental and Occupational Medicine,* Ed. W. N. Rom. Boston: Little, Brown, 1983.

REFERENCES

1. Ashford, N. A. "Risk Assessment and the Design of Policy for Worker Protection." *Am. J. Ind. Med.* 3(3), 242, 1982.
2. Polakoff, P. L. "Worker and Health Inseparable in the 80's." *Conference Proceedings.* NIH Publication No. 81—2293, p. 7, November 1980.
3. Bingham, E. "Winning the Battle of Occupational Disease." *Am. J. Ind. Med.* 3(4), 361, 1982.

Index

INDEX